VOLUME ONE HUNDRED AND FIVE

ADVANCES IN BOTANICAL RESEARCH

Oxidative Stress Responses In Plants

ADVANCES IN BOTANICAL RESEARCH

Series Editor

Jean-Pierre Jacquot
Membre de l'Institut Universitaire de France, Unité Mixte de Recherche INRA, UHP 1136 "Interaction Arbres Microorganismes", Université de Lorraine, Faculté des Sciences, Vandoeuvre, France

VOLUME ONE HUNDRED AND FIVE

ADVANCES IN BOTANICAL RESEARCH

Oxidative Stress Responses In Plants

Edited by

FRANK VAN BREUSEGEM

Department of Plant Biotechnology and Bioinformatics, Ghent University and VIB Center of Plant Systems Biology, Ghent, Belgium

RON MITTLER

The Division of Plant Sciences and Technology and Interdisciplinary Plant Group Christopher S. Bond Life Sciences Center, University of Missouri, Columbia, MO, United States

ELSEVIER

ACADEMIC PRESS

An imprint of Elsevier

Academic Press is an imprint of Elsevier
125 London Wall, London, EC2Y 5AS, United Kingdom
The Boulevard, Langford Lane, Kidlington, Oxford OX5 1GB, United Kingdom
50 Hampshire Street, 5th Floor, Cambridge, MA 02139, United States
525 B Street, Suite 1650, San Diego, CA 92101, United States

First edition 2023

ISBN: 978-0-323-91374-4
ISSN: 0065-2296

For information on all Academic Press publications
visit our website at https://www.elsevier.com/books-and-journals

Publisher: Zoe Kruze
Acquisitions Editor: Mariana L. Kuhl
Developmental Editor: Jhon Michael Peñano
Production Project Manager: James Selvam
Cover Designer: Vicky Pearson

Typeset by MPS Limited, India

Contents

Contributors

Jahed Ahmed
Louvain Institute of Biomolecular Science and Technology, Université catholique de Louvain, Louvain-la-Neuve, Belgium

Christophe Bailly
Sorbonne Université, IBPS, CNRS, UMR 7622 Biologie du Développement, Paris, France

Johannes Ballauff
Forest Botany and Tree Physiology, Georg-August University of Goettingen, Göttingen, Germany

François Chaumont
Louvain Institute of Biomolecular Science and Technology, Université catholique de Louvain, Louvain-la-Neuve, Belgium

Francisco J. Corpas
Group of Antioxidants, Free Radicals and Nitric Oxide in Biotechnology, Food and Agriculture, Department of Biochemistry, Cell and Molecular Biology of Plants, Estación Experimental del Zaidín, Spanish National Research Council (CSIC), Granada, Spain

Thomas Dresselhaus
Cell Biology and Plant Biochemistry, University of Regensburg, Regensburg, Germany

Christine H. Foyer
School of Biosciences, College of Life and Environmental Sciences, University of Birmingham, Edgbaston, United Kingdom

Luciano Freschi
Department of Botany, Institute of Biosciences, University of São Paulo (USP), São Paulo, Brazil

Stanisław Karpiński
Warsaw University of Life Sciences – SGGW, Poland

Chanhong Kim
Center for Plant Stress Biology, Center for Excellence in Molecular Plant Sciences, Chinese Academy of Sciences, Shanghai, P.R. China

Karl J. Kunert
Department of Plant and Soil Sciences, FABI, University of Pretoria, Pretoria, South Africa

Mengping Li
Center for Plant Stress Biology, Center for Excellence in Molecular Plant Sciences, Chinese Academy of Sciences, Shanghai, P.R. China

Amna Mhamdi
Center for Plant Systems Biology, VIB, and Department of Plant Biotechnology and Bioinformatics, Ghent University, Gent, Belgium

José M. Palma
Group of Antioxidants, Free Radicals and Nitric Oxide in Biotechnology, Food and Agriculture, Department of Biochemistry, Cell and Molecular Biology of Plants, Estación Experimental del Zaidín, Spanish National Research Council (CSIC), Granada, Spain

Andrea Polle
Forest Botany and Tree Physiology, Georg-August University of Goettingen, Göttingen, Germany

Maaria Rosenkranz
Research Unit Environmental Simulation, Institute of Biochemical Plant Pathology, Helmholtz Munich, Neuherberg, Germany

Jos H.M. Schippers
Department of Molecular Genetics, Leibniz Institute of Plant Genetics and Crop Plant Research (IPK), Gatersleben, Germany

Jörg-Peter Schnitzler
Research Unit Environmental Simulation, Institute of Biochemical Plant Pathology, Helmholtz Munich, Neuherberg, Germany

Huili Shi
Forest Botany and Tree Physiology, Georg-August University of Goettingen, Göttingen, Germany

Magdalena Szechyńska-Hebda
Warsaw University of Life Sciences – SGGW, Poland; Plant Breeding and Acclimatization Institute – National Research Institute, Radzików, Błonie, Poland

Liang-Zi Zhou
Cell Biology and Plant Biochemistry, University of Regensburg, Regensburg, Germany

Preface

Plant cells take advantage of a flexible, yet complex, machinery to modulate their response to adverse environmental stress conditions, thereby setting the stage for initiating defense strategies at various levels. In these, the accumulation of reactive oxygen species (ROS) is an inherent aspect. ROS are partially reduced (e.g., superoxide anions) or excited forms (e.g., singlet oxygen) of molecular oxygen. ROS are known to act as double-edged swords, as excessive levels of ROS can cause indiscriminate oxidation of nearly all biomolecules, including DNA/RNA, proteins, and lipids. Such events are referred to as oxidative stress. On the other side, at more moderate levels, ROS act as important signaling molecules that regulate various processes, such as development and responses of plants to biotic and/or abiotic environmental stresses. In this volume, we invited a suite of experts to share their insights in various aspects of ROS biology in plants.

In Chapter 1, Li and Kim provide a comprehensive overview of both the genesis and signaling characteristics of *singlet oxygen* ($^1O_2$), which is a nonradical ROS. Like many other ROS, this photosynthetic byproduct of plants can damage biological molecules, but interestingly it also plays an interdependent signaling role leading to stress acclimation and programmed cell death. Among the different ROS, *hydrogen peroxide* (H_2O_2) is the most efficient signaling molecule due to its relative stability and ability to migrate within and across cells and its impact on the redox state of the cell. In Chapter 2, Mhamdi elaborates on the processes involved in H_2O_2 generation and processing and the different ways by which H_2O_2 can act as a signal, including the recent discoveries of H_2O_2 sensors and receptors that link H_2O_2 to other signal transduction pathways. Low molecular weight antioxidants, such as *ascorbate* and *glutathione*, and antioxidant enzymes are present in most, if not all, plant cell compartments, where they play important roles in propagating or limiting ROS and redox signals. In Chapter 3, Kunert and Foyer review the synthesis and functions of ascorbate and glutathione, the evidence for interactions between these antioxidants in the ascorbate/glutathione cycle and the position of this cycle within the network of cellular redox systems.

In recent years it has become clear that plants also use ROS homeostasis to guide their *development*. In Chapter 4, Shippers presents an overview for the role of ROS gradients and homeostasis in the development of leaves from the shoot apical meristem, based mainly on the model plant species

Arabidopsis thaliana. Furthermore, this chapter summarizes current knowledge on the involvement of different ROS species and molecular components in the regulation of leaf development. Like leaf development, recent studies revealed that ROS also play a critical role in various aspects of *flowering* and *plant reproduction.* Zhou and Dresselhous provide in Chapter 5 an overview of the multiple functions of ROS signaling during plant reproductive development from germline initiation to double fertilization. They mainly focus on flowering plants and discuss the generation and elimination of ROS, and the induction and regulation of various downstream signaling pathways. Seed germination is a critical event in higher plant life because its success ensures the establishment of a new generation and survival of the species. In Chapter 6, Bailly summarizes the current knowledge on the role of ROS in *seed germination* and provides novel insights, such as the concepts of oxidative memory and oxidative networks, which bring a better understanding to the function of ROS in these key developmental processes. In Chapter 7, Corpas et al. provide a broad perspective of the relevance of ROS metabolism during the *ripening* of fleshy fruits, which usually involves physiological oxidative stress. In addition, they review how the application of exogenous molecules with signaling functions such as nitric oxide (NO), hydrogen sulfide (H_2S), or melatonin can contribute to regulating ROS metabolism and how these can exert beneficial effects during postharvest storage. Furthermore, Corpas et al. describe the antioxidant composition of fleshy fruits, recognized as a nutraceutical trait that contributes to human health when regularly consumed. In the next two chapters, the critical roles of ROS in plant interactions with the environment are described. In Chapter 8, Polle et al. review the roles of ROS and antioxidative systems in the fungal partner and plant host for the establishment of symbiosis and for the induction of local and systemic *mycorrhizal* resistance. In Chapter 9, Karpinski and Szechyńska-Hebda summarize the latest insights into the roles of ROS, in concert with electrical signaling, within the induction of a new acclimation phenomenon termed "network acquired acclimation" (NAA), which is necessary for the induction of systemic acquired acclimation (SAA) and systemic acquired resistance (SAR) within a community of plants. The function of H_2O_2 as a signal requires its diffusion through the plasma membrane or internal membranes, a process that is facilitated by the presence of channels named *aquaporins* (AQPs). In the final chapter (Chapter 10), Ahmed and Chaumont discuss the mechanisms by which aquaporins facilitate and regulate H_2O_2 membrane diffusion, and the relevance of their contribution to organellar, cellular, and cell-to-cell H_2O_2 signaling processes in the context of abiotic and biotic stresses.

Together, we believe this volume provides a detailed, comprehensive, and up-to-date overview of ROS biology in plants, and hope that it will be a useful reading material to both young researchers who are new in this field and more experienced scientists who are interested to learn about the critical roles of ROS in plant biology.

Ron Mittler[1], and Frank Van Breusegem[2]
[1]The Division of Plant Sciences and Technology and Interdisciplinary Plant Group, Christopher S. Bond Life Sciences Center, University of Missouri, Columbia, MO, United States
[2]Department of Plant Biotechnology and Bioinformatics, Ghent University and VIB Center of Plant Systems Biology, Ghent, Belgium

CHAPTER ONE

Singlet oxygen in plants: From genesis to signaling

Mengping Li, and Chanhong Kim*
Center for Plant Stress Biology, Center for Excellence in Molecular Plant Sciences, Chinese Academy of Sciences, Shanghai, P.R. China
*Corresponding author. e-mail address: chanhongkim@cemps.ac.cn

Contents

Abstract

Singlet oxygen ($^1O_2$) is a non-radical reactive oxygen species (ROS), a photosynthetic byproduct in plants. Like other chloroplastic radicals, such as superoxide and hydroxyl radicals, $^1O_2$ can damage biological molecules including nucleic acids, lipids, and proteins, exacerbating photoinhibition and oxidative stress. Since the photosystem II reaction center (PSII RC) is a primary place of $^1O_2$ production, PSII damage, quality control, and reassembly occur as light-driven default processes. In addition, the chloroplast has evolved systems to quench excess light energy and to directly detoxify $^1O_2$ as counteractive measures. However, upon the first publication of a research article elucidating the genetic basis of $^1O_2$-triggered stress responses by Klaus Apel's group in 2004, the classical view of $^1O_2$ as a toxic agent has dramatically changed. Since then, many scientists started the incredible journey of $^1O_2$-signaling research with more discoveries of $^1O_2$-overproducing mutants under certain conditions. This chapter summarizes the current knowledge of the biogenesis of

Advances in Botanical Research, Volume 105
ISSN 0065-2296, https://doi.org/10.1016/bs.abr.2022.08.023

photosynthetic and non-photosynthetic $^1O_2$, PSII damage and repair cycle, and the $^1O_2$-dependent oxidative modification of biomolecules and related retrograde signaling pathways in plants. In particular, we highlight recent progress in $^1O_2$ signaling pathways mediated by two spatially separated chloroplast $^1O_2$ sensors, β-carotene and EXECUTER proteins, as well as possible other branches of the $^1O_2$ signaling network, providing insights into the importance of $^1O_2$ in plant stress responses, such as acclimation and programmed cell death.

1. Introduction

The chloroplast, a light-harvesting energy-producing organelle, produces photosynthetic singlet oxygen ($^1O_2$) due to excess light energy absorption over photosynthetic capacity (Apel & Hirt, 2004; Dogra & Kim, 2019; Krieger-Liszkay, 2005; Triantaphylidès & Havaux, 2009). $^1O_2$ has long been considered a toxic reactive oxygen species (ROS) mainly produced in the chloroplast through direct energy transfer from the light-excited chlorophyll (Chl) to the ground-state triplet oxygen molecule ($^3O_2$) (Fischer, Hideg, & Krieger-Liszkay, 2013; Krieger-Liszkay, Fufezan, & Trebst, 2008; Santabarbara et al., 2007; van Mieghem et al., 1995). Since $^3O_2$ has two unpaired electrons on the two outer orbitals with parallel spin rotation, to interact with other molecules, it covalently accepts two electrons. However, once $^3O_2$ becomes excited by energy transferred by photosensitizers, these two unpaired electrons become paired with an unstable unparallel spin rotation, transforming the ground state $^3O_2$ to the excited state of $^1O_2$ (Aldred, Buck, & Vall, 2009; Apel & Hirt, 2004; Dogra & Kim, 2019). Relative to $^3O_2$, $^1O_2$ is unstable and thus tends to rapidly return to a ground state (less than 0.04 microseconds) by chemically interacting with or transferring the excitation energy to an adjacent molecule, rendering $^1O_2$ highly reactive and electrophilic, but with a limited diffusion distance (around 220 nm) and a short lifespan (Redmond & Kochevar, 2006; Wilkinson & Brummer, 1981). Since the light-harvesting photosynthetic complexes produce $^1O_2$, electron-rich molecules in the vicinity are prone to oxidation by $^1O_2$ through physical and chemical quenching processes (Davies, 2004; Gorman & Rodgers, 1992; Krieger-Liszkay et al., 2008; Triantaphylidès & Havaux, 2009; Triantaphylidès et al., 2008). Increasing $^1O_2$-dependent damages across the thylakoid membrane ultimately deteriorates chloroplast photosynthetic capacity and functional integrity. Thus, deciphering $^1O_2$ is of great interest to a broad spectrum of chloroplast scientists whose research is centered on photosynthesis, photodamage, or plant oxidative stress responses.

Its high reactivity and short lifespan are the major challenging factors in measuring $^1O_2$ levels in living cells. The relative $^1O_2$ level was initially determined through luminescence measurements at 1270 nm based on the unstable electron pair at the outer orbital with the unparallel spin rotation. However, this method is applicable with only isolated photosystem apparatus (Macpherson, Telfer, Barber, & Truscott, 1993). Similarly, measuring $^1O_2$ levels through spin-trapping assays is mainly achieved in vitro (Fufezan, Rutherford, & Krieger-Liszkay, 2002). Higher demand for in vivo $^1O_2$ measurement eventually led to the invention of fluorescent probes, including DanePy (3-(*N*-diethylaminoethyl)-*N*-dansyl) aminomethyl-2,5-dihydro-2,2,5,5-tetramethyl-1*H*-pyrrole) (Hideg, Ogawa, Kálai, & Hideg, 2001) and SOSG (singlet oxygen sensor green) (Flors et al., 2006). The core of these probes is a $^1O_2$-trapping moiety and a chromophore. The intramolecular energetics of these fluorescent probes change upon reacting with $^1O_2$, affecting the chromophores' maximum excitation/emission wavelengths (Fischer et al., 2013). In this regard, SOSG has been used to detect $^1O_2$ in various studies (Cazzaniga, Li, Niyogi, Bassi, & Dall'Osto, 2012; Chen, Xing, Tan, Tang, & He, 2003; Dall'Osto, Cazzaniga, Havaux, & Bassi, 2010; Flors et al., 2006; Hideg, 2008). In the absence of $^1O_2$, SOSG emits a weak blue fluorescence with emission peaks at 395 nm and 416 nm when excited at 372 nm and 393 nm, respectively (Flors et al., 2006). However, once interacting with $^1O_2$, SOSG exhibits a strong green fluorescence with an excitation peak at 504 nm and an emission peak at 525 nm. This $^1O_2$-driven oxidation of the SOSG chromophore allows the quantification of $^1O_2$ by monitoring its altered fluorescence emission (Flors et al., 2006). However, limited penetration and diffusion efficiency of fluorescent probes and the presence of light-absorbing and auto-fluorescing chlorophyll molecules interfere with the accurate measurement of $^1O_2$ in plants. Recent studies also showed that SOSG and the $^1O_2$-mediated oxidation intermediate SOSG endoperoxide produce $^1O_2$ as photosensitizers, additionally interfering with the accuracy of the analysis in vivo (Gollmer et al., 2011; Ragas, Jimenez-Banzo, Sanchez-Garcia, Batllori, & Nonell, 2009). As polyunsaturated fatty acids (PUFAs) with carbon-carbon double bond(s) are prone to $^1O_2$ oxidation, the relative level of lipid peroxidation has also been used to indicate the related level of $^1O_2$ (Triantaphylidès et al., 2008).

Besides $^1O_2$, chloroplasts simultaneously generate chemically distinct photosynthetic ROS, such as superoxide radical (O_2^-), hydrogen peroxide

(H_2O_2), and hydroxyl radical (HO•), further limiting the research of individual ROS and their impact on different physiological processes. This problem has been, at least in part, resolved by discovering a conditional *fluorescent* (*flu*) mutant in *Arabidopsis thaliana* (Arabidopsis) generating $^1O_2$ in chloroplasts upon a dark-to-light shift (Meskauskiene et al., 2001). The genetic basis of $^1O_2$-triggered stress responses in the *flu* mutant has changed the classical view of $^1O_2$ from being a toxic to a signaling molecule (Wagner et al., 2004). Identification and characterization of more Arabidopsis mutants overproducing $^1O_2$ in chloroplasts under certain conditions revealed $^1O_2$ to be more multifaceted and intertwined with other stress signaling pathways. Undoubtedly, exploring such Arabidopsis mutants has proven to be a milestone in $^1O_2$ research, paving the way for their use in dissecting $^1O_2$-triggered retrograde signaling pathways and plant stress responses. Here, we summarize $^1O_2$ research from old to new understandings and provide a perspective for future research.

1.1 $^1O_2$ Genesis

1.1.1 Photosynthetic $^1O_2$

1.1.1.1 $^1O_2$ generation in PSII

In a light-harvesting antenna complex (LHC), Chl molecules absorb light energy exceeding its use for carbon assimilation through the Calvin-Benson cycle. This inefficient utilization of light energy inevitably produces destructive ROS. As protective measures, excess light energy from the excited singlet state of Chl (^{1}Chl) is harmlessly dissipated as heat via a process called non-photochemical quenching (NPQ) and re-emitted as Chl fluorescence (Demmig-Adams & Adams, 2000; Muller, Li, & Niyogi, 2001) (Fig. 1A). NPQ is a transient and reversible system that can mitigate PSII damage under fluctuating light intensity (Li et al., 2000). Three mechanisms support NPQ, including feedback de-excitation quenching (qE), photoinhibition (qI), and state transitions (qT) (Holt, Fleming, & Niyogi, 2004; Horton & Hague, 1988; Triantaphylidès & Havaux, 2009). In fact, qE can quench ^{1}Chl up to 80% (Bassi & Caffarri, 2000; Demmig-Adams et al., 1996; Holt et al., 2004; Li et al., 2000). In vascular plants, a decrease in pH in the thylakoid lumen along with the protonation of PsbS (to two lumen-faced glutamate residues) in PSII and the subsequent activation of zeaxanthin (one of the carotenoids) synthesis via the xanthophyll cycle are required for activating qE under excess light (Briantais, Vernotte, Picaud, & Krause, 1979; Demmig-Adams, 1990; Li et al., 2004; Li, Muller-Moule, Gilmore, & Niyogi, 2002; Yamamoto, Nakayama, & Chichester,

1962). Zeaxanthin can directly quench ^{1}Chl or induce the conformational change of the PSII antenna from light-harvesting into an energy dissipation state (Frank, Bautista, Josue, & Young, 2000; Gilmore, 1997; Horton, Ruban, & Wentworth, 2000; Ruban, Johnson, & Duffy, 2012). Accordingly, NPQ is largely absent in the zeaxanthin biosynthesis-impaired *npq1* mutant (Havaux & Niyogi, 1999). Likewise, the *npq4* mutant lacking PsbS also is unable to dissipate excess energy, encountering increased oxidative damage (Golan, Muller-Moule, & Niyogi, 2006; Havaux & Niyogi, 1999; Li et al., 2000). Otherwise, unquenched ^{1}Chl (lifespan: $\sim 10^{-8}$ s) undergoes a decay process resulting in the formation of the triplet excited state (^{3}Chl) with an increased lifespan ($\sim 10^{-3}$ s) (Fig. 1A). Although xanthophyll carotenoids, such as lutein and zeaxanthin, can quench ^{3}Chl in LHC (Dall'Osto et al., 2006, 2012; Triantaphylidès & Havaux, 2009), approximately 5% of ^{3}Chl remains unquenched (Mozzo, Dall'Osto, Hienerwadel, Bassi, & Croce, 2008; Triantaphylidès & Havaux, 2009). With long-lasting excess light, the overwhelmed quenching system further increases the ^{3}Chl level, which generates $^1O_2$ by the energy transfer to the inert form of stable $^3O_2$ (Li, Wakao, Fischer, & Niyogi, 2009; Santabarbara et al., 2007) (Fig. 1A).

Like Chls in LHC, P680 Chl molecules in the PSII reaction centers (PSII RC) become excited by light energy delivered from LHC, referred to as a single state (^{1}P680) (Fig. 1A). The subsequent charge separation between ^{1}P680 and the first electron carrier intermediate pheophytin (Pheo) forms a ^{1}P680$^+$ Pheo$^-$ radical pair, enabling the electron transfer to plastoquinone acceptors and then to PSI (Adir, Zer, Shochat, & Ohad, 2003; Vass et al., 1992). The oxidized P680$^+$ gains an electron released by the water-splitting reactions of the oxygen-evolving complex and returns to its ground state. However, when the electron transfer chain becomes over-reduced and cannot accept any further electrons under stress conditions, the primary radical pair ^{1}P680$^+$ Pheo$^-$ reemerges by the charge recombination reaction, which subsequently decays into ^{3}P680$^+$ Pheo$^-$ via an intersystem crossing (Durrant, Giorgi, Barber, Klug, & Porter, 1990; Krieger-Liszkay, 2005) (Fig. 1A). The ^{3}P680$^+$ Pheo$^-$ also can be directly formed by ^{1}P680$^+$ QA$^-$ (Krieger-Liszkay, 2005). ^{3}P680 then generates $^1O_2$ by transferring excitation energy to $^3O_2$ (Adir et al., 2003; Krieger-Liszkay, 2005) (Fig. 1A). Although two β-carotene molecules are attached to the D1/D2 proteins in the core of PSII RC, they cannot quench ^{3}P680 because their location is not close enough (Ferreira, Iverson, Maghlaoui, Barber, & Iwata, 2004). Efficient quenching of excited Chl molecules by carotenoids can only occur within the van der Waals distance of 3.6 Å (Fischer et al., 2013; Telfer, 2002). However, β-carotene, located in the PSII RC, is too distant (18–20 Å) to directly quench ^{3}P680 (Braslavsky & Holzwarth, 2012; Trebst, 2003). Instead, β-carotene scavenges $^1O_2$ generated by ^{3}P680 in the PSII RC and thereby protects PSII RC (van Gorkom & Schelvis, 1993) (Fig. 1A).

1.1.1.2 $^1O_2$ generation in the cytochrome b_6f complex and PSI-LHCI

In addition to LHC and PSII RC, the cytochrome b_6f complex (Cyt b_6f) and PSI were also reported to generate $^1O_2$ in the sub-thylakoid membrane fractions isolated from spinach leaves (Suh, Kim, & Jung, 2000). The formation of $^1O_2$ via the isolated Cyt b_6f was shown by spin-trapping assay upon illumination. Here, the Rieske iron (Fe)-sulfur (S) center of Cyt b_6f was thought to be a photo-reactive chromophore (Suh et al., 2000).

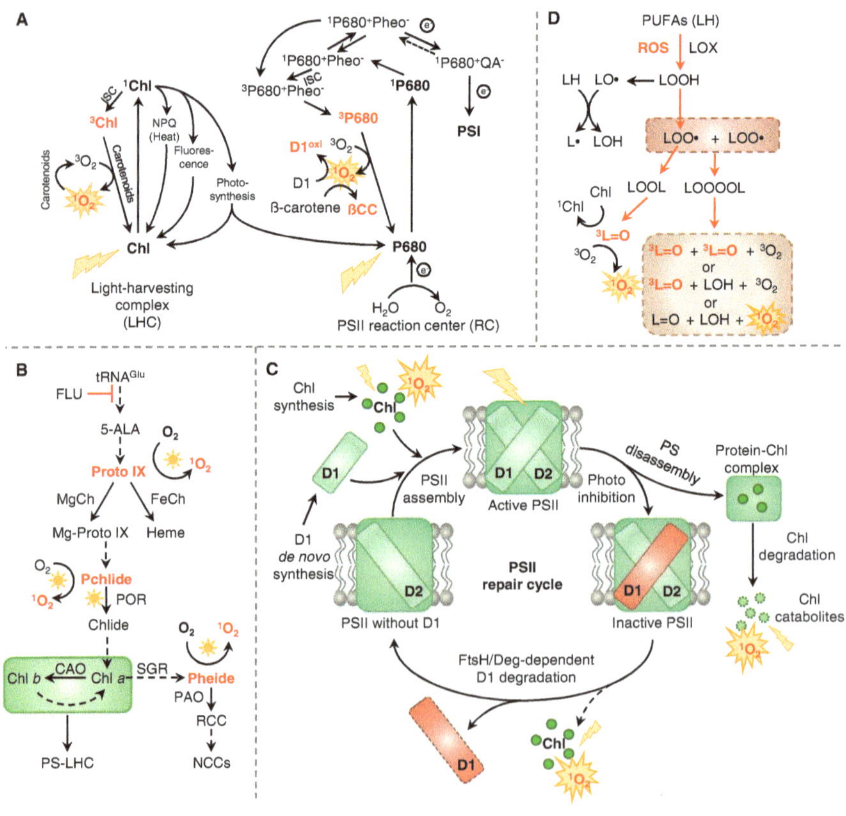

(caption on next page)

Fig. 1 **Distinct sources of $^1O_2$ generation. (A) Photosynthetic $^1O_2$ generation.** Upon absorption of light energy, Chl molecules in the PSII LHC become excited ^{1}Chl, enabling the conversion of excited energy into an electrochemical potential via charge separation in the PSII RC. Due to the inefficient utilization of light energy, part of the excess energy is re-emitted as fluorescence or dissipated as heat via non-photochemical quenching (NPQ), returning the ^{1}Chl to its ground state. Otherwise, ^{1}Chl spontaneously decays to its long-lived state (^{3}Chl) through intersystem crossing (ISC, spin changes in the molecular orbitals). Carotenoids in the LHC readily quench ^{3}Chl. The unquenched ^{3}Chl can transfer the excited energy to the ground state of $^3O_2$ to generate $^1O_2$. Carotenoids can also quench LHC-generated $^1O_2$. Similarly, $^1O_2$ is also formed by ^{3}P680 when the electron (e^-) transfer chain becomes over-reduced under stress conditions. **(B) $^1O_2$ production by tetrapyrrole intermediates and catabolites.** Tetrapyrrole biosynthesis is initiated with glutamyl-tRNA (tRNAGlu), which converts to the universal precursor 5-aminolevulinic acid (ALA). Then, several enzymatic reactions produce closed macrocycle tetrapyrroles, including protoporphyrin IX (Proto IX). From Proto IX, the pathway is divided into Chl and heme branches with Mg^{2+} and Fe^{2+} insertion via magnesium chelatase (MgCh) and ferrochelatase (FeCh), respectively. In the dark, the FLU protein negatively regulates the accumulation of Pchlide by repressing the synthesis of ALA in darkness. In plants, Chl *a* oxygenase (CAO) converts Chl *a* to Chl *b*, and Chl *a* and Chl *b* can be interconverted, essential for the stable accumulation of LHC. During leaf senescence, the central Mg^{2+} from Chl *a* is released by magnesium dechelatase SGR to produce pheophorbide a (Pheide). Next, pheophorbide *a* oxygenase PAO opens the chlorin ring to produce the red Chl catabolite (RCC), which is subsequently reduced into fluorescent Chl catabolites, followed by the export from chloroplasts to the vacuole where they undergo isomerization into non-fluorescent Chl catabolites (NCCs). Most free tetrapyrrole intermediates and Chl catabolites with the macrocycle structure act as photosensitizers upon exposure to light, thereby generating $^1O_2$. **(C) $^1O_2$ generation during the disassembly of pigment-protein complexes.** D1 protein in the PSII RC is susceptible to photodamage, constantly activating the PSII repair cycle to replace damaged D1 with freshly synthesized D1. The PSII repair cycle includes the partial disassembly of the inactive PSII complex, the FtsH- and Deg-dependent proteolysis of damaged D1, insertion of newly synthesized D1, and reassembly of the functional PSII. Degradation of damaged D1 and reassembly of PSII require the coordinated release and re-insertion of pigments (Tian et al., 2021; Wang & Grimm, 2021). During leaf senescence, the disassembly of photosystems is also associated with the release and degradation of Chl. Free Chl and Chl catabolites can act as photosensitizers, producing $^1O_2$ upon light exposure. **(D) $^1O_2$ formation via lipid peroxidation.** Lipoxygenase (LOX)-dependent enzymatic or ROS-dependent non-enzymatic oxidation of PUFAs (LH) generates lipid hydroperoxide (LOOH). With metal ions, it is reduced to lipid alkoxyl radical (LO•). The reaction of LO• with another LH generates a lipid radical (L•) and lipid hydroxide (LOH). The oxidation of LOOH forms lipid hydroperoxides (LOO•), spontaneously producing cyclic endoperoxide (LOOL) and linear tetroxide (LOOOOL). The decomposition of LOOOOL gives rise to different molecules, which include ketone (L=O) and oxygen molecules in either exited (colored by red) or ground state, and LOH based on the Russell mechanism. The thermal cleavage of LOOL is also known as the source of ^{3}L=O, which forms $^1O_2$ and ^{1}Chl by transferring excitation energy to $^3O_2$ and Chl, respectively.

Indeed, Cyt *b6f*-derived $^1O_2$ reinforced PSII photoinhibition when Cyt *b6f* was co-incubated with PSII in vitro. Accordingly, the application of the $^1O_2$ quencher azide anion repressed the Cyt b_6f effect on PSII photoinhibition. In contrast, heavy water, which extends the $^1O_2$ lifetime, enhanced the impact of Cyt b_6f on PSII inactivation. Another study in marine green alga *Bryopsis corticulans* showed that ^{3}Chl *a* formed by the excited Chl *a* molecule in Cyt b_6f can generate $^1O_2$, and the high enzymatic activity of Cyt b_6f complex might be crucial for the efficient ^{1}Chl *a* deactivation, preventing $^1O_2$ formation (Ma et al., 2009). Apparently, $^1O_2$ was detected in purified Chl *a*-containing but Rieske-depleted Cyt b_6f complexes through electron paramagnetic resonance spectroscopy under high light, suggesting that in fact the Chl *a* molecule instead of the Fe-S center might be responsible for the light-induced $^1O_2$ production (Sang et al., 2010). Besides, the increasing light intensity proportionally increases $^1O_2$ levels in the PSI-LHCI super-complexes isolated from Arabidopsis leaves (Cazzaniga et al., 2012). Moreover, in isolated chloroplasts from spinach leaves, PSI photoinhibition induced by repetitive short-pulse illumination was partially suppressed by the $^1O_2$ scavenger 2,2,5,7,8-pentamethyl-6-chromanol, suggesting that $^1O_2$ contributes to PSI photoinhibition as well (Takagi, Takumi, Hashiguchi, Sejima, & Miyake, 2016). However, whether $^1O_2$ can be generated in vivo either as a byproduct of photosynthesis or specifically under photooxidative stress in the Cyt *b6f* complex and PSI-LHC1 super-complexes still needs to be further verified.

1.1.1.3 $^1O_2$ generation by Chl precursors and catabolites

Macrocyclic tetrapyrroles and carotenoid intermediates such as tetra-*cis*-lycopene act as photosensitizers (Fischer et al., 2013; Han, Sakuraba, Koh, & Paek, 2012). The Arabidopsis *flu* mutant, deficient in repressing the Mg^{2+} branch of tetrapyrrole biosynthesis, accumulates an excess amount of free protochlorophyllide (Pchlide) in the dark, thereby generating $^1O_2$ upon exposure to light with the concurrent activation of $^1O_2$ signaling (see section "EX1-dependent signaling") (Lee, Kim, Landgraf, & Apel, 2007; Op den Camp et al., 2003) (Fig. 1B). The loss of *CAROTENOID ISOMERASE* (*CRTISO*) in the rice *zebra2* (*z2*) mutant leads to the accumulation of tetra-*cis*-lycopene (CRTISO substrate) during the dark periods under short-day conditions. This tetra-*cis*-lycopene accumulation in *z2* is positively correlated with $^1O_2$ generation during the light period and consequently alters the expression of $^1O_2$-responsive and cell death-related genes (Han et al., 2012). Under light, exogenous treatment of herbicides, such as acifluorfen and oxadiazon, blocks the protoporphyrinogen oxidase and leads to protoporphyrin IX (Proto IX) accumulation and high levels of $^1O_2$ (Becerril & Duke, 1989).

Disassembly of the protein-pigment complexes during leaf senescence or rapid photosystem turnover under high light might also lead to a release of free Chl and Chl catabolites (Beisel et al., 2010; Fischer et al., 2013; Wang & Grimm, 2021) (Fig. 1C). Excitation of Chl catabolites, especially pheophorbide a (Pheide), produces $^1O_2$ and induces the hypersensitive response (HR)-type cell death in both Arabidopsis and rice (Jiang, Chen, Li, Xu, & Wu, 2011; Mur et al., 2010) (Fig. 1B). In Arabidopsis, the expression of a chloroplast-targeted protein STAYGREEN (SGR), which is required for the LHC disassembly and Chl catabolism (Fig. 1B), is significantly increased 6 h after incubation with avirulent pathogen *Pseudomonas syringae pv tomato* (*Pst*) *avrRpm1* (Mur et al., 2010). The upregulation of SGR leads to an enhanced Pheide accumulation, $^1O_2$ generation, and accelerated HR-type cell death, while RNAi *SGR* plants exhibit a delayed cell death in response to *Pst avrRmp1*. Rice transgenic plants overexpressing an SGR homologue also show increased levels of Pheide, $^1O_2$ generation, and foliar cell death at the two-leaf stage when grown under natural conditions (Jiang et al., 2011). Similarly, the Arabidopsis mutant *accelerated cell death 1* (*acd1*), lacking the functional Chl catabolite reductase PHEOPHORBIDE A OXYGENASE (PAO), accumulates Pheide and $^1O_2$ with augmented cell death following *Pst avrRpm1* infection (Mur et al., 2010; Pruzinska et al., 2007). All these findings suggest that the Chl catabolism-derived $^1O_2$ might contribute to the immune responses toward pathogen infection by enhancing HR responses.

To avoid photooxidative damages caused by Chl metabolite/catabolite-derived $^1O_2$, various enzymes and auxiliary factors function in preventing the release of their free forms (i.e., photosensitizers) (Wang & Grimm, 2021). Several LIGHT-HARVESTING COMPLEX (LHC)-LIKE (LIL) proteins with chlorophyll-binding motifs belonging to EARLY LIGHT-INDUCIBLE PROTEINs (ELIPs) protect chloroplasts from oxidative damage and photoinhibition under various oxidative stress conditions, such as excess light, blue light, UV exposure, and cold (Adamska & Kloppstech, 1994; Adamska, Kloppstech, & Ohad, 1992; Adamska, Ohad, & Kloppstech, 1992; Heddad & Adamska, 2002). In the microalga *Chlamydomonas reinhardtii* (Chlamydomonas), ELIP3 regulates the cellular redox status and protects the photosystem under photooxidative stress conditions (Lee, Lee, Han, & Kim, 2020). Therefore,

ELIP3 knockdown renders Chlamydomonas more susceptible to high-light and cold stress, whereas *ELIP3* overexpression confers resistance (Lee et al., 2020). LIL3 is involved in late Chl biosynthesis, physically interacting with GERANYLGERANYL REDUCTASE (CHLP) during de-etiolation of maize seedlings (Reisinger, Plöscher, & Eichacker, 2008; Tanaka et al., 2010). Besides, LIL3 also interacts with and stabilizes protochlorophyllide oxidoreductases (PORs) in Arabidopsis (Hey et al., 2017). Thus, LIL3 and CHLP may form a protein complex close to the Chl insertion site of LHC. This event may decrease exposure to phototoxicity driven by free Chl or its precursors during photosystem assembly or under light stress conditions. Besides, the recombinant water-soluble Chl-binding protein (WSCP) of cauliflower (*Brassica oleracea* var. botrytis) was shown to inhibit $^1O_2$ generation independent of carotenoids (Schmidt, Fufezan, Krieger-Liszkay, Satoh, & Paulsen, 2003). Later, the crystal structure of the native tetrameric WSCP-Chl complex purified from *Lepidium virginicum* revealed that Chl molecules are tightly packed in the core of the tetramer (Horigome et al., 2007). The watertight cavity of the tetrameric WSCP-Chl complex might shield Chl molecules from $^3O_2$ contact, thereby preventing $^1O_2$ generation (Horigome et al., 2007). In line with this, transgenic tobacco plants overexpressing cauliflower *WSCP* exhibited lower ROS levels and enhanced resistance to high light (Damaraju, Schlede, Eckhardt, Lokstein, & Grimm, 2011). In addition, cyanobacterial Chl-binding proteins, namely small CAB-like proteins (SCPs), function in preventing $^1O_2$ generation in PSII, as evidenced by the fact that the *Synechocystis sp.* PCC 6803 mutant lacking SCPs exhibits elevated $^1O_2$ levels and protein oxidation upon exposure to high light (Sinha, Komenda, Knoppova, Sedlarova, & Pospisil, 2012).

1.2 Non-photosynthetic $^1O_2$

Similar to photosynthetic $^1O_2$ generation, the triplet-to-triplet energy transfer from the excited-state carbonyl ($^3C{=}O$) in lipids and proteins to $^3O_2$ appears to form $^1O_2$ (Cilento & Adam, 1995; Fedorova, Trofimov, Vasil'ev, & Veprintsev, 2007; Pathak, Prasad, & Pospisil, 2017). The decomposition of tetrahydroperoxides (C–OOOO–C) of lipids and proteins also produces $^1O_2$ or $^3C{=}O$ through the Russell mechanism (Cilento & Adam, 1995; Fedorova et al., 2007; Russell, 1957). In plants, the lipid peroxidation and its peroxyl radical products facilitate $^1O_2$ generation. Mechanical damage, such as wounding, releases PUFAs from the plasma membrane, thereby accumulating PUFAs at the wounding site

(Roach et al., 2015). PUFAs (hereafter LH) then undergo lipoxygenase (LOX)-mediated enzymatic or ROS-mediated non-enzymatic oxidation, generating hydroperoxyl PUFAs (also called lipid hydroperoxide, LOOH) (de Bruxelles & Roberts, 2001) (Fig. 1D). With transition metal ions, LOOH can be reduced to lipid alkoxyl radical (LO•). LO• can be further converted to lipid radical (L•) and lipid hydroxide (LOH) by reacting with another LH. Under oxidation-prone conditions, LOOH is oxidized into lipid hydroperoxides (LOO•). The cyclization of LOO• spontaneously forms high-energy intermediates such as cyclic endoperoxide (LOOL) and linear tetroxide (LOOOOL), which may directly generate excited-state carbonyl (^{3}L=O) or $^1O_2$ through disproportionation reactions based on the Russell mechanism (Russell, 1957). Thermal cleavage of LOOL also produces ^{3}L=O, which can transfer excitation energy to $^3O_2$ and Chl to form $^1O_2$ and ^{1}Chl, respectively (Fig. 1D). In fact, wounding in dark-adapted Arabidopsis leaves and germinating soybean leads to the production of $^1O_2$ (Chen et al., 2003). Interestingly, after crushing the cotyledons of Arabidopsis seedlings with a hemostat in the dark, the SOSG probe reveals $^1O_2$ generation in the systemic root tissue within one to three minutes, indicating a systemic wound-induced $^1O_2$ generation in the root (Mor et al., 2014). However, the genuine source of $^1O_2$ burst in roots remains unknown. $^1O_2$ was also detected in the root tip of drought- and bacteria elicitor peptide flagelline 22-treated Arabidopsis seedlings in the dark (Mor et al., 2014). Similarly, another study showed that polyethylene glycol (PEG)-induced osmotic stress rapidly elicits $^1O_2$ from the epidermal layers to the whole root and induces cell death (Chen & Fluhr, 2018). The application of lipoxygenase inhibitor salicylhydroxamic acid considerably inhibits the $^1O_2$ production and cell death, suggesting that the LOO• produced by lipoxygenase from PUFA could be the source of $^1O_2$ under the osmotic stress (Chen & Fluhr, 2018).

Another source of non-photosynthetic $^1O_2$ generation is the defense compound phytoalexin (Flors & Nonell, 2006; Triantaphylidès & Havaux, 2009). Microbe infection triggers the synthesis and accumulation of phytoalexins in various plant tissues to confer resistance. Phytoalexins with phenalenone-like chromophores can serve as $^1O_2$-generating phototoxins to kill pathogens (Flors & Nonell, 2006). Nevertheless, some species synthesize different phytoanticipins and secondary metabolites with photosensitizing properties, thereby deploying $^1O_2$ to enhance their antimicrobial activity (Pedras & Yaya, 2015). For instance, *Apiaceae* and

Hypericum spp produce psoralenes and polycyclic quinone hypericin as phototoxins, respectively (Flors & Nonell, 2006).

2. $^1O_2$-dependent oxidative modification of target molecules

There are two main strategies to detoxify $^1O_2$ upon its generation: either through energy transfer or through chemical reactions termed physical quenching and chemical quenching, respectively. Through physical quenching, $^1O_2$ directly transfers the excited energy to quenchers to return to the ground state. As a result, the quencher dissipates the energy, for example, as heat. With the multiple conjugated double bonds, carotenoids are considered the most efficient compounds that physically quench ^{3}Chl and $^1O_2$ produced in LHC. Besides, $^1O_2$ can also be physically quenched by tocopherols, plastoquinones, and flavonoids (Dogra & Kim, 2019; Li & Kim, 2022; Triantaphylidès & Havaux, 2009). In chemical quenching, quenchers become oxidized by $^1O_2$, resulting in their oxidative modifications. Due to its potent electrophilicity, $^1O_2$ readily oxidizes biomolecules containing double bonds, leading to the formation of hydroperoxides or endoperoxides (Dogra, Rochaix, & Kim, 2018; Triantaphylidès & Havaux, 2009). Carotenoids, ascorbate, glutathione, vitamin B6, PUFAs, some specific amino acid residues [such as tryptophan (Trp), tyrosine (Tyr), phenylalanine (Phe), histidine (His), methionine (Met), and cysteine (Cys)], and guanine, are known as the main chemical quenchers of $^1O_2$ (Dogra & Kim, 2019; Li & Kim, 2022; Triantaphylidès & Havaux, 2009). Upon illumination or interaction with HO•, hydroperoxides generated by $^1O_2$ initiate free radical chain reactions, further aggravating the $^1O_2$-dependent cellular damages. Due to the limited diffusion distance, $^1O_2$ will primarily react with biomolecules in its vicinity, highlighting the importance of $^1O_2$ production sites in finding its targets. Under long-lasting or adverse stress conditions, the rate of $^1O_2$ generation exceeds the scavenging capacity of chloroplasts and results in an increased PSII damage and consequently decreased photosynthetic efficiency (Dogra & Kim, 2019; Hideg, Kálai, Hideg, & Vass, 1998).

2.1 β-Carotene

$^1O_2$ oxidizes β-carotene in PSII RC (Ramel, Birtic, Cuine, et al., 2012; Telfer, Dhami, Bishop, Phillips, & Barber, 1994; Trebst, 2003). For this, $^1O_2$ oxidizes each double bond of the β-carotene polyene chain and

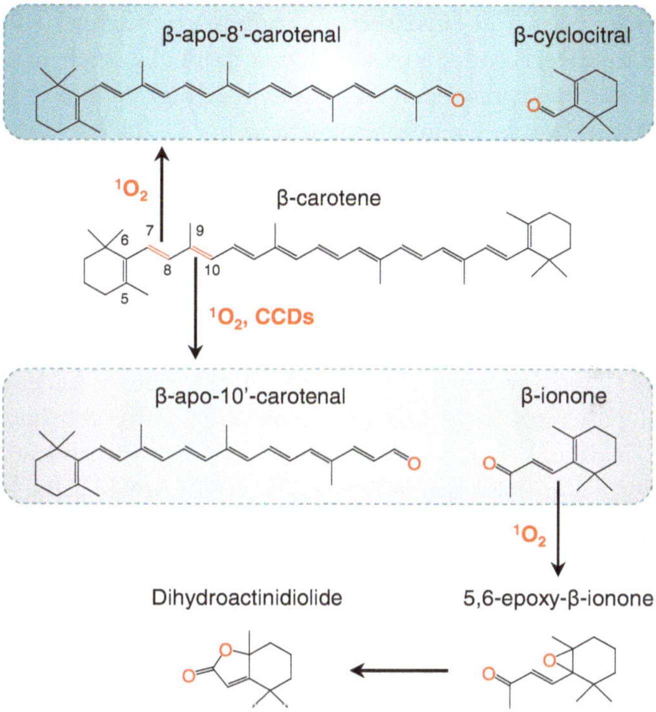

Fig. 2 Oxidative modifications of β-carotene. The oxidative cleavage of β-carotene polyene chain double bond at position 7′,8′ by $^1O_2$ produces β-apo-8′-carotenal and β-cyclocitral. The oxidative cleavage at positions 9′,10′ by either $^1O_2$ or carotenoid cleavage dioxygenases (CCDs) forms β-ionone and β-apo-10′-carotenal. β-ionone can undergo further $^1O_2$-dependent oxidation to produce 5′,6′-epoxy-β-ionone and then forms dihydroactinidiolide (dhA).

generates a variety of oxidized derivatives (D'Alessandro & Havaux, 2019; Havaux, 2014; Ramel, Birtic, Cuine, et al., 2012; Stratton, Schaefer, & Liebler, 1993) (Fig. 2). β-carotene endoperoxides, which are formed via the oxidative cleavage of polyene chain double bond at positions 7′,8′, are considered to be $^1O_2$-specific oxidation products since other oxidized carotenoid products can also be enzymatically produced by carotenoid cleavage dioxygenases (CCD) (Auldridge, McCarty, & Klee, 2006; Ramel, Birtic, Cuine, et al., 2012). Cleavage of the double bond at positions 9′,10′ generates β-ionone, which can undergo further $^1O_2$-dependent oxidation to produce lactone dihydroactinidiolide (dhA) (Havaux, 2014). Particularly, β-carotene endoperoxides, including the short-chain volatile compound β-cyclocitral (βCC), are found to accumulate in high light-exposed leaves with the concurrent turnover of β-carotene (Beisel et al., 2010;

Ramel, Birtic, Cuine, et al., 2012). In contrast to β-carotene, it is nearly impossible to detect endoperoxides of xanthophylls, such as lutein and zeaxanthin, and the turnover of xanthophylls in high light-exposed leaves, indicating the selective chemical quenching of $^1O_2$ by β-carotene (Beisel et al., 2010; Ramel, Birtic, Cuine, et al., 2012).

2.2 PSII core proteins and PSII quality control

Excess photosynthetic $^1O_2$ spontaneously oxidizes other nearby targets, especially the D1 protein in PSII RC (Nixon, Michoux, Yu, Boehm, & Komenda, 2010; Vass & Cser, 2009). Irreversible D1 damage hinders the electron transfer function of D1 and thus impairs PSII activity (Aro, Virgin, & Andersson, 1993). Other PSII proteins such as D2, CP43, and CP47 may also undergo $^1O_2$-dependent oxidations (Dogra & Kim, 2019). The light-dependent phosphorylation, which is mainly catalyzed by STATE TRANSITION KINASE 8 (STN8) on those PSII RC proteins, promotes thylakoid membrane unfolding and triggers the release of PSII monomers from the PSII mega-complexes (Bonardi et al., 2005; Goral et al., 2010; Herbstová, Tietz, Kinzel, Turkina, & Kirchhoff, 2012; Tikkanen, Nurmi, Kangasjärvi, & Aro, 2008; Vainonen, Hansson, & Vener, 2005). Re-localization of THYLAKOID FORMATION 1 (THF1) from the stroma to thylakoid membranes promotes the timely disassembly of PSII mega-complexes during dark-induced senescence and high-light acclimation in Arabidopsis (Huang et al., 2013). PSII monomers with photodamaged D1 migrate from the grana core (GC) to the non-appressed region of grana called the grana margin (GM) (Theis & Schroda, 2016). CP43 and the oxygen-evolving complex (OEC) are subsequently disassembled from PSII to generate a CP43-free PSII monomer, stimulating D1 proteolysis (Aro et al., 2005; Theis & Schroda, 2016). In the GM, the hetero-hexameric FtsH metalloprotease and lumenal Deg proteases function in the degradation of damaged D1 protein (Haussuhl, Andersson, & Adamska, 2001; Lindahl et al., 2000). Since the phosphorylation on the N-terminal threonine hinders the access of FtsH to photodamaged D1, the dephosphorylation by PP2C-like phosphatase and PP2A-like phosphatase is critical for D1 degradation (Kato & Sakamoto, 2014; Rokka, Aro, Herrmann, Andersson, & Vener, 2000; Vener, Rokka, Fulgosi, Andersson, & Herrmann, 2000). In addition, P680 molecules in PSII RC might be released to enable effective D1 degradation (Fig. 1C). Release and subsequent degradation of Chl are also considered a precondition for proteolysis of Chl-binding proteins during foliar senescence (Nickelsen & Rengstl, 2013). D1 degradation is followed by the insertion of nascent D1 from the thylakoid-bound ribosomes, perhaps together with recycled or newly synthesized chlorophyll molecules (Dogra & Kim, 2019; Kato & Sakamoto, 2009) (Fig. 1C). Under stress conditions, the impaired repair process may release free Chl molecules, which became the source of $^1O_2$ in the GM. Controlled D1 replacement and Chl insertion may minimize accidental $^1O_2$ generation in the GM. Similarly, high light induces several LHC family proteins to transiently bind Chl, which might also prevent $^1O_2$ production and PSII damage (Hutin et al., 2003; Sinha et al., 2012).

Previous studies suggested that $^1O_2$ may compromise PSII activity by damaging the D1 protein and inhibiting chloroplast protein synthesis through direct oxidation of the Cys residues of the elongation factor G (EF-G), an essential component of chloroplast protein translation (Kojima et al., 2007; Nishiyama, Allakhverdiev, & Murata, 2011). Likewise, elongation factor Tu (EF-Tu) was shown to undergo oxidation upon exposure to high-light stress, compromising PSII repair (Jimbo et al., 2018).

2.3 Polyunsaturated fatty acids

PUFAs are prone to oxidation because of their unsaturated double bond(s) (Davies, Griller, Ingold, Lindsay, & Walton, 1981; El-Aal, 2012) (Fig. 3). Accordingly, PUFAs' peroxidation increases with the number of unsaturated bonds in the lipid chain. Briefly, PUFAs (LH) can be oxidized by free radicals, such as $O_2^{\cdot-}$ and HO• (Type I), and $^1O_2$ (Type II) via non-enzymic oxidation pathways, generating LOOH (Di Mascio et al., 2019; El-Aal, 2012; Hanaa Ali Hassan Mostafa Abd, 2012; Van Dyck, 2010). Type I oxidation reactions generate 9′- and 13′-hydroperoxides, while the $^1O_2$-mediated Type II oxidations result in the formation of 9′-, 10′-, 12′-, and 13′-hydroperoxides. Thus, the 10′- and 12′- hydroperoxides are considered as markers of $^1O_2$-specific oxidation (Girotti, 1990; Terao, Hirota, Kawakatsu, & Matsushita, 1981). In Arabidopsis, around 0.1% of total foliar PUFAs are oxidized by $^1O_2$ in vivo under normal growth conditions (Triantaphylidès et al., 2008). The intermediate LOO• from Type I peroxidation can take a hydrogen ion from the neighboring LH, generating a highly reactive L• and propagating the lipid peroxidation chain reactions (Waldeck & Stocker, 1996). On the other hand, LOOH produced by $^1O_2$ via Type II oxidation can undergo decomposition to radicals (LO• and LOO•) in the presence of excited sensitizers, contributing to the Type I lipid peroxidation chain reactions (Stratton & Liebler, 1997). LOOH can further be converted into reactive carbonyl species (RCS), including 13-keto-octadecatrienoic acid (13-KOTE) and 13-keto-octadecadienoic acid (13-KODE), as shown in Fig. 3 (Mano, Biswas, &

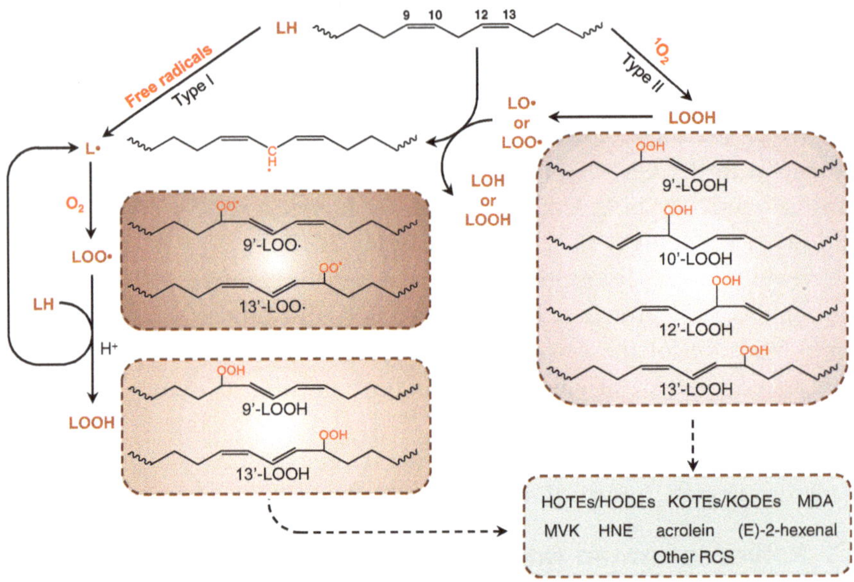

Fig. 3 Non-enzymatic oxidation of PUFAs and their derivatives. In the Type I oxidations, free radicals such as O_2^- and HO• extract the allylic hydrogen (H^+) of PUFA, forming a carbon-centered lipid radical (L•). The lipid radical rapidly reacts with O_2 to produce 9′- or 13′-lipid peroxy radicals (LOO•), which spontaneously extract an H^+ from another LH, generating a new L• along with 9′- and 13′-LOOH. The $^1O_2$-mediated Type II oxidation results in the formation of 9′-, 10′-, 12′-, and 13′-LOOH. The Type II-derived LOOH produces lipid hydroxy radical (LO•) or LOO• in the presence of excited sensitizers, contributing to the Type I lipid peroxidation chain reactions. Alternatively, LOOH can be detoxified into hydroxy fatty acid (LOH) isomers, such as 13-hydroxy octadecatrienoic acid (13-HOTE) and 13-hydroxy octadecadienoic acid (13-HODE), or further processed into various reactive carbonyl species (RCS), including 13-keto-octadecatrienoic acid (13-KOTE) and 13-keto-octadecadienoic acid (13-KODE), malonaldehyde (MDA), methyl vinyl ketone (MVK), 4-hydroxy-(E)-2-nonenal (HNE), acrolein, and (E)-2-hexenal.

Sugimoto, 2019; Mueller & Berger, 2009). RCS include a large number of electrophilic biological compounds, which contain α,β-unsaturated carbonyl groups (Farmer & Davoine, 2007; Mano et al., 2019). Since RCS can readily react with electron-rich atoms in biomolecules via nucleophilic Michael additions, they are believed to act as critical cell-damaging agents in plants (Farmer & Davoine, 2007; Liebler, 2008; Mano et al., 2019). Peroxidases and reductases actively detoxify LOOH to mitigate PUFA oxidations, resulting in the formation of hydroxy fatty acid (LOH) isomers, such as 13-hydroxy octadecatrienoic acid (13-HOTE) or 13-hydroxy octadecadienoic acid (13-HODE) (Fischer et al., 2013) (Fig. 3).

2.4 Oxidative modifications of the Trp residue

Due to their high concentration relative to other macromolecules, proteins are considered one of the primary cellular targets of $^1O_2$ in cells (Davies, 2003). $^1O_2$-protein interactions cause oxidative modification, conformational changes, misfolding, dysfunction, peptide backbone fragmentation, protein aggregation, and proteolysis (Davies, 2003; Di Mascio et al., 2019; Dogra & Kim, 2019; Li & Kim, 2022). $^1O_2$ rapidly reacts with certain amino acid residues by cycloaddition to aromatic rings and by addition to sulfur centers, which consequently produces peroxidic species (Davies, 2003; Di Mascio et al., 2019; Michaeli & Feitelson, 1994). The side chain (R group) of Trp, Try, Phe, His, Met, and Cys are prone to $^1O_2$ oxidation (Davies, 2003; Di Mascio et al., 2019; Dogra & Kim, 2019; Li & Kim, 2022; Michaeli & Feitelson, 1994). Among them, His and Trp are highly responsive to $^1O_2$ at physiological pH (Davies, 2005; Ehrenshaft, Zhao, Andley, Mason, & Roberts, 2011). While Trp can physically and chemically quench $^1O_2$, the reactions of $^1O_2$ with other amino acids occur only through chemical reactions (Matheson, Etheridge, Kratowich, & Lee, 1975; Michaeli & Feitelson, 1994). The $^1O_2$-driven oxidation of the Tyr residue delivers endoperoxide intermediates, which subsequently undergo ring-opening events of the affected indole side chain (Agon, Bubb, Wright, Hawkins, & Davies, 2006; Gracanin, Hawkins, Pattison, & Davies, 2009; Jin, Leitich, & von Sonntag, 1995).

The $^1O_2$-dependent oxidation of the Trp residue drives the formation of a dioxetane on the heterocyclic indole ring across the C2-C3 bond or a hydroperoxide at the C3 position (Davies, 2003; Gracanin et al., 2009). These intermediates are subsequently decomposed to *N*-formylkynurenine (NFK) and kynurenine via cleavage at the C2-C3 bond (Fig. 4). NFK and kynurenine were reported to function as photosensitizers, further promoting $^1O_2$ generation in response to a prolonged light exposure in vitro (Ehrenshaft et al., 2011; Gracanin et al., 2009). Since Trp residues have been implicated in various biological processes, including protein-protein interaction and anchorage of peripheral membrane proteins on the surface of the lipid bilayer, the modification might modulate Trp-linked biological processes. In photosynthetic tissues, exposing the thylakoid membrane to light led to the $^1O_2$ generation and oxidative Trp modifications in the light-harvesting complex II (LHCII) (Rinalducci et al., 2005). NFKs in the LHCII trimer were suggested to promote LHCII monomerization and turnover (Yang, Paulsen, & Andersson, 2000). Previous studies showed

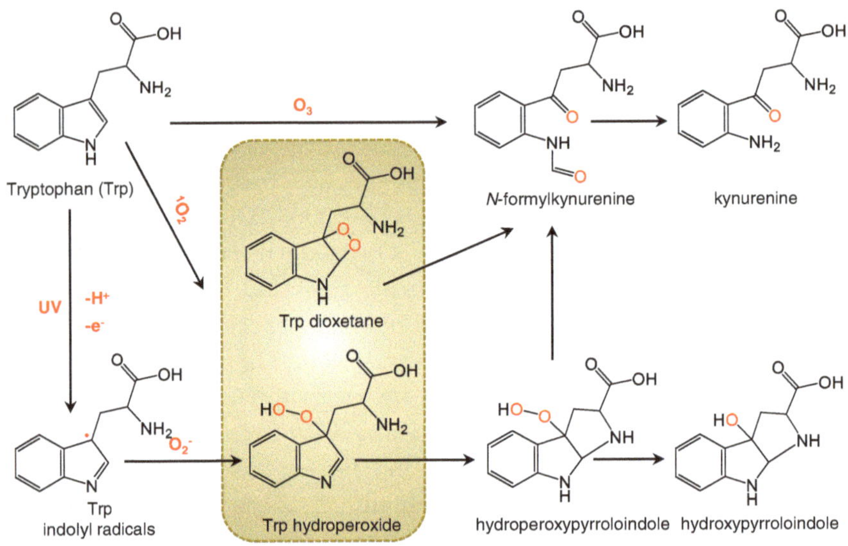

Fig. 4 Oxidative modifications of Trp. Trp can directly absorb UV light radiation and undergo photoionization to form Trp radicals, which subsequently oxidize into Trp hydroperoxide with O_2^-. The $^1O_2$-dependent oxidation of Trp produces dioxetane or hydroperoxide. Then the cyclization of Trp hydroperoxide generates hydroperoxypyrroloindole. Subsequent cleavage at the C2-C3 bond in the ring of Trp dioxetane and hydroperoxypyrroloindole forms *N*-formylkynurenine (NFK), which subsequently is decomposed to kynurenine. The reaction of Trp with ozone (O_3) also forms NFK. Hydroperoxypyrroloindole can be further reduced to hydroxypyrroloindole.

that the Trp residues in D1 and CP43 undergo NFK modifications in response to the oxidative stress (Dreaden Kasson, Rexroth, & Barry, 2012; Dreaden, Chen, Rexroth, & Barry, 2011). Notably, the levels of light-induced NFK modifications in D1 and CP43 appeared to be positively correlated with the levels of photoinhibition in PSII (Dreaden et al., 2011; Dreaden Kasson et al., 2012; Kasson & Barry, 2012). Accordingly, NFK modifications in D1 and CP43 might also activate the PSII repair cycle. It is important to note that UV, ozone, and O_2^- also lead to Trp modifications, as shown in Fig. 4 (Ehrenshaft, Deterding, & Mason, 2015). The mass spectrometry-based analysis of protein oxidation recently showed that only specific Trp residues in PSII core proteins (D1, D2, CP43, and CP47) are prone to oxidation (Dogra, Li, Singh, Li, & Kim, 2019).

3. $^1O_2$-triggered retrograde signaling pathways

In addition to cytotoxic effects caused by the high reactivity of $^1O_2$ with essential biomolecules, including lipids and proteins, $^1O_2$-triggered oxidative stress also activates specific signaling pathways priming either local and systemic acclimation or programmed cell death (PCD) in Arabidopsis (Carmody et al., 2016; Dogra & Kim, 2019; Dogra et al., 2018; Li & Kim, 2022; Mano et al., 2019). $^1O_2$ released in chloroplasts induces transcriptional changes in specific groups of nuclear genes through chloroplast-to-nucleus retrograde signaling (Dogra et al., 2018; Op den Camp et al., 2003; Triantaphylidès et al., 2008). Due to its high reactivity and short life span, $^1O_2$ is thought to trigger signaling by targeting other oxidation-prone biomolecules. These $^1O_2$ sensor molecules might reside in close vicinity to the $^1O_2$ production site. The discovery of two Arabidopsis mutants, *flu* and *chlorina 1* (*ch1*), which overproduce $^1O_2$ in chloroplasts under certain conditions, enabled scientists to reveal two $^1O_2$ sensors, β-carotene and EX proteins, located in the grana core and grana margin, respectively (Dogra et al., 2022; Ramel, Birtic, Ginies, et al., 2012; Triantaphylidès et al., 2008; Wagner et al., 2004; Wang et al., 2016).

3.1 β-Cyclocitral

The loss of CHLOROPHYLL A OXYGENASE (CAO) leads to Chl *b* deficiency in the Arabidopsis *ch1* mutant (Espineda, Linford, Devine, & Brusslan, 1999). As Chl *b* plays a critical role in assembling PSII LHCs, *ch1* mutant shows a reduced NPQ, resulting in enhanced levels of $^1O_2$ from the PSII RC (Dall'Osto et al., 2010; Triantaphylidès et al., 2008). High light impairs foliar $^1O_2$ homeostasis in *ch1* with a rapid induction of $^1O_2$-responsive genes (SORGs) in the nucleus and increased non-enzymatic lipid peroxidation in chloroplasts, leading to the jasmonic acid (JA)-dependent PCD by the activation of OXIDATIVE SIGNAL INDUCIBLE1 (OXI1) kinase (Ramel et al., 2013). Thus, the *OXI1* mutation drastically attenuates PCD in both WT and *ch1* without changing $^1O_2$ levels under high light, suggesting that OXI1 acts as an essential component of the $^1O_2$-triggered JA-dependent signaling pathway linked to high light-induced PCD (Shumbe et al., 2016). Moreover, *ch1* mutant plants pretreated with moderate light stress appear to induce a distinct group of nuclear genes, enabling them to readily acclimate to a subsequent challenge of lethal high-light intensity (Ramel et al., 2013). This acclimation response is found to be mediated by the volatile β-carotene derivative βCC (Ramel et al., 2013). $^1O_2$-caused non-enzymatic oxidation of β-carotene in PSII RC generates volatile variants, including βCC, dhA, and β-ionone, proportionally to the level of the PSII photoinhibition (Ramel, Birtic, Ginies, et al., 2012; Shumbe et al., 2016). Among them, βCC and dhA are found to induce subsets of SORGs, whose functions are implicated in either stress or detoxification (D'Alessandro & Havaux, 2019; Ramel, Birtic, Ginies, et al., 2012; Shumbe, Bott, & Havaux, 2014). βCC pretreatment confers tolerance to both WT and *ch1* plants by enhancing detoxification

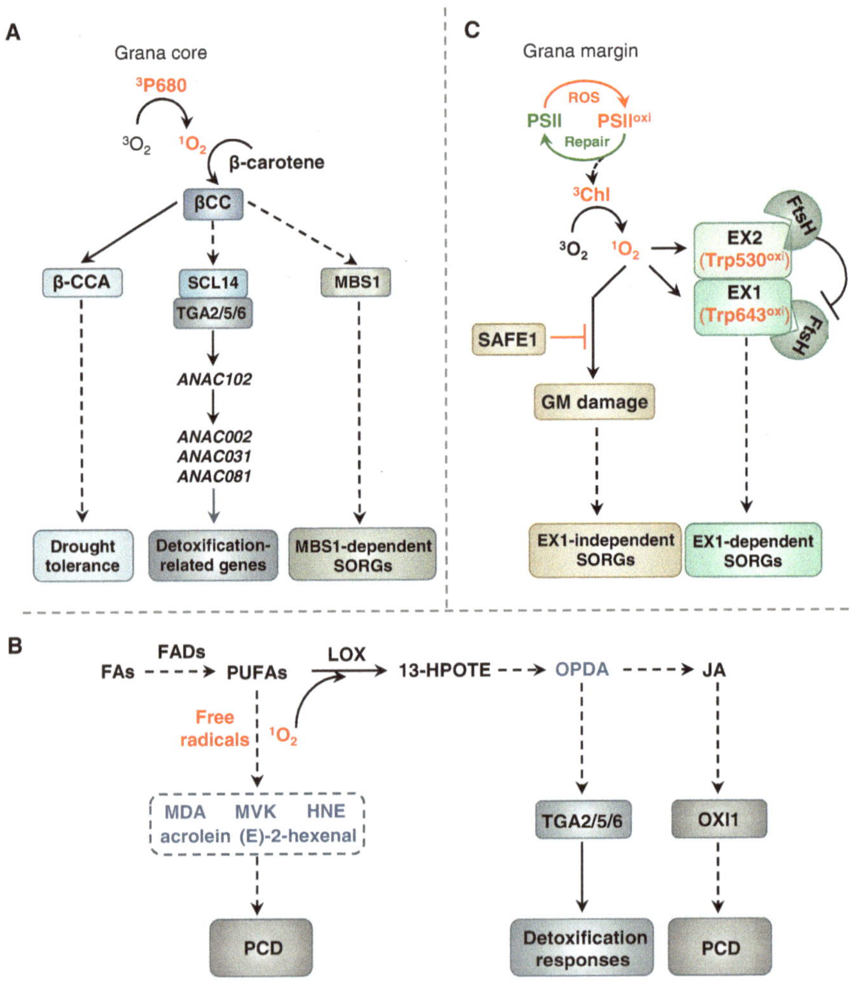

(caption on next page)

responses, thereby suppressing lipid peroxidation and PCD upon being challenged by a lethal dose of light intensity (Ramel et al., 2013; Ramel, Birtic, Ginies, et al., 2012). Released in response to excess light, βCC rapidly upregulates the GRAS protein transcription factor SCARECROW LIKE 14 (SCL14), which interacts with TGAII transcription factors, including TGA2, TGA5, and TGA6 (D'Alessandro, Ksas, & Havaux, 2018) (Fig. 5A). The SCL14-TGA II complex then promotes the expression of the *ANAC102* transcription factor and its target downstream genes, *ANAC002*, *ANAC031*, and *ANAC081*, facilitating the expression of a large set of their target genes encoding detoxifying enzymes and thereby boosting the cellular detoxification responses (D'Alessandro et al., 2018; Fode, Siemsen, Thurow, Weigel, & Gatz, 2008) (Fig. 5A). Indeed, Arabidopsis transgenic lines overexpressing *SCL14* show more tolerance to excessive light and drought stress, while the *scl14*

Fig. 5 **$^1O_2$-triggered retrograde signaling pathways. (A) βCC-mediated $^1O_2$-signaling pathways.** The excess light-induced $^1O_2$ produced by the excited triplet state of P680 (^{3}P680) in the grana core oxidizes β-carotene, generating the volatile oxidative derivate βCC. βCC induces GRAS protein SCL14 to interact with TGAII family transcription factors, including TGA2, TGA5, and TGA6. Such interaction activates downstream NAC family transcription factors (ANAC102, ANAC002, ANAC031, and ANAC081), facilitating the expression of detoxification-related genes. βCC also independently induces $^1O_2$-responsive genes (SORGs) via zinc finger protein METHYLENE BLUE SENSITIVITY 1 (MBS1). The oxidized form of βCC, β-cyclocitric acid (β-CCA), is also found to induce drought tolerance through a yet unknown mechanism. **(B) RCS-mediated $^1O_2$-signaling pathways.** FATTY ACID DESATURASEs (FADs) are responsible for the formation of PUFAs from fatty acids (FA). Under oxidative stress conditions, free radicals or $^1O_2$ can directly induce the non-enzymatic oxidation of PUFAs, giving rise to various reactive carbonyl species (RCS) such as acrolein, methyl vinyl ketone (MVK), 4-hydroxy-(E)-2-nonenal (HNE), (E)-2-hexenal, and MDA. RCS cause oxidative damage and induce PCD. On the other hand, $^1O_2$ also promotes the lipoxygenase (LOX)-dependent oxidation of PUFAs, generating 13-hydroperoxy-octadecatrienoic acid (13-HPOTE), a precursor of 12-oxo-phytodienoic acid (OPDA). OPDA also belongs to RCS and is known to activate detoxification responses through TGA2/5/6 transcription factors. OPDA can be further converted to JA, which is required to induce PCD via OXI1 kinase under high-light stress. RCS are indicated by blue. **(C) EXs- and SAFE1-mediated $^1O_2$-signaling pathways.** In the grana margin, $^1O_2$ generated by free Chl released during the PSII repair or by free tetrapyrroles can oxidize EX1-Trp643 residue (Trp643^{oxi}), leading to FtsH-dependent EX1 proteolysis and eliciting retrograde signaling to induce the expression of EX1-dependent SORGs. The EX1 paralog EX2 also undergoes $^1O_2$-dependent Trp530 oxidation (Trp530^{oxi}) and FtsH-dependent turnover, which attenuates $^1O_2$ signaling by decelerating EX1-Trp643 oxidation and subsequent EX1 degradation. Another chloroplast protein, SAFEGUARD 1 (SAFE1), protects grana margin proteins from oxidative damages caused by $^1O_2$. The loss of SAFE1 induces the expression of EX1-independent SORGs.

mutant becomes more susceptible, unable to respond to the βCC pretreatment (D'Alessandro et al., 2018; D'Alessandro, Mizokami, Legeret, & Havaux, 2019). βCC also induces the expression and the cytosol-to-nucleus re-localization of the small zinc finger protein METHYLENE BLUE SENSITIVITY 1 (MBS1), required for the full expression of SORGs under both normal and light stress conditions (Shao, Duan, & Bock, 2013; Shumbe et al., 2017) (Fig. 5A). Arabidopsis *mbs1* is also insensitive to βCC, thus susceptible to light stress (Shumbe et al., 2017). However, the βCC-induced detoxification responses remain unchanged in *mbs1*, but the mode of action of MBS1 remains unclear (D'Alessandro et al., 2018). Besides, the water-soluble form of βCC, β-cyclocitric acid (β-CCA), also appears to act as a signaling molecule, triggering nuclear gene expression changes and conferring drought tolerance in Arabidopsis (D'Alessandro et al., 2019) (Fig. 5A). The mechanism underlying β-CCA-induced drought tolerance remains unknown, but it seems to be independent of MBS1 and JA signaling, suggesting that several branches of βCC signaling are responsive to various environmental stress (D'Alessandro et al., 2019).

3.2 Reactive carbonyl species

Some RCS produced by $^1O_2$-dependent lipid peroxidation can also act as putative transmitters of $^1O_2$ signals. Direct oxidation by $^1O_2$ or the LOX-dependent enzymatic oxidation of PUFAs leads to RCS formation, including 12-oxo-phytodienoic acid (OPDA), phytoprostanes, aldehydes, and ketones (Blée, 1998; Flors et al., 2006; Matsui, 2006; Schaller, 2001). In Arabidopsis, LOX-dependent lipid oxidation can be induced by $^1O_2$ through the rapid transcriptional accumulation of two chloroplast 13-LOXs enzymes, LOX2 and LOX3 (Mochizuki, Sugimoto, Koeduka, & Matsui, 2016; Viswanath et al., 2020). The enzymatic oxidation product 13-hydroperoxy-octadecatrienoic acid (13-HPOTE) is the precursor of 12-oxo-phytodienoic acid (OPDA), which subsequently converts to JA in the peroxisome (Glauser et al., 2009; Wasternack & Hause, 2013) (Fig. 5B). As mentioned above, JA induces PCD by OXI1 kinase in the *ch1* mutant under high-light stress conditions (Ramel et al., 2013). 13-HPOTE can also be reduced to 13-hydroxy octadecatrienoic acid (13-HOTE) or converted to keto-forms, such as 13-KOTE (Fischer et al., 2013; Mueller & Berger, 2009). While OPDA and 13-KOTE are RCS (electrophilic), JA does not belong to RCS due to the saturated ketone group. OPDA induces a set of SORGs, distinct from those induced by JA in *ch1* in response to high light (Gutierrez, Gonzalez-Perez, Garcia-Garcia, Lorenzo, & Arellano, 2011; Taki et al., 2005). Like βCC, OPDA-mediated signaling activates detoxification and defense responses through TGA2/5/6 transcription factors (Gutierrez et al., 2011; Stotz, Mueller, Zoeller, Mueller, & Berger, 2013; Taki et al., 2005) (Fig. 5B). In Chlamydomonas, excess light also increases certain RCS, including 2-propenal (also called acrolein), hexenal, and malonaldehyde (MDA) (Roach, Baur, Stöggl, & Krieger-Liszkay, 2017). Pretreating Chlamydomonas cells with a sub-lethal dose of $^1O_2$ confers an acclimation by a substantial upregulation of detoxification-related genes, such as *Glutathione Peroxidase 5* (*GPX5*/*GPXH*) and *Glutathione-S Transferase 1* (*GSTS1*) (Ledford, Chin, & Niyogi, 2007). Interestingly, this acclimation seems independent of βCC accumulation, since the βCC level was not increased in light-stressed Chlamydomonas (Roach et al., 2017). The bZIP transcription factor Singlet Oxygen Resistant 1 (SOR1) drives the expression of many oxidative stress-responsive and detoxification-related genes, including *GPXH* and *GSTS1* in Chlamydomonas (Fischer et al., 2012). All these findings suggest an essential role of RCS in $^1O_2$ signaling and acclimation responses.

Some RCS cause cellular toxicity and trigger PCD in plants (Biswas & Mano, 2015; Kai et al., 2012; Li et al., 2020) (Fig. 5B). Exogenous treatment with PUFA oxidation-derived RCS (non-enzymatic oxidation of PUFAs), such as acrolein, methyl vinyl ketone (MVK), 4-hydroxy-(E)-2-nonenal (HNE), (E)-2-hexenal, and MDA, is known to reduce PSII fluorescence, inhibit photosynthesis, inactivate multiple enzymes in the Calvin-Benson cycle, and cause oxidative damage on leaf tissue (Alméras et al., 2003; Mano, Miyatake, Hiraoka, & Tamoi, 2009). Treatment with low concentrations (0.2 mM) of acrolein and HNE causes PCD in tobacco Bright Yellow-2 (BY-2) cells and in roots of both tobacco and Arabidopsis plants (Biswas & Mano, 2015). Interestingly, the reduced level of C18:3 PUFAs, which are the targets of ROS, inhibits oxidative damage and enhances thermotolerance in *Cyclamen persicum* (Kai et al., 2012). Genetically modified fatty acid composition through the inactivation of the plastid-localized ω-3 FATTY ACID DESATURASE (FAD) 7 leads to a specific decrease in C18:3 PUFAs in Cyclamen. In contrast to WT plants, which showed elevated accumulation of acrolein, MVK, and (E)-2-hexenal with foliar lesions on leaves, the *FAD7* knock-down lines exhibited no apparent upregulation of those RCS and reduced foliar lesions under heat stress. Similarly, the compromised production of C16:1 (or C16:3) suppresses $^1O_2$-dependent PCD in Arabidopsis (Li et al., 2020). The Arabidopsis mutant *crumpled leaf* (*crl*) deficient in chloroplast protein import continually overproduces $^1O_2$ in chloroplasts with a concurrent upregulation of SORGs and the emergence of lesions under normal growth conditions (Asano, Yoshioka, Kurei, Sakamoto, & Machida, 2004; Li et al., 2020; Šimková et al., 2012; Wang, Fang, et al., 2020). *crl*-induced foliar cell death seems to be induced by PUFA oxidation-derived RCS signaling but is independent of neither stress hormones, such as salicylic acid (SA) and JA, nor EXECUTER 1/ EXECUTER 2 (EX) proteins (Li et al., 2020). Genetic inactivation of chloroplast-localized FATTY ACID DESATURASE 5 (FAD5), catalyzing the conversion of palmitic acid (16:0) to palmitoleic acid (16:1), decreased both C16:1 and C16:3 PUFA levels in *crl fad5* with a complete loss of cell death. The loss of FAD5 might presumably attenuate RCS levels in *crl fad5*, thereby impairing the expression of SORGs and abrogating cell death in *crl*. These studies strongly suggest that RCS-mediated signaling can induce PCD, but the potential role of RCS-associated signaling molecules and the downstream components regulating transcriptional changes need to be further investigated.

3.3 Executer1, Executer2, and Safeguard1

Unlike *ch1* chloroplasts, which generate $^1O_2$ in the PSII RC under high light, chloroplasts in *flu* generate $^1O_2$ in thylakoid membranes after a dark-to-light transition (Meskauskiene et al., 2001; Ramel et al., 2013; Wang et al., 2016). The nuclear-encoded chloroplast protein FLU negatively regulates Chl biosynthesis in the dark, preventing the accumulation of free Pchlide, a potent photosensitizer. This impaired regulation by *flu* results in the accumulation of free Pchlide in the dark because the subsequent reduction step from Pchlide to Chlide requires light and light-dependent POR enzymes (Meskauskiene et al., 2001) (Fig. 1B). Free Pchlide rapidly generates $^1O_2$ upon illumination by transferring the absorbed light energy to $^3O_2$ (Redmond & Kochevar, 2006), the reason for the tight regulation of Pchlide levels in the dark. In the GM, FLU forms a complex with Mg^{2+}-protoporphyrin IX monomethyl ester cyclase CFH27 and POR enzymes (Kauss, Bischof, Steiner, Apel, & Meskauskiene, 2012; Wang et al., 2016). Upon binding free Pchlide to POR enzymes, the FLU-CHL27-POR-Pchlide tetrameric complex physically interacts with and inhibits the glutamyl-tRNA reductase (GluTR) (Hou, Pang, Hedtke, & Grimm, 2021; Kauss et al., 2012). Since GluTR catalyzes the formation of 5-Aminolevulinic acid (ALA), the universal precursor of tetrapyrroles (Czarnecki & Grimm, 2012), FLU complex-GluTR interaction efficiently represses Pchlide accumulation in the dark. Therefore in *flu,* Pchlide over-accumulates in the dark and immediately produces $^1O_2$ upon re-illumination, resulting in the upregulation of SORGs and lipid peroxidation that promotes the synthesis of oxylipins such as JA and OPDA along with the other hormones SA and ethylene (Danon, Miersch, Felix, Opden Camp, & Apel, 2005; Ochsenbein et al., 2006; Op den Camp et al., 2003). This consequently leads to cell death in seedlings and growth inhibition in mature plants (Op den Camp et al., 2003; Wagner et al., 2004). The *flu* mutant is indistinguishable from WT plants when grown under continuous light conditions, classifying *flu* as a conditional mutant generating $^1O_2$ upon a dark-to-light shift. Besides, $^1O_2$ levels increase proportionally to the length of the dark period due to the correlated accumulation of Pchlide (Meskauskiene et al., 2001; Op den Camp et al., 2003). Therefore, *flu* mutants have been used as a bio-tool for investigating the $^1O_2$ dose-dependency in $^1O_2$ signaling and related physiological responses, chloroplast damage, and stress responses in a noninvasive manner (Meskauskiene et al., 2001; Op den Camp et al., 2003; Rentel et al., 2004). Exploring this conditional mutant has enabled scientists to find the nuclear-encoded chloroplast $^1O_2$ sensor EX1 protein, which mediates $^1O_2$ signaling in *flu* (Dogra, Li, et al., 2019; Lee et al., 2007; Wagner et al., 2004) (Fig. 5C). Inactivation of EX1 thus considerably abolishes the $^1O_2$-induced stress responses in *flu* (Lee et al., 2007; Wagner et al., 2004). Similarly, EX2, a paralog of EX1, also serves as a modulator of EX1-mediated $^1O_2$ signaling in both WT and the *flu* mutant (Dogra et al., 2022; Lee et al., 2007). Treating WT seedlings with norflurazon (NF), a carotenoid biosynthesis inhibitor, also enhances $^1O_2$ production due to the increased photosensitizing activity of tetrapyrroles, lipid peroxidation, and cell death in cotyledons and emerging true leaves, whereas those responses are largely suppressed in *ex1 ex2* seedlings (Kim & Apel, 2013). EX1 and EX2 also participate in the systemic expression of SORGs by modulating apoplastic ROS signaling involved in intercellular communication, conferring a systemic acclimation to high-light stress (Carmody et al., 2016). However, nearly no impact of EX1 and EX2 proteins in the βCC-induced chloroplast signaling and stress responses under high light is monitored (Ramel et al., 2013), indicating the existence of two distinct $^1O_2$-signaling pathways mediated by EX1/EX2 proteins and βCC in chloroplasts. The finding that EX1 and EX2 exclusively reside in GM would explain the spatial separation of these two signaling pathways. Note that EX1 and EX2 reside away from the primary site of $^1O_2$ generation, i.e., the PSII RC in GC where the volatile βCC is released (Dogra et al., 2018, 2022; Wang et al., 2016). Due to the short lifespan, $^1O_2$ generated at PSII in the GC would not reach the GM. Therefore, GM can be an alternative site for

$^1O_2$ generation, where EX1 and EX2 mediate $^1O_2$ signaling (Fig. 5C). EX1 and EX2 interact with proteins required for PSII quality control and reassembly, including PSII core proteins, Chl synthesis enzymes (especially POR enzymes), the PSII quality control-associated FtsH protease, and a protein elongation factor (Dogra et al., 2022; Dogra, Li, et al., 2019; Wang et al., 2016). Thus, EX1 is thought to sense $^1O_2$ generated by free Chl molecules or intermediate tetrapyrroles released accidentally during the PSII repair process. In line with this, one latest finding showed that chlorophyll catabolic enzymes are enriched in GM to ensure chlorophyll breakdown (Fukura, Tanaka, Tanaka, & Ito, 2021). Such a system may involve $^1O_2$ homeostasis in GM.

Recent works partly elucidated a working model of the EX1/EX2-mediated $^1O_2$ signaling (Dogra et al., 2022; Dogra, Li, et al., 2019; Wang et al., 2016). In *flu*, the conditional release of $^1O_2$ in chloroplasts results in the light-dependent oxidative post-translational modification (Oxi-PTM) of a tryptophan residue (Trp643) located at the C-terminal $^1O_2$ sensor domain (previously annotated as the domain of unknown function 3506, DUF3506) of EX1, as shown in Fig. 5C (Dogra, Li, et al., 2019). Trp643 modification promotes EX1 proteolysis by the membrane-bound FtsH protease implicated in PSII repair (Dogra, Li, et al., 2019; Wang et al., 2016). Either the replacement of Trp643 with $^1O_2$-insensitive but hydrophobic residues, such as alanine (Ala) and leucine (Leu), or inactivation of the major subunit of the FtsH protease, FtsH2, completely abrogates $^1O_2$-triggered stress responses in the *flu* mutant (Dogra, Li, et al., 2019). These results indicate that EX1 acts as a $^1O_2$ sensor in the GM and that EX1 proteolysis may promote the release of an actual signaling molecule to elicit $^1O_2$ responses. Dogra et al. (2022) further demonstrated that Trp530 oxidation in EX2 modulates EX1-mediated $^1O_2$ signaling. EX2-Trp530 oxidation mitigates the level of EX1-Trp643 oxidation, thereby attenuating $^1O_2$ signaling with increased EX1 stability (Fig. 5C). Consistent with this result, *EX2* knockout reinforces EX1-dependent $^1O_2$ signaling with enhanced EX1-Trp643 oxidation and EX1 proteolysis, coinciding with the increased expression of SORGs and cell death in *ex2 flu* versus *flu*. Moreover, *EX2* overexpression confers a greater tolerance to photoinhibitory conditions that aggravate PSII photoinhibition and the $^1O_2$ burst, while the loss of EX2 renders seedlings more susceptible. A phylogenetic study provides an important clue that EX2 emerged in land plants through genome duplication of EX1, suggesting a safeguarding role of EX2 towards the $^1O_2$ sensitivity of EX1. This paralog of EX1 may finely modulate $^1O_2$-triggered EX1-dependent cell death (Tano & Woodson, 2022).

A recent forward genetic study aiming to find EX1/EX2-independent $^1O_2$-signaling components unveiled an intriguing chloroplast protein, dubbed SAFEGUARD 1 (SAFE1) (Wang, Leister, et al., 2020) (Fig. 5C). SAFE1 localizes in the stroma and undergoes degradation upon $^1O_2$ exposure. The inactivation of SAFE1 in *flu ex1* reactivates cell death, indicating that SAFE1 is required to inhibit EX1-independent $^1O_2$ signaling. It was further suggested that SAFE1 might protect grana margin-associated proteins from oxidative damage and degradation in response to $^1O_2$ (Fig. 5C), as evident by the augmented turnover of GM proteins such as POR, CURVATURE THYLAKOID 1 (CURT1), Mg^{2+}-CHELATASE SUBUNIT I (CHLI), and Mg^{2+}-PROTOPORPHYRIN IX METHYL TRANSFERASE (CHLM) in *flu ex1 safe1* after the release of $^1O_2$. Moreover, more plastoglobules (oxidative stress markers) are observed in *flu ex1 safe1* compared with *flu* and *flu ex1* upon the release of $^1O_2$ (Wang, Leister, et al., 2020). However, the underlying mechanism of the SAFE1-dependent pathway in antagonizing $^1O_2$-signaling remains to be elucidated.

In *flu*, a $^1O_2$ burst results in chloroplast leakage, manifested by the release of stromal proteins to the cytosol, followed by vacuole rupture and cell death (Kim et al., 2012). This leakage and related phenotypes were abrogated by the *EX1* mutation, indicating an EX1 dependency. Similar phenotypes were also observed in the *ferrochelatase 2* (*fc2*) mutant owing to accumulated free protoporphyrin IX (Proto IX), as FC2 catalyzes the formation of heme by inserting ferrous ion (Fe^{2+}) into Proto IX (Scharfenberg et al., 2015; Woodson et al., 2015) (Fig. 1B). Like free Pchlide, free Proto IX also acts as a photosensitizer generating $^1O_2$ after absorption of light energy (Kennedy & Pottier, 1992) (Fig. 1B). Indeed, similar to cell death phenotypes observed in *flu*, the *fc2* mutant exhibits rapid chloroplast leakage and vacuole rupture upon release of $^1O_2$ via accumulated free Proto IX. However, despite the comparable cellular phenotypes, the subsequent genetic study found that EX1 is dispensable for the cell death response in *fc2* (Woodson et al., 2015), revealing an additional $^1O_2$-signaling pathway. A plant U-box 4 (PUB4) E3 ubiquitin ligase appears to ubiquitinate chloroplast envelope proteins and promotes chloroplast degradation upon $^1O_2$ accumulation in *fc2* (Woodson et al., 2015). Given the dual localization of FC2 in both envelope and thylakoid membranes (Roper & Smith, 1997), it is reasonable to suggest that $^1O_2$ produced in the envelope membrane rather than in the thylakoid membrane may induce EX1-independent cell death responses. Further exploration of the precise sub-plastidial localization of FC2 and of the spatial accumulation of free Proto IX may provide a clue for deciphering how the $^1O_2$ signaling pathway induces cell death.

3.4 Chloroplast UPR and $^1O_2$ signaling

An earlier study revealed that the endoplasmic reticulum (ER)-mediated unfolded protein response (erUPR) is another crucial player in the $^1O_2$-triggered PCD (Beaugelin, Chevalier, D'Alessandro, Ksas, & Havaux, 2020). erUPR is a stress response activated upon the accumulation of aggregates of unfolded and misfolded proteins in the ER lumen (called ER stress), enabling the restoration of ER functions while eliciting PCD if the adaptation is unsuccessful (Read & Schröder, 2021). Interestingly, $^1O_2$ generation in the *ch1* mutant under moderate light or excess light conditions appears to activate erUPR by inducing the expression of nucleus-encoded genes, such as *bZIP60*, *bZIP28*, and *bZIP17*, involved in two erUPR branches (Beaugelin et al., 2020). erUPR modulates plant resistance to high-light stress to varying degrees. For instance, moderate induction of erUPR genes (including *bZIP60*) enhances tolerance to photooxidative stress, whereas overexpression of erUPR genes leads to cell death in *ch1* under high-light stress. Notably, some erUPR-related genes induced in *ch1* under high light are also responsive to $^1O_2$ generated in *flu* after a dark-to-light transition (Op den Camp et al., 2003). It was proposed that the chloroplast might directly contact ER to relay the $^1O_2$ signaling. For example, chloroplasts can be associated with ER through stromules, i.e., tubular extensions of the chloroplast, for inter-organellar communication or by the exchange of metabolites (Brunkard, Runkel, & Zambryski, 2015). The ER also has specific membrane extensions to connect to the chloroplast (Andersson, Goksör, & Sandelius, 2007). However, signaling molecules involved in ER-chloroplast communication remain to be identified. Besides, erURP is known to be regulated by SA in plants (Poor, Czekus, Tari, & Ordog, 2019). Thus, the potential involvement of SA in $^1O_2$-triggered erUPR is worthy of further investigation.

Similarly to erUPR, dysfunctional chloroplasts resulting from impaired chloroplast proteostasis also induce a chloroplast unfolded/misfolded/damaged protein response (cpUPR) by upregulating a suite of nuclear genes involved in protein quality control to compensate for the chloroplast defect (Dogra, Duan, Lee, & Kim, 2019; Kessler & Longoni, 2019). For instance, impaired PSII quality control in the Arabidopsis *yellow variegated 2* (*var2*) mutant, lacking the FtsH2 metalloprotease, upregulates a group of nuclear genes involved in protein quality control and ROS detoxification in chloroplasts, compensating the *var2*-induced deficiency in proteostasis (Dogra, Duan, et al., 2019).

Likewise, Chlamydomonas and Arabidopsis mutants with reduced stromal Clp protease activity are also found to induce cpUPR (Llamas, Pulido, & Rodriguez-Concepcion, 2017; Ramundo et al., 2014). Moreover, an enhanced SA level and related signaling were observed in *var2*, linking impaired proteostasis or increased photodamage in chloroplasts to the activation of the SA signaling (Dogra, Duan, et al., 2019).

4. Concluding remarks

Photosynthetic $^1O_2$ is produced within thylakoid membranes, whereas lipid peroxyl radicals, excited-state carbonyls generated by the LOX-dependent peroxidation of PUFAs, and phytoalexins produce non-photosynthetic $^1O_2$. In any case, $^1O_2$ tends to oxidize unsaturated double bond-containing biomolecules, including carotenoids, specific amino acid residues in proteins, and PUFAs, exacerbating oxidative stress in plants. However, the characterization of Arabidopsis mutants, which conditionally produce $^1O_2$ in chloroplasts, has changed the orthodox view of $^1O_2$ from a toxin to a signaling-involved molecule. Such a signaling role of chloroplast-derived $^1O_2$ appears to prime acclimation or cell death, primarily controlled by $^1O_2$ levels (from low to high) and production sites (GC or GM). The latter case also enabled scientists to find different $^1O_2$ sensors, β-carotene in GC and EX1/EX2 proteins in GM. The $^1O_2$-triggered oxidation of β-carotene and EX1/EX2 proteins is integral in initiating specific retrograde signaling pathways. The water-soluble apocarotenoid β-cyclocitric acid also triggers signaling and confers drought tolerance in Arabidopsis. Although initial $^1O_2$ perception mechanisms are largely unveiled, the downstream signaling components of these signaling cascades remain to be revealed.

Finding receptor components of βCC and β-cyclocitric, for instance, will significantly advance our understanding of the β-carotene-dependent signaling. Likewise, it is yet unclear how the oxidation of EX1/EX2 and its subsequent degradation induces retrograde signaling. Uncovering a genuine signaling molecule associated with EX1/EX2 proteins would be the next goal for further dissecting the EX1/EX2 signaling pathway. As shown for EX1/EX2 proteins, recent studies also revealed Trp oxidation in other PSII core proteins (Dogra, Li, et al., 2019; Duan et al., 2019). Since EX1/EX2 undergoes $^1O_2$-triggered turnover by FtsH protease implicated in PSII repair, one can also suggest that PSII oxidation would initiate PSII quality control, including their turnover by FtsH protease. It is also possible that Trp oxidation in membrane proteins may act as a degron signal through the conformation change of targeted proteins. Trp oxidation in different proteins might trigger distinct signaling independently of β-carotene or EX1/EX2 proteins. Indeed, as discussed above, recent studies revealed more $^1O_2$ signaling pathways through forward and reverse genetic approaches. In this regard, Oxi-PTM of Trp residues in photosynthesis-associated proteins opens exciting research avenues for investigating the biological significance of Trp oxidation in plant oxidative stress responses and possible engineering of photosynthetic apparatus to gain resilience against $^1O_2$.

Various RCS derived from the peroxidation of PUFAs seem to participate in plant stress responses in an $^1O_2$ dose-dependent manner. Medium production of $^1O_2$ tends to trigger the LOX-dependent enzymatic peroxidation of PUFAs, generating RCS including OPDA to activate detoxification responses or PCD, while high levels of $^1O_2$ induce non-enzymatic peroxidation of PUFAs, resulting in oxidative damage and cell death. Considering that LOX-dependent peroxidation of PUFAs could be an alternative source of $^1O_2$, non-photosynthetic $^1O_2$ signaling can be activated via PUFA oxidation. Especially $^1O_2$ signaling in the absence of light would be an intriguing next challenge to investigate. Identifying bioactive RCS molecules is also interesting to link their activity to specific physiological responses. Besides, the chloroplast $^1O_2$ generation is always accompanied by the rapid accumulation of defense phytohormones, such as SA and JA. Indeed, some $^1O_2$-induced genes encode proteins involved in defense hormone-dependent signaling pathways, indicating a potential crosstalk among $^1O_2$-, RCS-, and defense hormone-mediated signaling in plant stress responses.

Although tremendous efforts have been made to elucidate $^1O_2$ signaling pathways in plants over the past decade, our current understanding remains largely fragmented. Future research aimed at elucidating the source and generation mechanisms of $^1O_2$, signaling molecules, and receptors would significantly fill the gaps still present in these signaling pathways. The knowledge obtained using model plants must be explored in other essential crops to produce sustainable crops and maintain food security against upcoming environmental changes.

Acknowledgments

We apologize to all authors whose work could not be cited here due to space constraints. This work was supported by the Strategic Priority Research Program from the Chinese Academy of Sciences (Grant No. XDB27040102), the 100-Talent Program of the Chinese Academy of Sciences, and the National Natural Science Foundation of China (NSFC) (Grant No. 31871397) to C.K.

References

Adamska, I., & Kloppstech, K. (1994). Low temperature increases the abundance of early light-inducible transcript under light stress conditions. *The Journal of Biological Chemistry, 269*, 30221–30226.

Adamska, I., Kloppstech, K., & Ohad, I. (1992). UV light stress induces the synthesis of the early light-inducible protein and prevents its degradation. *The Journal of Biological Chemistry, 267*, 24732–24737.

Adamska, I., Ohad, I., & Kloppstech, K. (1992). Synthesis of the early light-inducible protein is controlled by blue light and related to light stress. *Proceedings of the National Academy of Sciences, 89*, 2610–2613.

Adir, N., Zer, H., Shochat, S., & Ohad, I. (2003). Photoinhibition – A historical perspective. *Photosynthesis Research, 76*, 343–370.

Agon, V. V., Bubb, W. A., Wright, A., Hawkins, C. L., & Davies, M. J. (2006). Sensitizer-mediated photooxidation of histidine residues: Evidence for the formation of reactive side-chain peroxides. *Free Radical Biology and Medicine, 40*, 698–710.

Aldred, E. M., Buck, C., & Vall, K. (2009). Chapter 7 – Free radicals. In E. M. Aldred, C. Buck, & K. Vall (Eds.). *Pharmacology* (pp. 41–52). Edinburgh: Churchill Livingstone.

Alméras, E., Stolz, S., Vollenweider, S., Reymond, P., Mène-Saffrané, L., & Farmer, E. E. (2003). Reactive electrophile species activate defense gene expression in Arabidopsis. *The Plant Journal, 34*, 205–216.

Andersson, M. X., Goksör, M., & Sandelius, A. S. (2007). Optical manipulation reveals strong attracting forces at membrane contact sites between endoplasmic reticulum and chloroplasts. *The Journal of Biological Chemistry, 282*, 1170–1174.

Apel, K., & Hirt, H. (2004). Reactive oxygen species: Metabolism, oxidative stress, and signal transduction. *Annual Review of Plant Biology, 55*, 373–399.

Aro, E. M., Suorsa, M., Rokka, A., Allahverdiyeva, Y., Paakkarinen, V., Saleem, A., ... Rintamaki, E. (2005). Dynamics of photosystem II: A proteomic approach to thylakoid protein complexes. *Journal of Experimental Botany, 56*, 347–356.

Aro, E. M., Virgin, I., & Andersson, B. (1993). Photoinhibition of Photosystem II. Inactivation, protein damage and turnover. *Biochimica et Biophysica Acta, 1143*, 113–134.

Asano, T., Yoshioka, Y., Kurei, S., Sakamoto, W., & Machida, Y. (2004). A mutation of the CRUMPLED LEAF gene that encodes a protein localized in the outer envelope membrane of plastids affects the pattern of cell division, cell differentiation, and plastid division in Arabidopsis. *The Plant Journal: For Cell and Molecular Biology, 38*, 448–459.

Auldridge, M. E., McCarty, D. R., & Klee, H. J. (2006). Plant carotenoid cleavage oxygenases and their apocarotenoid products. *Current Opinion in Plant Biology, 9*, 315–321.

Bassi, R., & Caffarri, S. (2000). Lhc proteins and the regulation of photosynthetic light harvesting function by xanthophylls. *Photosynthesis Research, 64*, 243–256.

Beaugelin, I., Chevalier, A., D'Alessandro, S., Ksas, B., & Havaux, M. (2020). Endoplasmic reticulum-mediated unfolded protein response is an integral part of singlet oxygen signalling in plants. *The Plant Journal: For Cell and Molecular Biology, 102*, 1266–1280.

Becerril, J. M., & Duke, S. O. (1989). Protoporphyrin IX content correlates with activity of photobleaching herbicides. *Plant Physiology, 90*, 1175–1181.

Beisel, K. G., Jahnke, S., Hofmann, D., Köppchen, S., Schurr, U., & Matsubara, S. (2010). Continuous turnover of carotenes and chlorophyll a in mature leaves of arabidopsis revealed by $^{14}CO_2$ pulse-chase labeling. *Plant Physiology, 152*, 2188–2199.

Biswas, M. S., & Mano, Ji. (2015). Lipid peroxide-derived short-chain carbonyls mediate hydrogen peroxide-induced and salt-induced programmed cell death in plants. *Plant Physiology, 168*, 885–898.

Blée, E. (1998). Phytooxylipins and plant defense reactions. *Progress in Lipid Research, 37*, 33–72.

Bonardi, V., Pesaresi, P., Becker, T., Schleiff, E., Wagner, R., Pfannschmidt, T., ... Leister, D. (2005). Photosystem II core phosphorylation and photosynthetic acclimation require two different protein kinases. *Nature, 437*, 1179–1182.
Braslavsky, S., & Holzwarth, A. (2012). Role of carotenoids in photosystem II (PSII) reaction centers. *International Journal of Thermophysics, 33*.
Briantais, J. M., Vernotte, C., Picaud, M., & Krause, G. H. (1979). A quantitative study of the slow decline of chlorophyll a fluorescence in isolated chloroplasts. *Biochimica et Biophysica Acta, 548*, 128–138.
Brunkard, J. O., Runkel, A. M., & Zambryski, P. C. (2015). Chloroplasts extend stromules independently and in response to internal redox signals. *Proceedings of the National Academy of Sciences of the United States of America, 112*, 10044–10049.
Carmody, M., Crisp, P. A., d'Alessandro, S., Ganguly, D., Gordon, M., Havaux, M., ... Pogson, B. J. (2016). Uncoupling high light responses from singlet oxygen retrograde signaling and spatial-temporal systemic acquired acclimation. *Plant Physiology, 171*, 1734–1749.
Cazzaniga, S., Li, Z., Niyogi, K. K., Bassi, R., & Dall'Osto, L. (2012). The Arabidopsis szl1 mutant reveals a critical role of beta-carotene in photosystem I photoprotection. *Plant Physiology, 159*, 1745–1758.
Chen, T., & Fluhr, R. (2018). Singlet oxygen plays an essential role in the root's response to osmotic stress. *Plant Physiology, 177*, 1717–1727.
Chen, W. L., Xing, D., Tan, S., Tang, Y., & He, Y. (2003). Imaging of ultra-weak biochemiluminescence and singlet oxygen generation in germinating soybean in response to wounding. *Luminescence: The Journal of Biological and Chemical Luminescence, 18*, 37–41.
Cilento, G., & Adam, W. (1995). From free radicals to electronically excited species. *Free Radical Biology and Medicine, 19*, 103–114.
Czarnecki, O., & Grimm, B. (2012). Post-translational control of tetrapyrrole biosynthesis in plants, algae, and cyanobacteria. *Journal of Experimental Botany, 63*, 1675–1687.
D'Alessandro, S., & Havaux, M. (2019). Sensing beta-carotene oxidation in photosystem II to master plant stress tolerance. *The New Phytologist, 223*, 1776–1783.
D'Alessandro, S., Ksas, B., & Havaux, M. (2018). Decoding β-cyclocitral-mediated retrograde signaling reveals the role of a detoxification response in plant tolerance to photooxidative stress. *The Plant Cell, 30*, 2495–2511.
D'Alessandro, S., Mizokami, Y., Legeret, B., & Havaux, M. (2019). The apocarotenoid beta-cyclocitric acid elicits drought tolerance in plants. *iScience, 19*, 461–473.
Dall'Osto, L., Cazzaniga, S., Havaux, M., & Bassi, R. (2010). Enhanced photoprotection by protein-bound vs free xanthophyll pools: A comparative analysis of chlorophyll b and xanthophyll biosynthesis mutants. *Molecular Plant, 3*, 576–593.
Dall'Osto, L., Holt, N. E., Kaligotla, S., Fuciman, M., Cazzaniga, S., Carbonera, D., ... Bassi, R. (2012). Zeaxanthin protects plant photosynthesis by modulating chlorophyll triplet yield in specific light-harvesting antenna subunits. *The Journal of Biological Chemistry, 287*, 41820–41834.
Dall'Osto, L., Lico, C., Alric, J., Giuliano, G., Havaux, M., & Bassi, R. (2006). Lutein is needed for efficient chlorophyll triplet quenching in the major LHCII antenna complex of higher plants and effective photoprotection in vivo under strong light. *BMC Plant Biology, 6*, 32.
Damaraju, S., Schlede, S., Eckhardt, U., Lokstein, H., & Grimm, B. (2011). Functions of the water soluble chlorophyll-binding protein in plants. *Journal of Plant Physiology, 168*, 1444–1451.
Danon, A., Miersch, O., Felix, G., Opden Camp, R. G. L., & Apel, K. (2005). Concurrent activation of cell death-regulating signaling pathways by singlet oxygen in Arabidopsis thaliana. *The Plant Journal, 41*, 68–80.
Davies, A. G., Griller, D., Ingold, K. U., Lindsay, D. A., & Walton, J. C. (1981). An electron spin resonance study of pentadienyl and related radicals: Homolytic fission of

cyclobut-2-enylmethyl radicals. *Journal of the Chemical Society, Perkin Transactions, 2*, 633–641.

Davies, M. J. (2003). Singlet oxygen-mediated damage to proteins and its consequences. *Biochemical and Biophysical Research Communications, 305*, 761–770.

Davies, M. J. (2004). Reactive species formed on proteins exposed to singlet oxygen. *Photochemical & Photobiological Sciences: Official Journal of the European Photochemistry Association and the European Society for Photobiology, 3*, 17–25.

Davies, M. J. (2005). The oxidative environment and protein damage. *Biochimica et Biophysica Acta, 1703*, 93–109.

de Bruxelles, G. L., & Roberts, M. R. (2001). Signals regulating multiple responses to wounding and herbivores. *Critical Reviews in Plant Sciences, 20*, 487–521.

Demmig-Adams, B. (1990). Carotenoids and photoprotection in plants: A role for the xanthophyll zeaxanthin. *Biochimica et Biophysica Acta (BBA) - Bioenergetics, 1020*, 1–24.

Demmig-Adams, B., Adams Iii, W. W., Barker, D. H., Logan, B. A., Bowling, D. R., & Verhoeven, A. S. (1996). Using chlorophyll fluorescence to assess the fraction of absorbed light allocated to thermal dissipation of excess excitation. *Physiologia Plantarum, 98*, 253–264.

Demmig-Adams, B., & Adams, W. W., 3rd (2000). Harvesting sunlight safely. *Nature, 403*(371), 373–374.

Di Mascio, P., Martinez, G. R., Miyamoto, S., Ronsein, G. E., Medeiros, M. H. G., & Cadet, J. (2019). Singlet molecular oxygen reactions with nucleic acids, lipids, and proteins. *Chemical Reviews, 119*, 2043–2086.

Dogra, V., Duan, J., Lee, K. P., & Kim, C. (2019). Impaired PSII proteostasis triggers a UPR-like response in the var2 mutant of Arabidopsis. *Journal of Experimental Botany, 70*, 3075–3088.

Dogra, V., & Kim, C. (2019). Singlet oxygen metabolism: From genesis to signaling. *Frontiers in Plant Science, 10*, 1640.

Dogra, V., Li, M., Singh, S., Li, M., & Kim, C. (2019). Oxidative post-translational modification of EXECUTER1 is required for singlet oxygen sensing in plastids. *Nature Communications, 10*, 2834.

Dogra, V., Rochaix, J.-D., & Kim, C. (2018). Singlet oxygen-triggered chloroplast-to-nucleus retrograde signalling pathways: An emerging perspective. *Plant, Cell & Environment, 41*, 1727–1738.

Dogra, V., Singh, R. M., Li, M., Li, M., Singh, S., & Kim, C. (2022). EXECUTER2 modulates the EXECUTER1 signalosome through its singlet oxygen-dependent oxidation. *Molecular Plant, 15*, 438–453.

Dreaden, T. M., Chen, J., Rexroth, S., & Barry, B. A. (2011). N-formylkynurenine as a marker of high light stress in photosynthesis. *The Journal of Biological Chemistry, 286*, 22632–22641.

Dreaden Kasson, T. M., Rexroth, S., & Barry, B. A. (2012). Light-induced oxidative stress, N-formylkynurenine, and oxygenic photosynthesis. *PLoS One, 7*, e42220.

Duan, J., Lee, K. P., Dogra, V., Zhang, S., Liu, K., Caceres-Moreno, C., ... Kim, C. (2019). Impaired PSII proteostasis promotes retrograde signaling via salicylic acid. *Plant Physiology, 180*, 2182–2197.

Durrant, J. R., Giorgi, L. B., Barber, J., Klug, D. R., & Porter, G. (1990). Characterisation of triplet states in isolated Photosystem II reaction centres: Oxygen quenching as a mechanism for photodamage. *Biochimica et Biophysica Acta (BBA) - Bioenergetics, 1017*, 167–175.

Ehrenshaft, M., Deterding, L. J., & Mason, R. P. (2015). Tripping up Trp: Modification of protein tryptophan residues by reactive oxygen species, modes of detection, and biological consequences. *Free Radical Biology & Medicine, 89*, 220–228.

Ehrenshaft, M., Zhao, B., Andley, U. P., Mason, R. P., & Roberts, J. E. (2011). Immunological detection of N-formylkynurenine in porphyrin-mediated photooxided lens alpha-crystallin. *Photochemistry and Photobiology, 87*, 1321–1329.

El-Aal, H. A. H. M. A. (2012). Lipid Peroxidation End-Products as a Key of Oxidative Stress: Effect of Antioxidant on Their Production and Transfer of Free Radicals. In (Ed.), Lipid Peroxidation. IntechOpen. https://doi.org/10.5772/45944.

Espineda, C. E., Linford, A. S., Devine, D., & Brusslan, J. A. (1999). The AtCAO gene, encoding chlorophyll a oxygenase, is required for chlorophyll b synthesis in Arabidopsis thaliana. *Proceedings of the National Academy of Sciences, 96*, 10507–10511.

Farmer, E. E., & Davoine, C. (2007). Reactive electrophile species. *Current Opinion in Plant Biology, 10*, 380–386.

Fedorova, G., Trofimov, A., Vasil'ev, R., & Veprintsev, T. (2007). Peroxy-radical-mediated chemiluminescence: Mechanistic diversity and fundamentals for antioxidant assay. *Waldemar Adam ARKIVOC, 38*, 163–215.

Ferreira, K. N., Iverson, T. M., Maghlaoui, K., Barber, J., & Iwata, S. (2004). Architecture of the photosynthetic oxygen-evolving center. *Science (New York, N. Y.), 303*, 1831–1838.

Fischer, B. B., Hideg, E., & Krieger-Liszkay, A. (2013). Production, detection, and signaling of singlet oxygen in photosynthetic organisms. *Antioxidants & Redox Signaling, 18*, 2145–2162.

Fischer, B. B., Ledford, H. K., Wakao, S., Huang, S. G., Casero, D., Pellegrini, M., ... Niyogi, K. K. (2012). SINGLET OXYGEN RESISTANT 1 links reactive electrophile signaling to singlet oxygen acclimation in Chlamydomonas reinhardtii. *Proceedings of the National Academy of Sciences of the United States of America, 109*, E1302–E1311.

Flors, C., & Nonell, S. (2006). Light and singlet oxygen in plant defense against pathogens: Phototoxic phenalenone phytoalexins. *Accounts of Chemical Research, 39*, 293–300.

Flors, C., Fryer, M. J., Waring, J., Reeder, B., Bechtold, U., Mullineaux, P. M., ... Baker, N. R. (2006). Imaging the production of singlet oxygen in vivo using a new fluorescent sensor, Singlet Oxygen Sensor Green. *Journal of Experimental Botany, 57*, 1725–1734.

Fode, B., Siemsen, T., Thurow, C., Weigel, R., & Gatz, C. (2008). The Arabidopsis GRAS protein SCL14 interacts with class II TGA transcription factors and is essential for the activation of stress-inducible promoters. *The Plant Cell, 20*, 3122–3135.

Frank, H. A., Bautista, J. A., Josue, J. S., & Young, A. J. (2000). Mechanism of nonphotochemical quenching in green plants: Energies of the lowest excited singlet states of violaxanthin and zeaxanthin. *Biochemistry, 39*, 2831–2837.

Fufezan, C., Rutherford, A. W., & Krieger-Liszkay, A. (2002). Singlet oxygen production in herbicide-treated photosystem II. *FEBS Letters, 532*, 407–410.

Fukura, K., Tanaka, A., Tanaka, R., & Ito, H. (2021). Enrichment of chlorophyll catabolic enzymes in grana margins and their cooperation in catabolic reactions. *Journal of Plant Physiology, 266*, 153535.

Gilmore, A. M. (1997). Mechanistic aspects of xanthophyll cycle-dependent photoprotection in higher plant chloroplasts and leaves. *Physiologia Plantarum, 99*, 197–209.

Girotti, A. W. (1990). Photodynamic lipid peroxidation in biological systems. *Photochemistry and Photobiology, 51*, 497–509.

Glauser, G., Dubugnon, L., Mousavi, S. A., Rudaz, S., Wolfender, J. L., & Farmer, E. E. (2009). Velocity estimates for signal propagation leading to systemic jasmonic acid accumulation in wounded Arabidopsis. *The Journal of Biological Chemistry, 284*, 34506–34513.

Golan, T., Muller-Moule, P., & Niyogi, K. K. (2006). Photoprotection mutants of Arabidopsis thaliana acclimate to high light by increasing photosynthesis and specific antioxidants. *Plant, Cell & Environment, 29*, 879–887.

Gollmer, A., Arnbjerg, J., Blaikie, F. H., Pedersen, B. W., Breitenbach, T., Daasbjerg, K., ... Ogilby, P. R. (2011). Singlet Oxygen Sensor Green(R): Photochemical behavior in solution and in a mammalian cell. *Photochemistry and Photobiology, 87*, 671–679.

Goral, T. K., Johnson, M. P., Brain, A. P. R., Kirchhoff, H., Ruban, A. V., & Mullineaux, C. W. (2010). Visualizing the mobility and distribution of chlorophyll proteins in higher

plant thylakoid membranes: Effects of photoinhibition and protein phosphorylation. *The Plant Journal, 62*, 948–959.
Gorman, A. A., & Rodgers, M. A. (1992). Current perspectives of singlet oxygen detection in biological environments. *Journal of Photochemistry and Photobiology B, Biology, 14*, 159–176.
Gracanin, M., Hawkins, C. L., Pattison, D. I., & Davies, M. J. (2009). Singlet-oxygen-mediated amino acid and protein oxidation: Formation of tryptophan peroxides and decomposition products. *Free Radical Biology and Medicine, 47*, 92–102.
Gutierrez, J., Gonzalez-Perez, S., Garcia-Garcia, F., Lorenzo, O., & Arellano, J. B. (2011). Does singlet oxygen activate cell death in Arabidopsis cell suspension cultures?: Analysis of the early transcriptional defense responses to high light stress. *Plant Signaling & Behavior, 6*, 1937–1942.
Han, S.-H., Sakuraba, Y., Koh, H.-J., & Paek, N.-C. (2012). Leaf variegation in the rice zebra2 mutant is caused by photoperiodic accumulation of tetra-Cis-lycopene and singlet oxygen. *Molecules and Cells, 33*, 87–97.
Hanaa Ali Hassan Mostafa Abd, E. -A. (2012). Lipid peroxidation end-products as a key of oxidative stress: Effect of antioxidant on their production and transfer of free radicals.
Haussuhl, K., Andersson, B., & Adamska, I. (2001). A chloroplast DegP2 protease performs the primary cleavage of the photodamaged D1 protein in plant photosystem II. *The EMBO Journal, 20*, 713–722.
Havaux, M. (2014). Carotenoid oxidation products as stress signals in plants. *The Plant Journal, 79*, 597–606.
Havaux, M., & Niyogi, K. K. (1999). The violaxanthin cycle protects plants from photooxidative damage by more than one mechanism. *Proceedings of the National Academy of Sciences of the United States of America, 96*, 8762–8767.
Heddad, M., & Adamska, I. (2002). The evolution of light stress proteins in photosynthetic organisms. *Comparative and Functional Genomics, 3*, 504–510.
Herbstová, M., Tietz, S., Kinzel, C., Turkina, M. V., & Kirchhoff, H. (2012). Architectural switch in plant photosynthetic membranes induced by light stress. *Proceedings of the National Academy of Sciences of the United States of America, 109*, 20130–20135.
Hey, D., Rothbart, M., Herbst, J., Wang, P., Müller, J., Wittmann, D., ... Grimm, B. (2017). LIL3, a light-harvesting complex protein, links terpenoid and tetrapyrrole biosynthesis in Arabidopsis thaliana. *Plant Physiology, 174*, 1037–1050.
Hideg, É. (2008). A comparative study of fluorescent singlet oxygen probes in plant leaves. *Open Life Sciences, 3*, 273–284.
Hideg, É., Kálai, T., Hideg, K. L. N., & Vass, I. (1998). Photoinhibition of photosynthesis in vivo results in singlet oxygen production detection via nitroxide-induced fluorescence quenching in broad bean leaves. *Biochemistry, 37*(33), 11405–11411.
Hideg, É.É., Ogawa, K., Kálai, T., & Hideg, K. (2001). Singlet oxygen imaging in Arabidopsis thaliana leaves under photoinhibition by excess photosynthetically active radiation. *Physiologia Plantarum, 112*, 10–14.
Holt, N. E., Fleming, G. R., & Niyogi, K. K. (2004). Toward an understanding of the mechanism of nonphotochemical quenching in green plants. *Biochemistry, 43*, 8281–8289.
Horigome, D., Satoh, H., Itoh, N., Mitsunaga, K., Oonishi, I., Nakagawa, A., & Uchida, A. (2007). Structural mechanism and photoprotective function of water-soluble chlorophyll-binding protein. *The Journal of Biological Chemistry, 282*, 6525–6531.
Horton, P., & Hague, A. (1988). Studies on the induction of chlorophyll fluorescence in isolated barley protoplasts. IV. Resolution of non-photochemical quenching. *Biochimica et Biophysica Acta (BBA) - Bioenergetics, 932*, 107–115.
Horton, P., Ruban, A. V., & Wentworth, M. (2000). Allosteric regulation of the light-harvesting system of photosystem II. *Philosophical Transactions of the Royal Society of London. Series B, Biological Sciences, 355*, 1361–1370.

Hou, Z., Pang, X., Hedtke, B., & Grimm, B. (2021). In vivo functional analysis of the structural domains of FLUORESCENT (FLU). *The Plant Journal: For Cell and Molecular Biology, 107*, 360–376.
Huang, W., Chen, Q., Zhu, Y., Hu, F., Zhang, L., Ma, Z., ... Huang, J. (2013). Arabidopsis thylakoid formation 1 is a critical regulator for dynamics of PSII–LHCII complexes in leaf senescence and excess light. *Molecular Plant, 6*, 1673–1691.
Hutin, C., Nussaume, L., Moise, N., Moya, I., Kloppstech, K., & Havaux, M. (2003). Early light-induced proteins protect Arabidopsis from photooxidative stress. *Proceedings of the National Academy of Sciences of the United States of America, 100*, 4921–4926.
Jiang, H., Chen, Y., Li, M., Xu, X., & Wu, G. (2011). Overexpression of SGR results in oxidative stress and lesion-mimic cell death in rice seedlings. *Journal of Integrative Plant Biology, 53*, 375–387.
Jimbo, H., Yutthanasirikul, R., Nagano, T., Hisabori, T., Hihara, Y., & Nishiyama, Y. (2018). Oxidation of translation factor EF-Tu inhibits the repair of photosystem II. *Plant Physiology, 176*, 2691–2699.
Jin, F., Leitich, J., & von Sonntag, C. (1995). The photolysis ($\lambda = 254$ nm) of tyrosine in aqueous solutions in the absence and presence of oxygen. The reaction of tyrosine with singlet oxygen. *Journal of Photochemistry and Photobiology A: Chemistry, 92*, 147–153.
Kai, H., Hirashima, K., Matsuda, O., Ikegami, H., Winkelmann, T., Nakahara, T., & Iba, K. (2012). Thermotolerant cyclamen with reduced acrolein and methyl vinyl ketone. *Journal of Experimental Botany, 63*, 4143–4150.
Kasson, T. M., & Barry, B. A. (2012). Reactive oxygen and oxidative stress: N-formyl kynurenine in photosystem II and non-photosynthetic proteins. *Photosynthesis Research, 114*, 97–110.
Kato, Y., & Sakamoto, W. (2009). Protein quality control in chloroplasts: A current model of D1 protein degradation in the photosystem II repair cycle. *Journal of Biochemistry, 146*, 463–469.
Kato, Y., & Sakamoto, W. (2014). Phosphorylation of photosystem II core proteins prevents undesirable cleavage of D1 and contributes to the fine-tuned repair of photosystem II. *The Plant Journal: For Cell and Molecular Biology, 79*, 312–321.
Kauss, D., Bischof, S., Steiner, S., Apel, K., & Meskauskiene, R. (2012). FLU, a negative feedback regulator of tetrapyrrole biosynthesis, is physically linked to the final steps of the Mg(++)-branch of this pathway. *FEBS Letters, 586*, 211–216.
Kennedy, J. C., & Pottier, R. H. (1992). Endogenous protoporphyrin IX, a clinically useful photosensitizer for photodynamic therapy. *Journal of Photochemistry and Photobiology B, Biology, 14*, 275–292.
Kessler, F., & Longoni, P. (2019). How chloroplasts protect themselves from unfolded proteins. *eLife, 8*, e51430.
Kim, C., & Apel, K. (2013). $^1O_2$-mediated and EXECUTER-dependent retrograde plastid-to-nucleus signaling in norflurazon-treated seedlings of Arabidopsis thaliana. *Molecular Plant, 6*, 1580–1591.
Kim, C., Meskauskiene, R., Zhang, S., Lee, K. P., Lakshmanan Ashok, M., Blajecka, K., ... Apel, K. (2012). Chloroplasts of Arabidopsis are the source and a primary target of a plant-specific programmed cell death signaling pathway. *The Plant Cell, 24*, 3026–3039.
Kojima, K., Oshita, M., Nanjo, Y., Kasai, K., Tozawa, Y., Hayashi, H., & Nishiyama, Y. (2007). Oxidation of elongation factor G inhibits the synthesis of the D1 protein of photosystem II. *Molecular Microbiology, 65*, 936–947.
Krieger-Liszkay, A. (2005). Singlet oxygen production in photosynthesis. *Journal of Experimental Botany, 56*, 337–346.
Krieger-Liszkay, A., Fufezan, C., & Trebst, A. (2008). Singlet oxygen production in photosystem II and related protection mechanism. *Photosynthesis Research, 98*, 551–564.
Ledford, H. K., Chin, B. L., & Niyogi, K. K. (2007). Acclimation to singlet oxygen stress in Chlamydomonas reinhardtii. *Eukaryotic Cell, 6*, 919–930.

Lee, J. W., Lee, S. H., Han, J. W., & Kim, G. H. (2020). Early light-inducible protein (ELIP) can enhance resistance to cold-induced photooxidative stress in Chlamydomonas reinhardtii. *Frontiers in Physiology, 11*.

Lee, K. P., Kim, C., Landgraf, F., & Apel, K. (2007). EXECUTER1- and EXECUTER2-dependent transfer of stress-related signals from the plastid to the nucleus of Arabidopsis thaliana. *Proceedings of the National Academy of Sciences of the United States of America, 104*, 10270–10275.

Li, B., Fang, J., Singh, R. M., Zi, H., Lv, S., Liu, R., ... Kim, C. (2020). Fatty acid DESATURASE5 is required to induce autoimmune responses in gigantic chloroplast mutants of arabidopsis. *The Plant Cell, 32*, 3240–3255.

Li, M., & Kim, C. (2022). Chloroplast ROS and stress signaling. *Plant Communications, 3*, 100264.

Li, X.-P., Björkman, O., Shih, C., Grossman, A. R., Rosenquist, M., Jansson, S., & Niyogi, K. K. (2000). A pigment-binding protein essential for regulation of photosynthetic light harvesting. *Nature, 403*, 391–395.

Li, X. P., Gilmore, A. M., Caffarri, S., Bassi, R., Golan, T., Kramer, D., & Niyogi, K. K. (2004). Regulation of photosynthetic light harvesting involves intrathylakoid lumen pH sensing by the PsbS protein. *The Journal of Biological Chemistry, 279*, 22866–22874.

Li, X. P., Muller-Moule, P., Gilmore, A. M., & Niyogi, K. K. (2002). PsbS-dependent enhancement of feedback de-excitation protects photosystem II from photoinhibition. *Proceedings of the National Academy of Sciences of the United States of America, 99*, 15222–15227.

Li, Z., Wakao, S., Fischer, B. B., & Niyogi, K. K. (2009). Sensing and responding to excess light. *Annual Review of Plant Biology, 60*, 239–260.

Liebler, D. C. (2008). Protein damage by reactive electrophiles: Targets and consequences. *Chemical Research in Toxicology, 21*, 117–128.

Lindahl, M., Spetea, C., Hundal, T., Oppenheim, A. B., Adam, Z., & Andersson, B. (2000). The thylakoid FtsH protease plays a role in the light-induced turnover of the photosystem II D1 protein. *The Plant Cell, 12*, 419–431.

Llamas, E., Pulido, P., & Rodriguez-Concepcion, M. (2017). Interference with plastome gene expression and Clp protease activity in Arabidopsis triggers a chloroplast unfolded protein response to restore protein homeostasis. *PLoS Genetics, 13*, e1007022.

Ma, F., Chen, X. B., Sang, M., Wang, P., Zhang, J. P., Li, L. B., & Kuang, T. Y. (2009). Singlet oxygen formation and chlorophyll a triplet excited state deactivation in the cytochrome b6f complex from Bryopsis corticulans. *Photosynthesis Research, 100*, 19–28.

Macpherson, A. N., Telfer, A., Barber, J., & Truscott, T. G. (1993). Direct detection of singlet oxygen from isolated Photosystem II reaction centres. *Biochimica et Biophysica Acta (BBA) – Bioenergetics, 1143*, 301–309.

Mano, J., Biswas, M. S., & Sugimoto, K. (2019). Reactive carbonyl species: A missing link in ROS signaling. *Plants (Basel), 8*.

Mano, J. I., Miyatake, F., Hiraoka, E., & Tamoi, M. (2009). Evaluation of the toxicity of stress-related aldehydes to photosynthesis in chloroplasts. *Planta, 230*, 639–648.

Matheson, I. B., Etheridge, R. D., Kratowich, N. R., & Lee, J. (1975). The quenching of singlet oxygen by amino acids and proteins. *Photochemistry and Photobiology, 21*, 165–171.

Matsui, K. (2006). Green leaf volatiles: Hydroperoxide lyase pathway of oxylipin metabolism. *Current Opinion in Plant Biology, 9*, 274–280.

Meskauskiene, R., Nater, M., Goslings, D., Kessler, F., Opden Camp, R., & Apel, K. (2001). FLU: A negative regulator of chlorophyll biosynthesis in Arabidopsis thaliana. *Proceedings of the National Academy of Sciences of the United States of America, 98*, 12826–12831.

Michaeli, A., & Feitelson, J. (1994). Reactivity of singlet oxygen toward amino acids and peptides. *Photochemistry and Photobiology, 59*, 284–289.

Mochizuki, S., Sugimoto, K., Koeduka, T., & Matsui, K. (2016). Arabidopsis lipoxygenase 2 is essential for formation of green leaf volatiles and five-carbon volatiles. *FEBS Letters, 590*, 1017–1027.

Mor, A., Koh, E., Weiner, L., Rosenwasser, S., Sibony-Benyamini, H., & Fluhr, R. (2014). Singlet oxygen signatures are detected independent of light or chloroplasts in response to multiple stresses. *Plant Physiology, 165*, 249–261.

Mozzo, M., Dall'Osto, L., Hienerwadel, R., Bassi, R., & Croce, R. (2008). Photoprotection in the antenna complexes of photosystem II: Role of individual xanthophylls in chlorophyll triplet quenching. *The Journal of Biological Chemistry, 283*, 6184–6192.

Mueller, M. J., & Berger, S. (2009). Reactive electrophilic oxylipins: Pattern recognition and signalling. *Phytochemistry, 70*, 1511–1521.

Muller, P., Li, X. P., & Niyogi, K. K. (2001). Non-photochemical quenching. A response to excess light energy. *Plant Physiology, 125*, 1558–1566.

Mur, L. A., Aubry, S., Mondhe, M., Kingston-Smith, A., Gallagher, J., Timms-Taravella, E., ... Ougham, H. (2010). Accumulation of chlorophyll catabolites photosensitizes the hypersensitive response elicited by Pseudomonas syringae in Arabidopsis. *The New Phytologist, 188*, 161–174.

Nickelsen, J., & Rengstl, B. (2013). Photosystem II assembly: From cyanobacteria to plants. *Annual Review of Plant Biology, 64*, 609–635.

Nishiyama, Y., Allakhverdiev, S. I., & Murata, N. (2011). Protein synthesis is the primary target of reactive oxygen species in the photoinhibition of photosystem II. *Physiologia Plantarum, 142*, 35–46.

Nixon, P. J., Michoux, F., Yu, J., Boehm, M., & Komenda, J. (2010). Recent advances in understanding the assembly and repair of photosystem II. *Annals of Botany, 106*, 1–16.

Ochsenbein, C., Przybyla, D., Danon, A., Landgraf, F., Gobel, C., Imboden, A., ... Apel, K. (2006). The role of EDS1 (enhanced disease susceptibility) during singlet oxygen-mediated stress responses of Arabidopsis. *The Plant Journal: For Cell and Molecular Biology, 47*, 445–456.

Op den Camp, R. G., Przybyla, D., Ochsenbein, C., Laloi, C., Kim, C., Danon, A., ... Apel, K. (2003). Rapid induction of distinct stress responses after the release of singlet oxygen in Arabidopsis. *The Plant Cell, 15*, 2320–2332.

Pathak, V., Prasad, A., & Pospisil, P. (2017). Formation of singlet oxygen by decomposition of protein hydroperoxide in photosystem II. *PLoS One, 12*, e0181732.

Pedras, M. S. C., & Yaya, E. E. (2015). Plant Chemical Defenses: Are all constitutive antimicrobial metabolites phytoanticipins? *Natural Product Communications, 10* 1934578X1501000142.

Poor, P., Czekus, Z., Tari, I., & Ordog, A. (2019). The multifaceted roles of plant hormone salicylic acid in endoplasmic reticulum stress and unfolded protein response. *International Journal of Molecular Sciences, 20*.

Pruzinska, A., Anders, I., Aubry, S., Schenk, N., Tapernoux-Luthi, E., Muller, T., ... Hortensteiner, S. (2007). In vivo participation of red chlorophyll catabolite reductase in chlorophyll breakdown. *The Plant Cell, 19*, 369–387.

Ragas, X., Jimenez-Banzo, A., Sanchez-Garcia, D., Batllori, X., & Nonell, S. (2009). Singlet oxygen photosensitisation by the fluorescent probe Singlet Oxygen Sensor Green. *Chemical Communications (Camb)*, 2920–2922.

Ramel, F., Birtic, S., Cuine, S., Triantaphylides, C., Ravanat, J. L., & Havaux, M. (2012). Chemical quenching of singlet oxygen by carotenoids in plants. *Plant Physiology, 158*, 1267–1278.

Ramel, F., Birtic, S., Ginies, C., Soubigou-Taconnat, L., Triantaphylides, C., & Havaux, M. (2012). Carotenoid oxidation products are stress signals that mediate gene responses to singlet oxygen in plants. *Proceedings of the National Academy of Sciences of the United States of America, 109*, 5535–5540.

Ramel, F., Ksas, B., Akkari, E., Mialoundama, A. S., Monnet, F., Krieger-Liszkay, A., ... Havaux, M. (2013). Light-induced acclimation of the Arabidopsis chlorina1 mutant to singlet oxygen. *The Plant Cell, 25*, 1445–1462.

Ramundo, S., Casero, D., Muhlhaus, T., Hemme, D., Sommer, F., Crevecoeur, M., ... Rochaix, J. D. (2014). Conditional depletion of the chlamydomonas chloroplast ClpP protease activates nuclear genes involved in autophagy and plastid protein quality control. *The Plant Cell, 26*, 2201–2222.

Read, A., & Schröder, M. (2021). The unfolded protein response: An overview. *Biology (Basel), 10*, 384.

Redmond, R. W., & Kochevar, I. E. (2006). Spatially resolved cellular responses to singlet oxygen. *Photochemistry and Photobiology, 82*, 1178–1186.

Reisinger, V., Plöscher, M., & Eichacker, L. A. (2008). Lil3 assembles as chlorophyll-binding protein complex during deetiolation. *FEBS Letters, 582*, 1547–1551.

Rentel, M. C., Lecourieux, D., Ouaked, F., Usher, S. L., Petersen, L., Okamoto, H., ... Knight, M. R. (2004). OXI1 kinase is necessary for oxidative burst-mediated signalling in Arabidopsis. *Nature, 427*, 858–861.

Rinalducci, S., Campostrini, N., Antonioli, P., Righetti, P. G., Roepstorff, P., & Zolla, L. (2005). Formation of truncated proteins and high-molecular-mass aggregates upon soft illumination of photosynthetic proteins. *Journal of Proteome Research, 4*, 2327–2337.

Roach, T., Baur, T., Stöggl, W., & Krieger-Liszkay, A. (2017). Chlamydomonas reinhardtii responding to high light: a role for 2-propenal (acrolein). *Physiologia Plantarum, 161*, 75–87.

Roach, T., Colville, L., Beckett, R. P., Minibayeva, F. V., Havaux, M., & Kranner, I. (2015). A proposed interplay between peroxidase, amine oxidase and lipoxygenase in the wounding-induced oxidative burst in Pisum sativum seedlings. *Phytochemistry, 112*, 130–138.

Rokka, A., Aro, E. M., Herrmann, R. G., Andersson, B., & Vener, A. V. (2000). Dephosphorylation of photosystem II reaction center proteins in plant photosynthetic membranes as an immediate response to abrupt elevation of temperature. *Plant Physiology, 123*, 1525–1536.

Roper, J. M., & Smith, A. G. (1997). Molecular localisation of ferrochelatase in higher plant chloroplasts. *European Journal of Biochemistry/FEBS, 246*, 32–37.

Ruban, A. V., Johnson, M. P., & Duffy, C. D. (2012). The photoprotective molecular switch in the photosystem II antenna. *Biochimica et Biophysica Acta, 1817*, 167–181.

Russell, G. A. (1957). Deuterium-isotope effects in the autoxidation of aralkyl hydrocarbons. Mechanism of the interaction of PEroxy radicals1. *Journal of the American Chemical Society, 79*, 3871–3877.

Sang, M., Ma, F., Xie, J., Chen, X. B., Wang, K. B., Qin, X. C., ... Kuang, T. Y. (2010). High-light induced singlet oxygen formation in cytochrome b(6)f complex from Bryopsis corticulans as detected by EPR spectroscopy. *Biophysical Chemistry, 146*, 7–12.

Santabarbara, S., Agostini, G., Casazza, A. P., Syme, C. D., Heathcote, P., Bohles, F., ... Carbonera, D. (2007). Chlorophyll triplet states associated with Photosystem I and Photosystem II in thylakoids of the green alga Chlamydomonas reinhardtii. *Biochimica et Biophysica Acta, 1767*, 88–105.

Schaller, F. (2001). Enzymes of the biosynthesis of octadecanoid-derived signalling molecules. *Journal of Experimental Botany, 52*, 11–23.

Scharfenberg, M., Mittermayr, L., Von Roepenack-Lahaye, E., Schlicke, H., Grimm, B., Leister, D., & Kleine, T. (2015). Functional characterization of the two ferrochelatases in Arabidopsis thaliana. *Plant, Cell & Environment, 38*, 280–298.

Schmidt, K., Fufezan, C., Krieger-Liszkay, A., Satoh, H., & Paulsen, H. (2003). Recombinant water-soluble chlorophyll protein from Brassica oleracea var. Botrys binds various chlorophyll derivatives. *Biochemistry, 42*, 7427–7433.

Shao, N., Duan, G. Y., & Bock, R. (2013). A mediator of singlet oxygen responses in Chlamydomonas reinhardtii and Arabidopsis identified by a luciferase-based genetic screen in algal cells. *The Plant Cell, 25*, 4209–4226.

Shumbe, L., Bott, R., & Havaux, M. (2014). Dihydroactinidiolide, a high light-induced β-carotene derivative that can regulate gene expression and photoacclimation in Arabidopsis. *Molecular Plant, 7*, 1248–1251.

Shumbe, L., Chevalier, A., Legeret, B., Taconnat, L., Monnet, F., & Havaux, M. (2016). Singlet oxygen-induced cell death in arabidopsis under high-light stress is controlled by OXI1 kinase. *Plant Physiology, 170*, 1757–1771.

Shumbe, L., D'Alessandro, S., Shao, N., Chevalier, A., Ksas, B., Bock, R., & Havaux, M. (2017). Methylene blue sensitivity 1 (MBS1) is required for acclimation of Arabidopsis to singlet oxygen and acts downstream of β-cyclocitral. *Plant, Cell & Environment, 40*, 216–226.

Šimková, K., Kim, C., Gacek, K., Baruah, A., Laloi, C., & Apel, K. (2012). The chloroplast division mutant caa33 of Arabidopsis thaliana reveals the crucial impact of chloroplast homeostasis on stress acclimation and retrograde plastid-to-nucleus signaling. *The Plant Journal: For Cell and Molecular Biology, 69*, 701–712.

Sinha, R. K., Komenda, J., Knoppova, J., Sedlarova, M., & Pospisil, P. (2012). Small CAB-like proteins prevent formation of singlet oxygen in the damaged photosystem II complex of the cyanobacterium Synechocystis sp. PCC 6803. *Plant, Cell & Environment, 35*, 806–818.

Stotz, H. U., Mueller, S., Zoeller, M., Mueller, M. J., & Berger, S. (2013). TGA transcription factors and jasmonate-independent COI1 signalling regulate specific plant responses to reactive oxylipins. *Journal of Experimental Botany, 64*, 963–975.

Stratton, S. P., & Liebler, D. C. (1997). Determination of singlet oxygen-specific versus radical-mediated lipid peroxidation in photosensitized oxidation of lipid bilayers: Effect of beta-carotene and alpha-tocopherol. *Biochemistry, 36*, 12911–12920.

Stratton, S. P., Schaefer, W. H., & Liebler, D. C. (1993). Isolation and identification of singlet oxygen oxidation products of beta-carotene. *Chemical Research in Toxicology, 6*, 542–547.

Suh, H. J., Kim, C. S., & Jung, J. (2000). Cytochrome b6/f complex as an indigenous photodynamic generator of singlet oxygen in thylakoid membranes. *Photochemistry and Photobiology, 71*, 103–109.

Takagi, D., Takumi, S., Hashiguchi, M., Sejima, T., & Miyake, C. (2016). Superoxide and singlet oxygen produced within the thylakoid membranes both cause photosystem I photoinhibition. *Plant Physiology, 171*, 1626–1634.

Taki, N., Sasaki-Sekimoto, Y., Obayashi, T., Kikuta, A., Kobayashi, K., Ainai, T., ... Ohta, H. (2005). 12-oxo-phytodienoic acid triggers expression of a distinct set of genes and plays a role in wound-induced gene expression in Arabidopsis. *Plant Physiology, 139*, 1268–1283.

Tanaka, R., Rothbart, M., Oka, S., Takabayashi, A., Takahashi, K., Shibata, M., ... Tanaka, A. (2010). LIL3, a light-harvesting-like protein, plays an essential role in chlorophyll and tocopherol biosynthesis. *Proceedings of the National Academy of Sciences, 107*, 16721–16725.

Tano, D. W., & Woodson, J. D. (2022). Putting the brakes on chloroplast stress signaling. *Molecular Plant, 15*, 388–390.

Telfer, A. (2002). What is beta-carotene doing in the photosystem II reaction centre? *Philosophical Transactions of the Royal Society of London. Series B, Biological Sciences, 357*, 1431–1439 discussion 1439-1440, 1469-1470.

Telfer, A., Dhami, S., Bishop, S. M., Phillips, D., & Barber, J. (1994). beta-Carotene quenches singlet oxygen formed by isolated photosystem II reaction centers. *Biochemistry, 33*, 14469–14474.

Terao, J., Hirota, Y., Kawakatsu, M., & Matsushita, S. (1981). Structural analysis of hydroperoxides formed by oxidation of phosphatidylcholine with singlet oxygen. *Lipids, 16*, 427–432.

Theis, J., & Schroda, M. (2016). Revisiting the photosystem II repair cycle. *Plant Signaling & Behavior, 11*, e1218587.

Tian, Y. N., Zhong, R. H., Wei, J. B., Luo, H. H., Eyal, Y., Jin, H. L., ... Pang, X. Q. (2021). Arabidopsis CHLOROPHYLLASE 1 protects young leaves from long-term photodamage by facilitating FtsH-mediated D1 degradation in photosystem II repair. *Molecular Plant, 14*, 1149–1167.

Tikkanen, M., Nurmi, M., Kangasjärvi, S., & Aro, E.-M. (2008). Core protein phosphorylation facilitates the repair of photodamaged photosystem II at high light. *Biochimica et Biophysica Acta (BBA) – Bioenergetics, 1777*, 1432–1437.

Trebst, A. (2003). Function of β-carotene and tocopherol in photosystem II. *Zeitschrift für Naturforschung C, 58*, 609–620.

Triantaphylidès, C., & Havaux, M. (2009). Singlet oxygen in plants: Production, detoxification and signaling. *Trends in Plant Science, 14*, 219–228.

Triantaphylidès, C., Krischke, M., Hoeberichts, F. A., Ksas, B., Gresser, G., Havaux, M., ... Mueller, M. J. (2008). Singlet oxygen is the major reactive oxygen species involved in photooxidative damage to plants. *Plant Physiology, 148*, 960–968.

Vainonen, J. P., Hansson, M., & Vener, A. V. (2005). STN8 protein kinase in Arabidopsis thaliana is specific in phosphorylation of photosystem II core proteins. *The Journal of Biological Chemistry, 280*, 33679–33686.

Van Dyck, S. (2010). 3 – The impact of singlet oxygen on lipid oxidation in foods. In E. A. Decker (Ed.). *Oxidation in foods and beverages and antioxidant applications* (pp. 57–75). Woodhead Publishing.

van Gorkom, H. J., & Schelvis, J. P. M. (1993). Kok's oxygen clock: What makes it tick? The structure of P680 and consequences of its oxidizing power. *Photosynthesis Research, 38*, 297–301.

van Mieghem, F., Brettel, K., Hillmann, B., Kamlowski, A., Rutherford, A. W., & Schlodder, E. (1995). Charge recombination reactions in photosystem II. I. Yields, recombination pathways, and kinetics of the primary pair. *Biochemistry, 34*, 4798–4813.

Vass, I., & Cser, K. (2009). Janus-faced charge recombinations in photosystem II photoinhibition. *Trends in Plant Science, 14*, 200–205.

Vass, I., Styring, S., Hundal, T., Koivuniemi, A., Aro, E., & Andersson, B. (1992). Reversible and irreversible intermediates during photoinhibition of photosystem II: Stable reduced QA species promote chlorophyll triplet formation. *Proceedings of the National Academy of Sciences of the United States of America, 89*, 1408–1412.

Vener, A. V., Rokka, A., Fulgosi, H., Andersson, B., & Herrmann, R. G. (2000). A cyclophilin-regulated PP2A-like protein phosphatase in thylakoid membranes of plant chloroplasts. *Biochemistry, 39*, 2130.

Viswanath, K. K., Varakumar, P., Pamuru, R. R., Basha, S. J., Mehta, S., & Rao, A. D. (2020). Plant lipoxygenases and their role in plant physiology. *Journal of Plant Biology, 63*, 83–95.

Wagner, D., Przybyla, D., Opden Camp, R., Kim, C., Landgraf, F., Lee, K. P., ... Apel, K. (2004). The genetic basis of singlet oxygen-induced stress responses of Arabidopsis thaliana. *Science (New York, N. Y.), 306*, 1183–1185.

Waldeck, A. R., & Stocker, R. (1996). Radical-initiated lipid peroxidation in low density lipoproteins: Insights obtained from kinetic modeling. *Chemical Research in Toxicology, 9*, 954–964.

Wang, F., Fang, J., Guan, K., Luo, S., Dogra, V., Li, B., ... Kim, C. (2020). The Arabidopsis CRUMPLED LEAF protein, a homolog of the cyanobacterial bilin lyase, retains the bilin-binding pocket for a yet unknown function. *The Plant Journal: For Cell and Molecular Biology, 104*, 964–978.

Wang, L., Kim, C., Xu, X., Piskurewicz, U., Dogra, V., Singh, S., ... Apel, K. (2016). Singlet oxygen- and EXECUTER1-mediated signaling is initiated in grana margins and depends on the protease FtsH2. *Proceedings of the National Academy of Sciences of the United States of America, 113*, E3792–E3800.

Wang, L., Leister, D., Guan, L., Zheng, Y., Schneider, K., Lehmann, M., ... Kleine, T. (2020). The Arabidopsis SAFEGUARD1 suppresses singlet oxygen-induced stress responses by protecting grana margins. *Proceedings of the National Academy of Sciences of the United States of America, 117*, 6918–6927.

Wang, P., & Grimm, B. (2021). Connecting chlorophyll metabolism with accumulation of the photosynthetic apparatus. *Trends in Plant Science, 26*, 484–495.

Wasternack, C., & Hause, B. (2013). Jasmonates: Biosynthesis, perception, signal transduction and action in plant stress response, growth and development. An update to the 2007 review in Annals of Botany. *Annals of Botany, 111*, 1021–1058.

Wilkinson, F., & Brummer, J. G. (1981). Rate constants for the decay and reactions of the lowest electronically excited singlet state of molecular oxygen in solution. *Journal of Physical and Chemical Reference Data, 10*, 809–999.

Woodson, J. D., Joens, M. S., Sinson, A. B., Gilkerson, J., Salomé, P. A., Weigel, D., ... Chory, J. (2015). Ubiquitin facilitates a quality-control pathway that removes damaged chloroplasts. *Science (New York, N. Y.), 350*, 450–454.

Yamamoto, H. Y., Nakayama, T. O., & Chichester, C. O. (1962). Studies on the light and dark interconversions of leaf xanthophylls. *Archives of Biochemistry and Biophysics, 97*, 168–173.

Yang, D.-H., Paulsen, H., & Andersson, B. (2000). The N-terminal domain of the light-harvesting chlorophyll a/b-binding protein complex (LHCII) is essential for its acclimative proteolysis. *FEBS Letters, 466*, 385–388.

CHAPTER TWO

Hydrogen peroxide in plants

Amna Mhamdi*
Center for Plant Systems Biology, VIB, and Department of Plant Biotechnology and Bioinformatics, Ghent University, Gent, Belgium
*Corresponding author. e-mail address: amna.mhamdi@psb.vib-ugent.be

Contents

Abstract

Hydrogen peroxide (H_2O_2) is a molecule generated by plant cells during vital processes such as photosynthesis and photorespiration. This molecule is required for plant development and growth, as well as for coordinating appropriate responses to survive in a changing climate. Much attention has been dedicated to studying the functions of H_2O_2 in plant abiotic and biotic stress responses, mainly because of its role as a signal and its close interaction with phytohormones and redox-dependent signaling pathways. Like other reactive oxygen species (ROS), H_2O_2 has a higher energy state than oxygen which facilitate its interactions with cellular components, these include metabolites that are required to control the levels of H_2O_2 such as ascorbate and glutathione. The functions of H_2O_2 depend on its intracellular concentrations and these are regulated, among other factors, by enzymes that metabolize H_2O_2 to water and oxygen such as catalases and ascorbate peroxidases. H_2O_2 triggers signaling via oxidizing Methionine and Cysteine residues on different proteins, some acting as sensors that help to amplify and transduce H_2O_2 signals. In this Chapter, I provide a comprehensive overview of cellular components that are involved in H_2O_2 generation and processing. I also discuss the different ways by

Advances in Botanical Research, Volume 105
ISSN 0065-2296, https://doi.org/10.1016/bs.abr.2022.11.002

which H_2O_2 can act as a signal and summarize recent discoveries of H_2O_2 sensors and receptors that link H_2O_2 to other signal transduction pathways.

1. Introduction

What is hydrogen peroxide (H_2O_2)? It is a molecule derived from oxygen (O_2) and it is more reactive than O_2 itself (Foyer & Noctor, 2009; Mhamdi & Van Breusegem, 2018; Mittler, 2017). Every reactive oxygen species (ROS) has unique and distinct chemical properties, that makes it different is its ability to react with other molecule over its lifetime (Halliwell & Gutteridge, 2015). Singlet oxygen can oxidize lipids, proteins, and guanidine residues of DNA; superoxide ($O_2^{\cdot -}$) reacts with Fe-S proteins; and hydroxyl radical ($OH^{\cdot}$) is the most 'damaging' ROS due to its extreme reactivity and ability to oxidize almost all cell components (Mittler, 2017; Waszczak, Carmody, & Kangasjärvi, 2018). In contrast, H_2O_2 is fairly stable (more than 1 ms, so more than any other ROS) and, therefore, considered as an important signal.

Is H_2O_2 a signal? The ability of H_2O_2 to act as a signal is determined at the first level by the rates of its production and by its metabolism. More factors can determine how H_2O_2 signals are perceived, these include its ability to travel beyond the site of production and its interaction with other signals or signal transduction pathways (Mittler, Zandalinas, Fichman, & Van Breusegem, 2022; Noctor, Reichheld, & Foyer, 2018). H_2O_2 can interact with various cellular components, including metabolites and enzymes that play a role in keeping intracellular ROS under control (Noctor et al., 2012, 2018). H_2O_2 crosstalk with other metabolites including phytohormones such as salicylic acid (SA), jasmonic acid (JA), ethylene (ET), abscisic acid is well documented (Noctor, Lelarge-Trouverie, & Mhamdi, 2015; Waszczak et al., 2018).

Where is H_2O_2 produced at the subcellular level? All ROS are spontaneous products of plant metabolism (Foyer & Noctor, 2009; Mittler, 2017; Noctor et al., 2018). H_2O_2 is produced during respiration in the mitochondria, during photosynthesis in chloroplasts, and by apoplastic superoxide dismutases and NADPH oxidases (e.g., respiratory burst oxidase homologs, RBOHs), and at high rates during photorespiration in peroxisomes. This compartmentalization of ROS generation allows a tight control over H_2O_2 signaling between organelles and the nucleus (Mignolet et al., 2016; Noctor & Foyer, 2017; Phua, De Smet, Remacle, Chan, & Van Breusegem, 2021).

How is H_2O_2 metabolized? H_2O_2 can be processed by several enzymes including catalases (CATs) and ascorbate peroxidases (APXs), which are the main players involved in H_2O_2 metabolism. H_2O_2 as well as other peroxides can also be processed by enzymes that have peroxidase activities, these include glutathione S-transferases (GSTs), peroxiredoxins (PRXs), and thioredoxin (TRX)-dependent peroxiredoxins (Bela et al., 2015; Dietz, 2011; Dixon & Edwards, 2010; Iqbal, Yabuta, Takeda, Nakano, & Shigeoka, 2006).

Is H_2O_2 required for plant growth and survival? Like other ROS, H_2O_2 was long been studied for its roles in mediating responses to abiotic and biotic stress conditions. However, over the years, a number of studies have uncovered key roles for H_2O_2 during basic cellular processes such as the regulation of cell cycle and cell division and during various stages of plant development and growth, within seeds and meristems, leaves, roots and flowers (Mhamdi & Van Breusegem, 2018).

In the sections below, the basic chemistry and reactions involved in H_2O_2 generation, processing or signaling characteristics, via oxidation events, are described.

2. H_2O_2 generation in photosynthetic tissues

Hydrogen peroxide is generated at high rates in photosynthetic cells and this involves many organelles and subcellular sites (Exposito-Rodriguez, Laissue, Yvon-Durocher, Smirnoff, & Mullineaux, 2017; Noctor & Foyer, 2017; Phua et al., 2021). H_2O_2 results from a two-electron reduction of O_2. This reduction can occur chemically and involves interaction with metals or can be facilitated by enzymes, mainly oxidases, and dismutases. Oxidases catalyze the oxidation of other specific metabolites and simultaneously reduce O_2 to H_2O_2. Dismutation is a reaction in which the same molecule is both oxidized and reduced. Dismutases act on an oxygen molecule that already has an energy state higher than oxygen. In this particular context of H_2O_2 production, one molecule of superoxide is oxidized back to oxygen and another one is reduced to H_2O_2.

2.1 Glycolate oxidases

Photorespiration is triggered when ribulose-1,5-biphosphate carboxylase/oxygenase (RUBISCO) binds O_2 instead of CO_2 (Foyer et al., 2009;

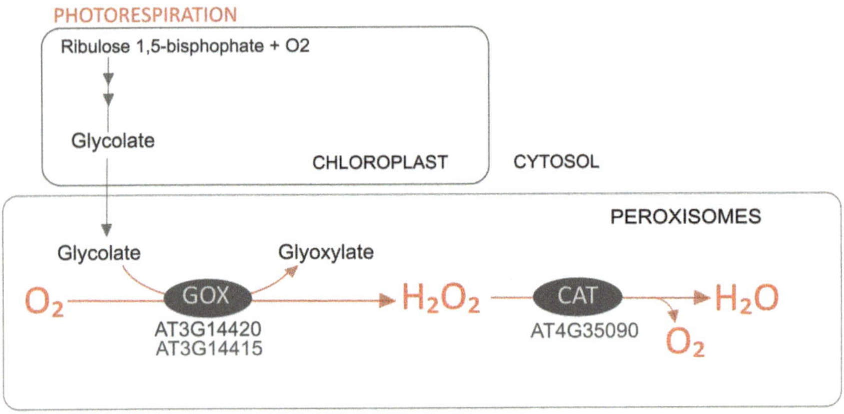

Fig. 1 Photorespiration production and metabolism of H_2O_2. The scheme shows an overview of the start of photorespiration in the chloroplasts and the link to glycolate oxidation by GOX. H_2O_2 produced in this condition is then metabolized by the highly abundant and highly active catalase. Major photorespiratory isoforms are encoded by 2 genes for GOX and by one gene for CAT.

Kangasjarvi, Neukermans, Li, Aro, & Noctor, 2012) This pathway consists of two interconnected cycles that functionally connect chloroplasts, peroxisomes, mitochondria, and the cytosol: (1) one cycle involved in glycolate carbon (C2) recycling and (2) one in ammonium recycling (Foyer & Noctor, 2009).

Oxidation of glycolate to glyoxylate is the first step of photorespiration that takes place in the peroxisome and is regarded as the major source of H_2O_2 (Fig. 1). Typical rates of H_2O_2 production via glycolate oxidation are estimated as 10,000 nmol.m^{-2}.s^{-1} of H_2O_2, while photosynthetic chloroplastic rates would be only 4030 nmol.m^{-2}.s^{-1} (Foyer & Noctor, 2003). This reaction, catalyzed by glycolate oxidase (GOX; Foyer & Noctor, 2003, 2009) can be particularly rapid in plants, because it is driven by high rates of glycolate produced in photorespiration. Probably, most of the H_2O_2 produced in peroxisomes is then metabolized by the peroxisome redox guardian, catalase 2, a photorespiratory enzyme that is also highly active and considered to be the first-line of defence against H_2O_2 produced by glycolate oxidation (Mhamdi, Noctor, & Baker, 2012).

Recombinant GOX proteins from C3 and C4 plant species share similar kinetic properties. In the literature, two main scenarios are described for the catalytic mechanism of α-hydroxy-acid oxidase family members (Dellero et al., 2015; Furuichi et al., 2008; Macheroux et al., 1993). The first mechanism involves a carbanion (a stepwise mechanism) and the second one involves a hydride transfer (a concerted but maybe asynchronous mechanism). Using isotope labeling, Dellero et al. (2015) show that, both in n Arabidopsis and Maize, (1) GOX reaction mechanism involves a hydride transfer mechanism, (2) that glycolate deprotonation is only partially rate-limiting, and (3) that the formation of the substrate-enzyme complex is also an important limiting factor.

Arabidopsis has five GOX genes, and it has been suggested that two of these encode the principal photorespiratory enzymes, GOX1 and GOX2 (Reumann et al., 2007). The relevance of their functions as major GOXs in photorespiration is also confirmed by transcriptomic analyses that show both isoforms to be highly expressed in leaves. The single *gox1* and *gox2* mutants retain 30%–45% of wild-type GOX activity and are not affected in photosynthetic capacities or growth and development suggesting that these are functionally redundant (Dellero et al., 2016). GOX1 and GOX2 reside in a tandem repeat on chromosome 3 and are separated by less than 1 kb, which makes it very difficult to obtain double mutants by crossing existing single mutant lines, and would explain why mutants affected in GOX activity were not isolated in photorespiratory mutants screens (Timm & Bauwe, 2013). However, when both GOX genes are mutated using microRNA technology, the double mutant retains only 5% of wild-type leaf GOX activity and exhibits a very strong growth inhibition in air (Dellero et al., 2016), reverted when by growth at a higher CO_2 environment (3000 ppm CO_2). But are GOX1 and GOX2 completely redundant? In a second-site suppressor screen for mutations that revert the photorespiratory phenotype of catalase mutants, GOX1 was identified, but not GOX2 (Kerchev et al., 2016). The introduction of the *gox2* mutation into the *cat2* background did not prevent the cell death phenotype or gene expression observed in the *cat2* mutants. However, the mutation of *gox1* was able to revert the *cat2* biochemical and physiological phenotypes, and thus revealed that GOX1 functions can be specific (Kerchev et al., 2016).

2.2 Superoxide dismutases

Chloroplasts and mitochondria generate H_2O_2 (from superoxide) via electron transport chains. In the chloroplast, superoxide production occurs at the site of photosystem I (PSI), together with the production of singlet oxygen at the site of PSII, these are considered to be highly active ROS sources in photosynthetic tissues (Foyer & Noctor, 2003; Mittler, 2002). By contrast to mammalian mitochondria, which have been described as the major source of ROS in animals, the contribution of plant mitochondria to

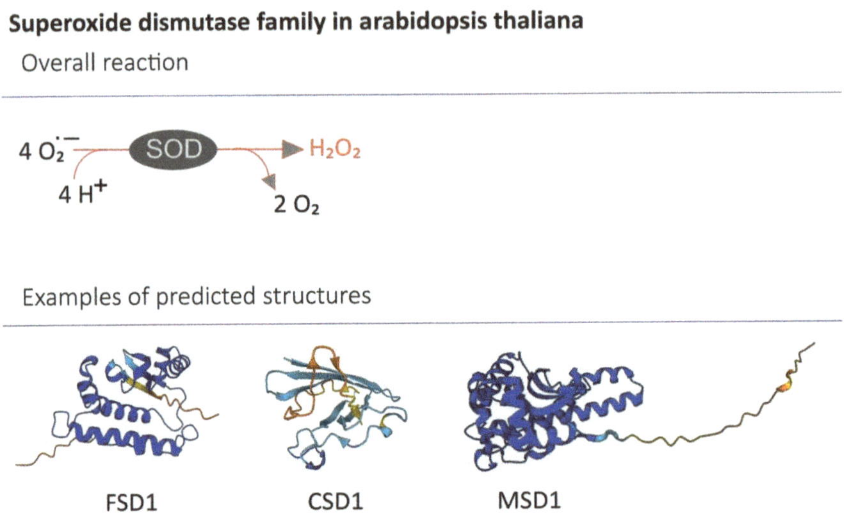

Genes and annotations

Gene id.	Annotation	Predicted localisation	Cofactor
At4g25100	FSD1	chloroplast stroma, cytosol, nucleus	Iron
At5g51100	FSD2	chloroplast thylakoids	Iron
At5g23310	FSD3	chloroplast thylakoids	Iron
At1g08830	CSD1	cytosol	copper/zinc
At2g28190	CSD2	chloroplast thylakoids	copper/zinc
At5g18100	CSD3	peroxisomes	copper/zinc
At3g10920	MSD1	mitochondria	manganese

Fig. 2 Overview of SOD gene family in Arabidopsis. The overall reaction of SOD is shown, known genes in Arabidopsis and the localization of their encoded proteins, as well as their specificity towards metal cofactors. Examples of SOD structures are predicted by Alphafold (alphafold.ebi.ac.uk).

ROS production is likely to be minor relative to the capacity of peroxisomes or chloroplasts (Foyer & Noctor, 2003). In mitochondria, superoxide is produced via the major NADH dehydrogenase and cytochrome bc1 (complexes I and III). Photo-oxidative stress environment stimulates the production of $O_2^{\cdot -}$ and consecutively H_2O_2 production (Asada, 2006; Foyer & Noctor, 2009; Foyer & Shigeoka, 2011; Mubarakshina et al., 2010; Pospisil, 2009).

Superoxide dismutase catalyzes the dismutation of $O_2^{\cdot -}$ to O_2 and H_2O_2 (Fig. 2). SODs use metal cofactors such as iron (Fe), manganese (Mn) and cupper/zinc (Cu/Zn). The Arabidopsis genome encode 8 genes, three genes encoding FeSOD (FSD1, FSD2, and FSD3), one encoding MnSOD (MSD1) and one gene encoding MnSOD-like, and three encoding Cu/ZnSOD (CSD1, CSD2, and CSD3) (Fig. 2) (Kliebenstein, Monde, & Last, 1998; Pilon, Ravet, & Tapken, 2011). SODs are distributed among different organelles like mitochondria (MSD1), peroxisomes (CSD3), and chloroplast stroma (FSD1), or chloroplast thylakoids (CSD2, FSD2, and FSD3), but are also found in the nucleus (FSD1) and the cytosol (CSD1, FSD1). This distribution in the nucleus and the cytosol ensures the metabolism of $O_2^{\cdot -}$ that escapes from other subcellular compartments (Dvořák et al., 2021; Kliebenstein et al., 1998; Melicher et al., 2022; Morgan et al., 2008; Myouga et al., 2008).

The genetic manipulation of some SODs triggers phenotypic changes, suggesting their important roles in plant development. Knock-out *fsd2* and *fsd3* mutants display chlorosis, abnormal chloroplast morphology and growth inhibition (Myouga et al., 2008). Although, *fsd1* mutant does not show obvious phenotypes or altered ROS levels in leaves (Myouga et al., 2008), overexpression of Arabidopsis FSD1 in *Zea mays* and *Nicotiana tabacum* caused increased tolerance against oxidative stress (Melicher et al., 2022; Van Breusegem et al., 1999; Van Camp, Capiau, Van Montagu, Inzé, & Slooten, 1996).

2.3 Other oxidases

When it comes to other cellular compartments that produce H_2O_2 directly via oxidases, peroxisomes are at the top of the list (Mhamdi et al., 2012). Peroxisomes host β-oxidation, a process involved in the degradation of fatty acids into acetyl-CoA. The pathway of β-oxidation degrades acyl-CoA substrates via oxidation at the C3 or β carbon to generate acetyl CoA and an acyl-CoA that is shortened by two carbons. In the first step of the β-oxidation H_2O_2 is generated through Acyl-CoA oxidases (ACX). These enzymes use FAD as a co-factor to produce FADH- which then donates a hydride ion to O_2 to generate H_2O_2. In Arabidopsis, six genes are described, and these differ by their specificity towards short, medium or long chains of lipids. This pathway is particularly relevant during seed germination and hormone synthesis, substrates of ACXs in Arabidopsis include indole butyric acid and the precursor of jasmonic acid (Baker, Graham, Holdsworth, Smith, & Theodoulou, 2006; Schilmiller, Koo, & Howe, 2007).

In plant peroxisomes, two other enzymes of the sulfur metabolism and the purine metabolism contribute to the production of H_2O_2 either directly or via $O_2^{\cdot -}$ (Mhamdi et al., 2012). Xanthine oxidoreductases (XOR) consist of xanthine oxidases and xanthine dehydrogenases. They catalyze the reaction of hypoxanthine to xanthine and then xanthine to urate (Corpas, de la Colina, Sanchez-Rasero, & Del Rio, 1997; Corpas et al., 2008; Smith & Atkins, 2002), the first reaction produces H_2O_2 and the second reaction produces $O2^{\cdot -}$. The peroxisomal XODs are described in legumes including castor beans and peas, however, in Arabidopsis, only a cytosolic xanthine dehydrogenase is reported (Corpas et al., 2008; Zarepour et al., 2010). Sulfite oxidase, a molybdoenzyme, oxidizes sulfite to sulfate, and produces H_2O_2 (Hansch et al., 2006). H_2O_2 is then rapidly metabolized by catalase, however in conditions of high sulfite levels, catalase is inhibited and the produced H_2O_2 assists in sulfite detoxification (Hansch et al., 2006).

An additional compartment that produces ROS, especially upon plant immunity or environmental stimuli, is the apoplast (Mhamdi & Van Breusegem, 2018). The NADPH oxidases (respiratory burst oxidase homologs, RBOH) are embedded in the plasma membrane and transfer electrons from cytoplasmic NADPH to apoplastic oxygen, forming $O_2^{\cdot -}$. Superoxide radicals are subsequently dismutated to H_2O_2 spontaneously or by SOD. Although there is evidence for apoplastic SOD, the type of SOD in the apoplast is not well characterized (Foreman et al., 2003; Jiménez-Quesada, Traverso, & Alché, 2016; Waszczak et al., 2018).

Polyamine oxidases (PAO) are further responsible for H_2O_2 production in the apoplast. These enzymes catabolism of polyamines spermidine and spermine, simultaneously generating H_2O_2. It seems that upon stress NADPH oxidases and PAO form a feed-forward loop to generate increased ROS levels (Gemes et al., 2016).

3. H_2O_2 processing pathways

H_2O_2 levels are tightly controlled to ensure their function as a signal. This is achieved by a concerted effort of antioxidant enzymes and metabolites. Catalases are enzymes that metabolize H_2O_2 and do not require reductants (Fig. 1) (Foyer et al., 2021; Mhamdi, Queval, et al., 2010; Mhamdi, Hager, et al., 2010; Mhamdi, Mauve, et al., 2010; Mhamdi et al., 2012). Almost all other enzymes do require the regeneration of reductants that get oxidized during the process of metabolizing H_2O_2. These pathways depend ultimately on NADPH regeneration either by photosynthesis or by specific NADP dehydrogenases. H_2O_2-metabolizing enzymes include (1) the ascorbate-glutathione pathway, (2) glutathione-dependent peroxidases and (3) TRX peroxidase pathways. Catalase and ascorbate peroxidase (APX) are considered to be H_2O_2-specific, however, other peroxidases (TRX or glutathione-dependent) may also metabolize organic peroxides as well as H_2O_2.

3.1 Catalases

Catalase exists in most organisms from bacteria to humans (Zamocky, Furtmüller, & Obinger, 2008). In plants, these enzymes catalyze the dismutation of H_2O_2 to H_2O and O_2 (Fig. 1). The Arabidopsis genome contains 3 genes encoding catalases named CAT1, CAT2, and CAT3. CAT1 and CAT3 are highly similar and are both located on chromosome 1, while CAT2 is located on chromosome 4 and has been found to encode the major isoform in Arabidopsis leaves and functions in photorespiration and maintaining leaf redox homeostasis (Fig. 1) (Mhamdi, Queval, et al., 2010; Mhamdi, Hager, et al., 2010; Mhamdi, Mauve, et al., 2010; Mhamdi et al., 2012; Queval et al., 2007; Yang, Mhamdi, & Noctor, 2019). Catalases can be classified based on organ and tissue expression into class I, class II, and class III enzymes (Mhamdi, Queval, et al., 2010; Mhamdi, Hager, et al., 2010; Mhamdi, Mauve, et al., 2010; Willekens, Inzé, Van Montagu, & Van Camp, 1995).

Catalase functions in plants have been investigated using photorespiratory transgenic lines and mutants (Chamnongpol et al., 1996, 1998; Kendall, Keys, Turner, Lea, & Miflin, 1983; Queval et al., 2007; Vandenabeele et al., 2004; Vanderauwera et al., 2005; Willekens et al., 1997). These analyses reported that loss of catalase function mimics plant responses to a variety of biotic and abiotic stress; this is mainly due to the fact that increased H_2O_2 production occurs via a physiologically relevant pathway, photorespiration, and triggers metabolic and transcriptional changes similar to those induced by biotic and abiotic stress (Mhamdi, Queval, et al., 2010). In the catalase-deficient lines, H_2O_2 triggers growth phenotypes, cell death and changes in gene expression and plant metabolism that are dependent on the growth conditions (Chaouch et al., 2010; Mhamdi, Queval, et al., 2010; Queval et al., 2007). For example, *cat2* mutants undergo cell death when grown in long days, but not in short days. These phenotypes are, however, accompanied by marked accumulation and oxidation of glutathione in both growth conditions (Mhamdi, Hager, et al., 2010; Queval et al., 2007). In recent years, catalase mutants were used as genetic tools to investigate oxidative stress signaling and the functions of other genes in response to H_2O_2. Introducing secondary mutations lacking ascorbate peroxidase 1 (APX1) or glutathione reductase 1 (GR1) abolished the cell death phenotype of *cat2* mutants (Mhamdi, Queval, et al., 2010; Mhamdi, Hager, et al., 2010; Mhamdi, Mauve, et al., 2010; Vanderauwera et al., 2011). These findings demonstrate that H_2O_2 signaling to cell death is more complex than merely H_2O_2 generation and that this requires additional factors and regulators. In this context studies of *cat2-associated* phenotypes in different photoperiod conditions show that SA-dependent pathogenesis responses are only activated in long days but not in short days (Chaouch et al., 2010; Queval et al., 2007). Photoperiod is a key factor that links H_2O_2 activating the SA responses, but also associated changes in the transcriptome (Queval, Neukermans, Vanderauwera, Van Breusegem, & Noctor, 2012). A protein phosphatase is identified as one of the factors that break the activation of the SA pathways in short days, the analysis of *cat2 pp2a-b′γ* double mutants demonstrated that PP2A-B′γ opposes the activation of the SA pathway and cell death in short days (Li, Mhamdi, Trotta, Kangasjärvi, & Noctor, 2014).

Catalases are long considered as peroxisome highly abundant proteins, though the exact import motif and mechanism remain to be established. More recently, our knowledge has shifted and the major photorespiratory isoform CAT2, is reported also to be localized in the nucleus, although CAT2 seems to be active in the nucleus, its function is yet to be defined (Al-Hajaya, Karpinska, Foyer, & Baker, 2022; Foyer et al., 2021).

3.2 The glutathione-ascorbate pathways

The glutathione ascorbate pathway is one of the most efficient players in H_2O_2 removal because of its specificity towards H_2O_2 relative to other organic peroxides. Fig. 3 shows glutathione, ascorbate, and NADPH and the key reactions or enzymes involved in H_2O_2 removal. Note that, these enzymes exist in almost all plant organelles. Although, ascorbate and glutathione are part of this highly sophisticated antioxidative system, their connection and function in an integrated manner (dependence) is not completely clear, changes in glutathione can occur without changes in ascorbate and both antioxidants can play specific functions in H_2O_2 metabolism but also in a development and growth context (few examples are cited below).

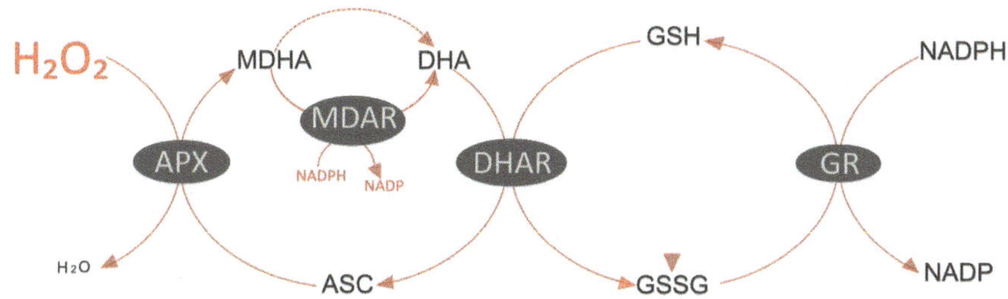

Fig. 3 Metabolites and enzymes associated with the ascorbate-glutathione cycle. The ascorbate and glutathione cycle. The scheme is showing glutathione, ascorbate and NAD(P)H and the key reactions and/or enzymes linking them to H_2O_2 metabolism. *ASC*, Ascorbate; *APX*, ascorbate peroxidase; *CAT*, catalase; *DHA*, dehydroascorbate; *DHAR*, dehydroascorbate reductase; *GR*, glutathione reductase; *GSH*, reduced glutathione; *GSSG*, glutathione disulfide; *MDHA*, monodehydroascorbate; *MDAR*, monodehydroascorbate reductase.

3.2.1 Glutathione and glutathione reductase

Glutathione is a tripeptide molecule found in most living organisms, and is composed of glutamate, cysteine and glycine (Noctor et al., 2012). As one of the most abundant low molecular-weight thiol compounds, glutathione maintains redox homeostasis under normal and stress conditions in higher plants. Glutathione is required for plant development, light signaling, metal sequestration, and pathogen responses (Ball et al., 2004; Cobbett, May, Howden, & Rolls, 1998; Mhamdi & Van Breusegem, 2018; Mhamdi, Hager, et al., 2010; Noctor et al., 2012; Parisy et al., 2007; Vernoux et al., 2000).

Glutathione is synthesized via two ATP-dependent reactions The first step is the ligation of cysteine and glutamate to form γ-glutamylcysteine and is catalyzed by γ-glutamylcysteine synthetase, also called GSH1. The Arabidopsis genome contains a single gene (At4g23100) encoding this protein, which is found only in the chloroplast (Wachter, Wolf, Steiniger, Bogs, & Rausch, 2005). The reaction is limiting for glutathione synthesis (Noctor et al., 1996), and T-DNA insertion mutants of GSH1 are not viable (Cairns, Pasternak, Wachter, Cobbett, & Meyer, 2006). Several mutants in this gene have been isolated in independent screens and have shown that plants can survive with 20%–30% of the wild-type glutathione levels (e.g., *cad2* and *pad2* alleles), but not when glutathione is depleted to 10% of the wild-type levels (Ball et al., 2004; Cobbett et al., 1998; Parisy et al., 2007; Vernoux et al., 2000). The GSH1 enzyme is composed of a homodimer linked by two disulfide bonds (Hothorn et al., 2006), and is redox-regulated, oxidation of the protein correlates with increased activity (Hicks et al., 2007).

The second step of glutathione synthesis involves glutathione synthetase, which is also encoded by a single gene in Arabidopsis (GSH2; At5g27380). However, the protein is located in both cytosol and chloroplast (Wachter et al., 2005). Knockouts of GSH2 are seedling-lethal, and accumulate the immediate precursor, γ-glutamylcysteine, to high levels. The expression of the protein in the cytosol and not the chloroplast is sufficient to restore the wild-type phenotype (Pasternak et al., 2008). Both enzymes were overexpressed in the chloroplast or the cytosol separately, or simultaneously in some plant species, Overexpression of GSH1 increases glutathione content to about four-fold relative to the control (Creissen et al., 1999; Noctor & Foyer, 1998; Noctor et al., 1996; Xiang, Werner, Christensen, & Oliver, 2001). Higher increases are obtained by using a bifunctional GSH1/GSH2 enzyme from Streptococcus thermophilus, which is insensitive to feedback inhibition by glutathione (Liedschulte, Wachter, Zhigang, & Rausch, 2010).

Glutathione is accumulated and oxidized in response to stress (Mhamdi, Hager, et al., 2010; Noctor et al., 2015; Queval et al., 2007, 2012). This accumulation is physiologically relevant and GSSG levels are often used as markers for oxidative stress and as an indirect measurement of H_2O_2 (Noctor et al., 2015, 2016). The changes in glutathione in this system occur without significant changes in ascorbate (Mhamdi, Hager, et al., 2010). Accumulation of glutathione is not a passive mechanism of defence but is of functional importance. Glutathione is required to transmit signals triggered by increased H_2O_2 and to link H_2O_2 signals to gene expression, cell death, and the activation of SA-dependent pathogenesis responses (Han, Chaouch, et al., 2013; Han, Mhamdi, Chaouch, & Noctor, 2013; Mhamdi et al., 2013).

Glutathione is oxidized by H_2O_2 to glutathione disulfide (GSSG) either chemically or via enzymes that use glutathione (reviewed in Rahantaniaina, Tuzet, Mhamdi, & Noctor, 2013). Glutathione is regenerated to its reduced form (GSH) and this is achieved by the NADPH-dependent glutathione reductase (GR) (Fig. 3). GR is a flavoprotein belonging to the pyridine nucleotide-disulfide oxidoreductase group, and thus contains a FAD group. As flavoproteins, GRs share similarities with other enzymes of this type, and regenerate GSH from GSSG using NADPH as an electron donor to the FAD prosthetic group and finally to a reactive disulfide bridge. This protein family also includes NADP-dependent thioredoxin reductase (NTR; discussed below). In Arabidopsis, two genes have been described and characterized. The loss of function of GR1, encoding the cytosol/peroxisome enzymes, causes only moderate GSSG accumulation in leaf tissue, in optimal growth conditions, and this is explained by the existence of a backup system with GSSG-reducing activity ensured by cytosolic NADPH-thioredoxin (TRX) systems (Marty et al., 2009).

Glutathione and GR1-dependent glutathione status are now established as signaling players and their role in H_2O_2 metabolism goes beyond their antioxidant functions. The first argument for this is that GR1 becomes more important in oxidative stress conditions and cannot be replaced either by GR2 or the cytosolic NTR-TRX system (Dghim et al., 2013; Mhamdi, Hager, et al., 2010). Moreover, the GR1-dependent glutathione status in the cytosol is required to link H_2O_2 signals to the activation of both SA and JA signaling pathways (Mhamdi, Hager, et al., 2010). This role seems to involve new yet unknown factors and does not seem to involve the known regulators of pathogenesis such as NPR1, NONEXPRESSER OF PR GENES 1.

3.2.2 Dehyroascorbate peroxidase

In contrast to GSSG reduction by GR, the reactions that are most important in converting GSH to GSSG are less clear and, potentially, more complex. Within the ascorbate glutathione pathway, one of the reactions that oxidize GSH to GSSG is catalyzed by dehydroascorbate reduactase (DHAR) (Fig. 3). DHARs are members of glutathione *S*-transferase (GST) superfamily with specific functions in ascorbate regeneration from dehydroascorbate, using glutathione. Unlike most GSTs, which are dimeric, DHARs are monomeric proteins (Dixon et al., 2002) with activities that are described to be localized in peroxisomes, mitochondria, cytosol, and chloroplast (Chew et al., 2003; Reumann et al., 2009). Arabidopsis has three annotated DHARs, encoded respectively by DHAR1 (At1g19570), DHAR2 (At1g75270), DHAR3 (At5g16710), while a fourth gene may encode another DHAR (Dixon & Edwards, 2010). Localization studies using fluorescent reporter proteins show that DHAR1 and DHAR2 are cytosolic proteins while DHAR3 is localized in chloroplasts, and that surprisingly there is no mitochondrial or peroxisomal DHARs (Rahantaniaina et al., 2017). DHAR2 plays a role in stress, Arabidopsis *dhar2* mutants are much more sensitive to ozone than wild-type (Tamaoki, Freeman, & Pilon-Smits, 2008; Yoshida et al., 2006). Furthermore, overexpression of DHAR confers tolerance to aluminium in transgenic tobacco plants (Yin et al., 2010). While DHARs play overlapping functions in growth and development, their function in oxidative stress and contributions to glutathione oxidation are specific (Rahantaniaina et al., 2017). The cytosolic DHAR activity (encoded DHAR1 and DHAR2) is particularly important in H_2O_2 metabolism that is linked to glutathione oxidation, and influences the H_2O_2-driven SA accumulation (Rahantaniaina et al., 2017). Noteworthy, glutathione oxidation by DHARs is not coupled to ascorbate depletion or ascorbate oxidation when H_2O_2 is generated in the peroxisomes. In conditions of high light stress and generation of H_2O_2 in chloroplasts, ascorbate accumulates and all DHAR isoforms play an important role (Terai et al., 2020). Thus, the relevance of DHAR activity is strictly dependent s on the ascorbate pool size and is required only to sustain high ascorbate accumulation (Terai et al., 2020).

3.2.3 Ascorbate and ascorbate peroxidase

Ascorbate, vitamin C, is the most abundant water-soluble antioxidant in plants and can participate chemically in the removal of many ROS (Foyer & Noctor, 2011; Smirnoff & Arnaud, 2019). Ascorbate is abundant in most cell organelles and its levels increase in some stress conditions and are required for stress perception (Smirnoff, Conklin, & Loewus, 2001; Zechmann, 2011). Ascorbate peroxidase (APX) has a higher affinity for H_2O_2 than catalase and catalyzes the reduction of H_2O_2 to H_2O by oxidizing ascorbate (ASC) to monodehydroascorbate (MDHA) (Asada, 1999; Shigeoka et al., 2002). In Arabidopsis, APX is encoded by nine genes and is found in the chloroplast, mitochondria, cytosol, and peroxisomes. The enzyme activity depends on ascorbate regeneration which occurs via the ascorbate-glutathione pathway (Fig. 3). ASC is regenerated from MDHA by MDHA reductase (MDHAR), and from dehydroascorbate (DHA) by DHAR (Fig. 3; see above) The first enzyme is NAD(P)H-dependent while the second, as described above, is glutathione-dependent. APXs, remain a powerful system to remove H_2O_2 generated during photosynthesis in the chloroplasts and are key to control signaling from and within this organelle (Dietz, 2016). Genetic studies provide evidence for the major role of cytosolic APXs in signaling. The expression of APXs, and in particular the cytosolic forms, is induced at the transcriptional level in different stresses such as high light (Ball et al., 2004; Karpinski, Escobar, Karpinska, Creissen, & Mullineaux, 1997; Shigeoka et al., 2002). Mutants for cytosolic APX1 exhibit oxidative stress phenotypes and decreased photosynthetic performance after exposure to high light (Davletova et al., 2005), suggesting a link between cytosolic and chloroplastic H_2O_2 pools. Moreover, *apx1* mutants show a more sensitive phenotype in response to a combination of heat and drought stress (Koussevitzky et al., 2008; Zandalinas et al., 2021). The growth and survival rates of *apx1* decline more rapidly relative to the wild-type when the stress factors increase, suggesting the importance of APX1 in stress acclimation (Zandalinas et al., 2021).

3.3 Peroxidases, peroxiredoxins and thioredoxins

Plants contain many genes that are predicted to encode peroxidases. Many of these are not considered to act as antioxidative peroxidases but rather to use H_2O_2 to oxidize other compounds (e.g., in cell wall cross-linking) or, indeed, to generate ROS. Other than these enzymes, plants contain peroxidases, whose main function is to control H_2O_2 levels (Dietz, 2016; Smirnoff & Arnaud, 2019). Given the high number and diversity of H_2O_2 -metabolizing systems, it remains unclear which are the most important and in which specific contexts of development and stress.

Peroxiredoxin (PRXs) form a multigenic family of proteins with the ability to reduce peroxides and H_2O_2 to the corresponding alcohol and H_2O, respectively. These enzymes can also reduce peroxynitrite to nitrite. Based on the number of cysteine residues and the catalytic mechanism, these proteins were subdivided into four sub-classes named 1-Cys PRX, 2-Cys PRX, Prx Q and type II PRX. 1-Cys PRXs contain only one cysteine in the catalytic site, 2-Cys PRXs and type II PRXs contain 2 cysteines and are therefore able to form an inter-molecular or intramolecular disulfide bridge during the reaction (Dietz, 2008, 2016). The regeneration of reduced PRX from their oxidized forms is achieved by glutaredoxins (GRX) for type II PRXs and by TRX or NADPH-thioredoxin reductases (NTR) for the other three classes (Dietz, 2003; Pérez-Ruiz & Cejudo, 2009; Rouhier & Jacquot, 2005).

Thioredoxins are very small proteins, around 12 kDa proteins, with disulfide oxidoreductase activities, and can be subdivided into six groups (Lemaire, 2004). Whereas TRX f, m, x and y are found in the chloroplast, TRXs o are mitochondrial and the large subgroup of TRX h is expressed, generally, in the cytosol (Meyer et al., 2008). TRX reduces the target protein by Cys thiol-disulfide exchange, yielding an oxidized TRX which is reduced by ferredoxin and ferredoxin-thioredoxin reductase (FTR) or by NADPH-dependent NTR. The genome of Arabidopsis contains three

NTR named NTRA (At2g17420), NTRB (At4g35460) and NTRC (A2tg41680). Using a recombinant protein from rice, NTRC was characterized as a bifunctional protein with both NADP-dependent TRX reductase and TRX domains: the loss of this isoform in Arabidopsis is associated with growth inhibition in optimal growth conditions and the mutants were hypersensitive to salt, methyl viologen, and drought (Kirchsteiger, Ferrandez, Pascual, Gonzalez, & Cejudo, 2012; Serrato, Pérez-Ruiz, Spínola, & Cejudo, 2004). NTRA and NTRB are expressed in the cytosol and mitochondria and are functionally redundant, combining both mutations causes only slight growth inhibition suggesting that cytosolic and mitochondrial systems are replaceable in Arabidopsis development (Reichheld et al., 2007). However, when these genes are knocked out together with GR1, the *ntra ntrb gr1*, the triple mutant does not produce viable pollen (Marty et al., 2009). Together with biochemical evidence that NTR-TRXh can reduce GSSG, it is now established that there is an overlap between cytosolic NADPH-NTR and NADPH-GR systems in maintaining GSH pools and redox homeostasis (Marty et al., 2009; Mhamdi, Hager, et al., 2010).

Glutathione peroxidases (GPXs) are non-heme peroxidases able to remove peroxides or lipid hydroperoxides and are monomeric proteins. Eight genes in Arabidopsis are annotated as GPX, and named GPX1 to GPX8 (Attacha et al., 2017; Bela et al., 2022). Although annotated as glutathione-dependent, and unlike animal cells, plant GPXs are likely to function with TRX rather than glutathione (Herbette et al., 2002; Iqbal et al., 2006). Their physiological importance has been associated with detoxification of lipid peroxides. GPXs are among enzymes responsive to oxidative stress; for example, GPX6 can be induced by H_2O_2 (Queval et al., 2007). The chloroplastic isoforms, GPX1 and GPX7, are involved in light acclimation, basal defense, and cell death (Chang et al., 2009). GPX3 has been described as both a stress transducer and scavenger in Arabidopsis, through interaction with the PP2C ABI2 (Miao et al., 2006). Peroxidases that depend chemically on glutathione are GSTs. Although their most studied role is in the removal of electrophilic xenobiotics by the formation of GS-conjugates (Dixon & Edwards, 2010), these enzymes can also act as peroxidases and are able to remove H_2O_2 (Dixon, Hawkins, Hussey, & Edwards, 2009).

3.4 NADPH regeneration

NADP(H) (nicotinamide adenine dinucleotide phosphate) is a small molecule that acts as a redox electron carrier; this molecule comprises nicotinamide, two phosphoryl groups, adenine and a ribose group. NADP differs from NAD by the presence of the phosphate attached to the carbon 2 of the adenosine ribose ring. The redox functions of NAD and NADP are due to the ability of the nicotinamide ring to accept or donate electrons. The oxidized forms are NAD or NADP, two-electron reductions of these molecules generate NADH and NADPH, respectively. In photosynthetic cells, a major origin of NADPH is the reduction of NADP by PSI during photosynthesis, a reaction catalyzed by the stromal enzyme, ferredoxin-NADP reductase (FNR). Noteworthy, most of this NADPH is supposed to be used by the Calvin cycle to fix CO_2. Alternatively, NADPH can be formed either by NADH kinase (NADHK) or by reduction of NADP+ by a variety of dehydrogenases. The best-characterized dehydrogenases are: isocitrate dehydrogenase (ICDH), glucose-6-phosphate dehydrogenase (G6PDH), malic enzyme (ME), and the non-phosphorylating glyceraldehyde-3-phosphate dehydrogenase (npGAPDH) (Fig. 4). The induction of activities, proteins or transcripts of the major NADP-dehydrogenases has been reported in response to oxidative stress (Leterrier, Del Rio, & Corpas, 2007; Marino, Gonzalez, Frendo, Puppo, & Arrese-Igor, 2007; Sarry et al., 2006; Valderrama et al., 2006), suggesting their importance in the production of NADPH as a key reductant to support H_2O_2 metabolism.

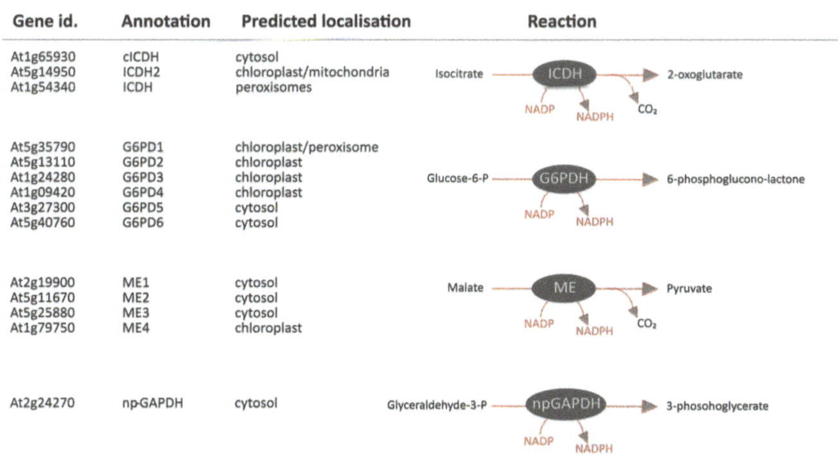

Fig. 4 Major arabidopsis dehydrogenases. The reaction catalyzed by each enzyme is displayed and this shows the link between NADPH regeneration and primary metabolism. *GAPDH,* Glyceraldehyde-3-phosphate dehydrogenase; *G6PDH,* glucose-6-phosphate dehydrogenase; *ICDH,* isocitrate dehydrogenase; *ME,* malic enzyme.

ICDH catalyzes the oxidative decarboxylation of isocitrate and produces 2-oxoglutarate (2-OG) while simultaneously converting NADP to the reduced form NADPH (Fig. 4). 2-OG produced in this reaction can be used for ammonia assimilation as a substrate of glutamate synthase, which constitutes with glutamine synthetase the main pathway of ammonia assimilation in plants (Scheible et al., 1997; Scheible, Krapp, & Stitt, 2000). Among ICDHs, the cytosolic cICDH appears to contribute to NADPH regeneration under oxidative stress (Mhamdi, Mauve, et al., 2010). Analysis of *icdh* mutants revealed that glutathione oxidation and accumulation were enhanced relative to the wild-type plants and that pathogenesis-related genes were constitutively induced. By contrast, the peroxisomal ICDH makes a very minor contribution to overall to redox homeostasis, neither glutathione oxidation, NADPH or cell death were different when *px-icdh* mutants were analyzed (Mhamdi & Noctor, 2015; Mhamdi, Mauve, et al., 2010).

The oxidative pentose phosphate pathway (OPPP) is coupled to NADPH production via G6PDH and 6-phosphogluconate dehydrogenase (6PGDH). The Arabidopsis genome contains six different isoforms of G6PDH (Table X), almost all of these are characterized at the biochemical level and exhibit different catalytic properties and redox sensitivity (Wakao & Benning, 2005). The cytosolic OPPP proteins were found to be insensitive to redox regulation while chloroplastic proteins are highly dependent on reversible redox modifications involving two cysteine residues that are highly conserved between plant species (Graeve, von Schaewen, & Scheibe, 1994; Wendt, Wenderoth, Tegeler, & von Schaewen, 2000). Altered activities of chloroplastic G6PDH in transgenic tobacco lines slightly increase the content of sucrose, fructose and glucose and impact the responses of these lines to methyl viologen (Debnam et al., 2004). The second reaction that produces NADPH in the OPPP is catalyzed by 6PGDH; plants have different cytosolic and chloroplastic isoforms of this enzyme (Averill, Bailey-Serres, & Kruger, 1998).

Malic enzyme (ME) catalyzes the oxidative decarboxylation of malate and releases pyruvate and CO_2 (Fig. 4) (Chang & Tong, 2003). Like ICDH, both NAD and NADP-dependent ME isoforms exist in plants (Maurino et al., 2009). While the NAD-dependent isoforms are especially mitochondrial and provide pyruvate required for the tricarboxylic acid cycle (Artus & Edwards, 1985), the NADP-dependent MEs are found in the chloroplast and cytosol (Drincovich, Casati, &

Andreo, 2001). The cytosolic C4 plants were found to be regulated by light and by mechanical injury suggesting the importance of these enzymes in both photosynthetic and non-photosynthetic functions (Lai, Tausta, & Nelson, 2002). Although these observations have highlighted some role of MEs in C4 plants, their physiological importance in C3 are not completely addressed. The cytosolic ME2 from Arabidopsis is activated by RPM1-INDUCED PROTEIN KINASE to fuel NADPH and to sustain RBOHD activity in plant immunity (Wu et al., 2022). Although in Arabidopsis, transcripts of *ME2* are induced by ozone and by oxidative stress, loss of ME2 function did not impact plant responses to H_2O_2, suggesting that even if gene expression induction is part of the oxidative stress, ME2 does not play an essential role in this context (Li, Mhamdi, Clement, Jolivet, & Noctor, 2013).

In plants, glyceraldehyde 3-phosphate can be converted to 3-phosphoglycerate by two different pathways. The first occurs via GAPDH coupled to phosphoglycerate kinase, generating both NADH and ATP, and forms the classical glycolytic reaction sequence. The second involves non-phosphorylating GAPDH, producing NADPH without ATP production. Although npGAPDH has long been known in plants (Kelly & Gibbs, 1973), its function is not well characterized. Functional analysis in Arabidopsis has showed that this enzyme is encoded by a single gene and localized in the cytosol (Rius, Casati, Iglesias-Diego, & Gomez-Casati, 2006). Different roles for npGAPDH have been proposed in plants. It could act as a 'bypass' to the classical glycolytic pathway (e.g., when ATP/ADP ratios are very high or phosphate is limiting) or it could act to generate NADPH in the cytosol. The second role may allow export of reductants from chloroplast, i.e., the transport of NADPH generated by FNR during photosynthesis from chloroplast into the cytosol in the form of triose-phosphate. To explore the functional importance of this enzyme, mutants from Arabidopsis have been studied (Rius et al., 2006). Interestingly, the mutants showed induction of phosphorylating cytosolic (NAD-dependent) GAPDH as well as an increase in total G6PDH activity and a significant increase in NADPH content relative to the control. Furthermore, slight inhibition of growth and ROS production in *np-gapdh* mutants was associated with down-regulation of several glycolytic and photosynthetic-associated genes (Rius et al., 2006). Moreover, npGAPDH seems to be required for NADPH production, phytohormones and defence metabolites synthesis when plants are grown in a higher CO_2 environment (Mhamdi & Noctor, 2016).

4. H_2O_2 sensing and signaling

Plants are more tolerant to H_2O_2 than animals (Foyer & Noctor, 2000). H_2O_2 signaling regulate every aspect of plant biology from gene expression and translation to enzyme chemistry. Hydrogen peroxide is not very reactive like other ROS namely $OH^{\cdot}$ and superoxide and this characteristic is crucial for its role in signaling. H_2O_2 does not 'damage' most of the cell components, including proteins, but it interacts in a specific manner with certain amino-acids in some proteins that could have regulatory roles. The oxidation of other cell components like nucleic acids and lipids is probably due to the non-enzymatic conversion of H_2O_2 to OH.

The importance of H_2O_2 as a signaling molecule is likely to be determined by 'factors' reflecting plant physiology and the redox state at a certain time and in a certain compartment. The first factor is the rate of H_2O_2 production, as discussed above, peroxisomes and then chloroplasts are major sites for H_2O_2 generation. The second is the rate of processing by antioxidative enzymes and pathways. The impact of H_2O_2 will be strongly influenced by the extent to which the potent antioxidative system allows its accumulation. Moreover, oxidative perturbation of components of the antioxidative system is likely to play a part in the initial transmission of H_2O_2 signals. H_2O_2 can diffuse and reach other compartments or even distal cells far from the site where they are produced. This can be very rapid and is facilitated by the action of membrane water channels such as aquaporins (Chapter four in this book). Long distance travel of ROS signals in the apoplast is known as the 'ROS wave' (Suzuki et al., 2013). The ROS wave is (1) dependent on RBOHs such as RBOHD and RBOHF, (2) an active ROS-dependent propagation of ROS signals that can occur in response stress and, (3) mediating local and systemic acclimation to different stresses (Fichman & Mittler, 2020, 2021; Mittler et al., 2022; Zandalinas, Sengupta, Burks, Azad, & Mittler, 2019; Zandalinas, Fichman, & Mittler, 2020; Zandalinas et al., 2021). One last factor that could influence the intensity of signaling is the ability of H_2O_2 to interact with other signals or signaling cascades such as those involving MAP kinases (Mittler et al., 2022).

Cysteine thiol groups are easily oxidized by H_2O_2 to sulfenic, sulfinic, and sulfonic acids (Fig. 5). These modifications can result in changes or loss of activity, particularly for enzymes with essential cysteine residues, although a gain of function has been described where oxidation of a cysteine residue can result in enzyme activation. Oxidation of plant

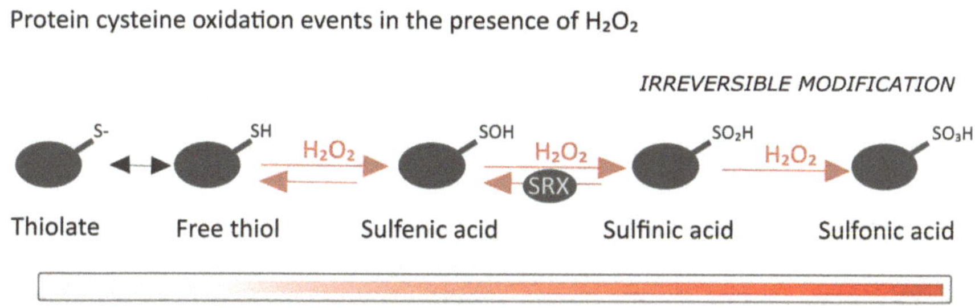

Fig. 5 Cysteine oxidation events induced by H_2O_2. Cysteine thiols (–SH) are targets of H_2O_2 and the first oxidation produces sulfenic acid. Sulfenic acid can be further oxidized by H_2O_2, to produce sulfinic acid and then sulfonic acid. Sulfinic acid has been reported to be reversed in some proteins by sulfiredoxin (SRX). The regeneration of SRX is dependent on The TRX-NTR system and NADPH. Sulfenic acid can react with protein or low-molecular-weight thiols to form inter- or intramolecular disulfides.

proteins induces in some cases changes in the subcellular localization and interactions with other proteins. Therefore, cysteine oxidation is regarded as a molecular switch regulating protein functions and chemistry (Waszczak et al., 2018). Recent methodological breakthroughs in trapping and labeling proteins that undergo cysteine oxidation in plants allowed the formation of an inventory of different cellular processes in which these modifications play a major role; these include RNA processing and amino acid metabolism (De Smet et al., 2019; Huang et al., 2019; Wei et al., 2020; Waszczak et al., 2014).

Cys is highly nucleophilic and therefore one of the amino acids most susceptible to oxidation (Roos & Messens, 2011) because of their highly reactive thiol group (–SH). Cys thiols are rather acidic and under physiological conditions, they often occur in the deprotonated thiolate (–S–) form (Conte & Carroll, 2013). The protein microenvironment, such as the presence of positively charged amino acid residues or hydrogens bonds, can significantly influence the reactivity of the Cys residue, by for example lowering the pKa value. The first oxidation step of a cysteine thiol is sulfenic acid (-SOH). Sulfenic acid is also a highly reactive species and is therefore considered to be an unstable residue. The presence of intramolecular hydrogen bonds in the environment, can, however, stabilize the sulfenic acid, reducing its reactivity (Conte & Carroll, 2013). The formation of sulfenic acid is a specific and reversible modification, making it a potent molecular switch. A second oxidation can further occur and sulfenic acid get oxidized to sulfinic ($–SO_2H$) and subsequently to sulfonic ($–SO_3H$) acid. Both are considered to be mainly irreversible modifications,

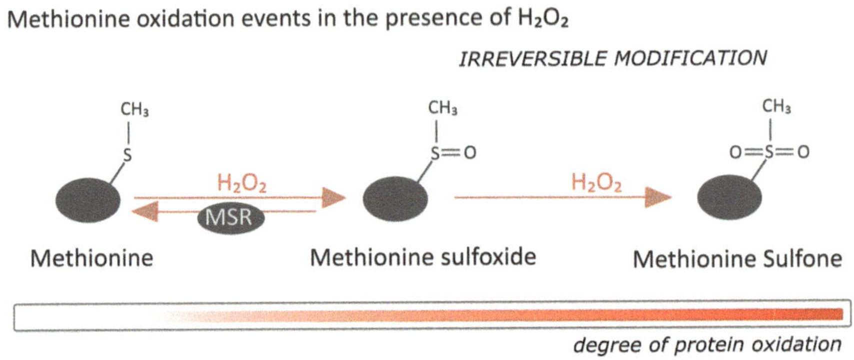

Fig. 6 Methionine oxidation events induced by H_2O_2. Methionine can be oxidized H_2O_2 to Methionine sulfoxide and Methionine sulfone. Only the first reaction is reversible and catalyzed by Methionine Sulfoxide reductase (MSR). The regeneration of MSR is dependent on the TRX-NTR system and NADPH.

although sulfinic acid can, in some cases, be reversed through the action of sulfiredoxins (Fig. 5). Alternatively, sulfenic acid can react with free thiols from proteins or low molecular weight thiols (such as glutathione) to form inter- or intramolecular disulfides. When it forms mixed-disulfides with glutathione, the protein is glutathionylated, which is often regarded as a post-translational modification that protects the protein from irreversible over-oxidation. Typically, glutathionylation is reduced by GRXs whereas protein-disulfides are reduced by TRXs.

In addition to cysteine thiols, H_2O_2 can also react with methionine residues (–S–CH_3) leading to the formation of methionine sulfoxide (–(SO–CH_3)), which can be further irreversibly oxidized to methionine sulfone (–(SO_2–CH_3)) (Fig. 6). Methionine sulfoxide can be reduced by methionine sulfoxide reductases (MSR), that use thioredoxin as electron donors (Fig. 6). Proteome-wide study of methionine oxidation in vivo revealed around 500 sites in 400 proteins that are affected by oxidative stress in catalase deficient plants. Exposure of Arabidopsis to photorespiratory stress resulted in the oxidation of methionine residues in approximately 400 proteins (Jacques et al., 2015), indicating that methionine oxidation is a dynamic modification under oxidative stress. However, compared with mechanisms involving cysteine oxidations the inventory of proteins that undergo methionine oxidation is rather scarce.

Intracellular H_2O_2 signals can be mediated by several pathways and involve transcription factors and other regulators (reviewed in He, Van Breusegem, & Mhamdi, 2018; Mittler et al., 2022). More recently, an

H_2O_2 sensor was reported from a genetic screen using plants expressing a calcium-level reporter, aequorin (Wu et al., 2020). HYDROGEN--PEROXIDE-INDUCED Ca^{2+} INCREASES (HPCA1), encodes a typical Receptor-like kinase (RLK) that has a couple of cysteine residues in the extracellular domain. In response to an increased apoplastic H_2O_2, cysteines are oxidized and form disulfide bonds which lead to autophosphorylation of the kinase domain and interaction with yet unknown calcium channels (Wu et al., 2020). However, the precise molecular mechanisms downstream of HPCA1 remain to be determined. HPCA1 was also identified as CANNOT RESPOND TO DMBQ 1 (CARD1) involved in the perception of quinones (Laohavisit et al., 2020). Sensing H_2O_2 levels at the extracellular site might be a strategy to initiate intracellular cellular transcriptional and metabolic reprogramming without the need to adjust intracellular H_2O_2. Interestingly, HPCA1 is also important to link RBOHD-dependent ROS signals and cell-to-cell signaling to stress acclimation (Fichman, Zandalinas, Peck, Luan, & Mittler, 2022). HPCA1, is required during the ROS wave, calcium wave, but not electric wave. In response to stress, HPCA1 is required for both local and systemic responses and acclimation to high light and in cell-to-cell communication and propagation of ROS signals in response to bacteria infection or salt stress, but not wounding (Fichman et al., 2022).

5. Conclusions and perspectives

H_2O_2, the more stable ROS, is produced by vital cellular processes and is involved in signaling stress conditions. H_2O_2 generation and metabolism in higher plants are very complex because most of the enzymes and proteins involved are produced by several hundred genes. Reverse genetics approaches and the development of chemical methods and genetic backgrounds to study oxidative stress has helped in elucidating the functions of some of these genes. However, the roles in H_2O_2 signaling or metabolism of many of these genes remain obscure. The diversity of ROS-producing and processing pathways is also highlighted by the different cellular compartments involved and the connections between these compartments are not always clear. Some of these organelles produce extensions during stress conditions, and these might be involved in H_2O_2 transport and in communications between organelles (Hirt, 2016; Mittler et al., 2022).

In vivo ROS imaging in cells and tissues has contributed significantly to many of the recent advances in understanding oxidative signaling. Of particular relevance are advances in H_2O_2 imaging using non-invasive and non-destructive genetically encoded probes that have helped to understand the dynamics of H_2O_2 diffusion between chloroplasts and the nucleus during stress (Exposito-Rodriguez et al., 2017). These technologies continue to gain success and are used within the community to understand organelles' communications (Ugalde et al., 2021) with the cytosol, the nucleus, and the apoplast, and could in the future help to reveal how H_2O_2 signals diffuse far from their sites of production. Another important question is the role of other 'oxidizing' compartments such as the vacuoles and the endoplasmic reticulum (Mittler et al., 2022). For instance, the vacuoles contain both antioxidants such as ascorbate and glutathione, and transporters for H_2O_2 and have been proposed to function as H_2O_2 sinks, while the ER and its close proximity to the nucleus and chloroplasts might be key to facilitating the exchange of H_2O_2 and other redox signals with the nucleus.

References

Al-Hajaya, Y., Karpinska, B., Foyer, C. H., & Baker, A. (2022). Nuclear and peroxisomal targeting of catalase. *Pant, Cell & Environment, 45*, 1096–1108.

Artus, N. N., & Edwards, G. E. (1985). NAD-malic enzyme from plants. *FEBS lett, 182*, 225–233.

Asada, K. (1999). The water–water cycle in chloroplasts: Scavenging of active oxygen and dissipation of excess photons. *Annual Review of Plant Physiology and Plant Molecular Biology, 50*, 601–639.

Asada, K. (2006). Production and scavenging of reactive oxygen species in chloroplasts and their functions. *Plant Physiology, 141*, 391–396.

Attacha, S., Solbach, D., Bela, K., Moseler, A., Wagner, S., Schwarzländer, M., et al. (2017). Glutathione peroxidase-like enzymes cover five distinct cell compartments and membrane surfaces in Arabidopsis thaliana. *Plant Cell Environment, 40*, 1281–1295.

Averill, R. H., Bailey-Serres, J., & Kruger, N. J. (1998). Co-operation between cytosolic and plastidic oxidative pentose phosphate pathways revealed by 6-phosphogluconate dehydrogenase-deficient genotypes of maize. *The Plant Journal, 14*, 449–457.

Baker, A., Graham, I. A., Holdsworth, M., Smith, S. M., & Theodoulou, F. L. (2006). *Trends Plant Sciences, 11*, 124–132.

Ball, L., Accotto, G., Bechtold, U., Creissen, G., Funck, D., Jimenez, A., et al. (2004). Evidence for a direct link between glutathione biosynthesis and stress defense gene expression in Arabidopsis. *The Plant Cell, 16*, 2448–2462.

Bela, K., Riyazuddin, R., & Csiszár, J. (2022). Plant Glutathione Peroxidases: Non-Heme Peroxidases with Large Functional Flexibility as a Core Component of ROS-Processing Mechanisms and Signalling. *Antioxidants (Basel, Switzerland), 11*(8), 1624. https://doi.org/10.3390/antiox11081624

Bela, K., Horváth, E., Gallé, Á., Szabados, L., Tari, I., & Csiszár, J. (2015). Plant glutathione peroxidases: Emerging role of the antioxidant enzymes in plant development and stress responses. *Journal of Plant Physiology, 176*, 192–201.

Cairns, N. G., Pasternak, M., Wachter, A., Cobbett, C. S., & Meyer, A. J. (2006). Maturation of Arabidopsis seeds is dependent on glutathione biosynthesis within the embryo. *Plant Physiology, 141*, 446–455.

Chamnongpol, S., Willekens, H., Langebartels, C., Van Montagu, M., Inzé, D., & Van Camp, W. (1996). Transgenic tobacco with a reduced catalase activity develops necrotic lesions and induces pathogenesis related expression under high light. *The Plant Journal, 10*, 491–503.

Chamnongpol, S., Willekens, H., Moeder, W., Langebartels, C., Sandermann, H., Van Montagu, M., et al. (1998). Defense activation and enhanced pathogen tolerance induced by H_2O_2 in transgenic tobacco. *Proceedings of the National Academy of Sciences of the United States of America, 95*, 5818–5823.

Chang, C. C., Slesak, I., Jordá, L., Sotnikov, A., Melzer, M., Miszalski, Z., et al. (2009). Arabidopsis chloroplastic glutathione peroxidases play a role in cross talk between photooxidative stress and immune responses. *Plant Physiology, 150*, 670–683.

Chang, G., & Tong, L. (2003). Structure and function of malic enzymes, a new class of oxidative decarboxylases. *Biochemistry, 42*, 12721–12733.

Chaouch, S., Queval, G., Vanderauwera, S., Mhamdi, A., Vandorpe, M., Langlois-Meurinne, M., et al. (2010). Peroxisomal hydrogen peroxide is coupled to biotic defense responses by ISOCHORISMATE SYNTHASE1 in a daylength-related manner. *Plant Physiology, 153*, 1692–1705.

Chew, O., Whelan, J., & Millar, A. H. (2003). Molecular definition of the ascorbate-glutathione cycle in Arabidopsis mitochondria reveals dual targeting of antioxidant defenses in plants. *The Journal of biological chemistry, 278*(47), 46869–46877. https://doi.org/10.1074/jbc.M307525200

Cobbett, C. S., May, M. J., Howden, R., & Rolls, B. (1998). The glutathione-deficient, cadmium-sensitive mutant, cad2-1, of Arabidopsis thaliana is deficient in γ-glutamylcysteine synthetase. *The Plant Journal, 16*, 73–78.

Conte, M. L., & Carroll, K. (2013). The Chemistry of Thiol Oxidation and Detection. In U. Jakob, & D. Reichmann (Eds.). *Oxidative Stress and Redox Regulation*. Dordrecht: Springer.

Corpas, F. J., de la Colina, C., Sanchez-Rasero, F., & Del Rio, L. A. (1997). *Journal of Plant Physiology, 151*, 246–250.

Corpas, F. J., Palma, J. M., Sandalio, L. M., Valderrama, R., Barroso, J. B., & Del Río, L. A. (2008). *Journal of Plant Physiology, 165*, 1319–1330

Creissen, G., Firmin, J., Fryer, M., Kular, B., Leyland, M., Reynolds, H., et al. (1999). Elevated glutathione biosynthetic capacity in the chloroplasts of transgenic tobacco paradoxically causes increased oxidative stress. *The Plant Cell, 11*, 1277–1291.

Davletova, S., Rizhsky, L., Liang, H., Shengqiang, Z., Oliver, D. J., Coutu, J., ... Mittler, R. (2005). Cytosolic ascorbate peroxidase 1 is a central component of the reactive oxygen gene network of Arabidopsis. *The Plant cell, 17*(1), 268–281. https://doi.org/10.1105/tpc.104.026971

De Smet, B., Willems, P., Fernandez-Fernandez, A. D., Alseekh, S., Fernie, A. R., Messens, J., et al. (2019). In vivo detection of protein cysteine sulfenylation in plastids. *The Plant Journal, 97*, 765–778.

Debnam, P. M., Fernie, A. R., Leisse, A., Golding, A., Bowsher, C. G., Grimshaw, C., et al. (2004). Altered activity of the P2 isoform of plastidic glucose 6-phosphatedehydrogenase in tobacco (Nicotiana tabacum cv. Samsun) causes changes in carbohydrate metabolism and response to oxidative stress in leaves. *The Plant Journal, 38*, 49–59.

Dellero, Y., Mauve, C., Boex-Fontvieille, E., Flesch, V., Jossier, M., Tcherkez, G., et al. (2015). Experimental evidence for a hydride transfer mechanism in plant glycolate oxidase catalysis. *Journal of Biological Chemistry, 290*, 1689–1698.

Dellero, Y., Jossier, M., Glab, N., Oury, C., Tcherkez, G., & Hodges, M. (2016). Decreased glycolate oxidase activity leads to altered carbon allocation and leaf senescence after a transfer from high CO_2 to ambient air in Arabidopsis thaliana. *Journal Experimental Botany, 67*, 3149–3163.

Dghim, A. A., Mhamdi, A., Vaultier, M. N., Hasenfratz-Sauder, M. P., Le Thiec, D., Dizengremel, P., et al. (2013). Analysis of cytosolic isocitrate dehydrogenase and glutathione reductase 1 in photoperiodinfluenced responses to ozone using Arabidopsis knockout mutants. *Plant, Cell & Environment, 36*, 1981–1991.

Dietz, K. J. (2003). Plant peroxiredoxins. *Annual Review of Plant Biology, 54*, 93–107.

Dietz, K. J. (2008). Redox signal integration: From stimulus to networks and genes. *Physiologia Plantarum, 133*, 459–468.

Dietz, K. J. (2011). Peroxiredoxins in plants and cyanobacteria. *Antioxidant and Redox Signaling, 15*, 1129–1159.

Dietz, K. J. (2016). Thiol-based peroxidases and ascorbate peroxidases: Why plants rely on multiple peroxidase systems in the photosynthesizing chloroplast? *Molecules and Cells, 39*, 20–25.

Dixon, D. P., & Edwards, R. (2010). Glutathione transferases. *The Arabidopsis Book, 8*, e0131. https://doi.org/10.1199/tab.0131

Dixon, D. P., Lapthorn, A., & Edwards, R. (2002). Plant glutathione transferases. *Genome biology, 3*(3), REVIEWS3004. https://doi.org/10.1186/gb-2002-3-3-reviews3004

Dixon, D. P., Hawkins, T., Hussey, P. J., & Edwards, R. (2009). Enzyme activities and subcellular localization of members of the Arabidopsis glutathione transferase superfamily. *Journal Experimental Botany, 60*, 1207–1218.

Drincovich, M. F., Casati, P., & Andreo, C. S. (2001). NADP-malic enzyme from plants: A ubiquitous enzyme involved in diferent metabolic pathways. *FEBS Letters, 490*, 1–6.

Dvořák, P., Krasylenko, Y., Ovečka, M., Basheer, J., Zapletalová, V., Šamaj, J., et al. (2021). In vivo light-sheet microscopy resolves localisation patterns of FSD1, a superoxide dismutase with function in root development and osmoprotection. *Plant, Cell & Environment, 44*, 68–87.

Exposito-Rodriguez, M., Laissue, P. P., Yvon-Durocher, G., Smirnoff, N., & Mullineaux, P. M. (2017). Photosynthesis-dependent H_2O_2 transfer from chloroplasts to nuclei provides a high-light signalling mechanism. *Nature Communications, 8*, 49.

Fichman, Y., & Mittler, R. (2020). Rapid systemic signaling during abiotic and biotic stresses: Is the ROS wave master of all trades? *The Plant Journal, 102*, 887–896.

Fichman, Y., & Mittler, R. (2021). Integration of electric, calcium, reactive oxygen species and hydraulic signals during rapid systemic signaling in plants. *The Plant Journal, 107*, 7–20.

Fichman, Y., Zandalinas, S. I., Peck, S., Luan, S., & Mittler, R. (2022). HPCA1 is required for systemic reactive oxygen species and calcium cell-to-cell signaling and plant acclimation to stress. *The Plant Cell, 34*, 4453–4471.

Foreman, J., Demidchik, V., Bothwell, J. H. F., Mylona, P., Miedema, H., Torres, M. A., et al. (2003). Reactive oxygen species produced by NADPH oxidase regulate plant cell growth. *Nature, 422*, 442–446.

Foyer, C. H., & Noctor, G. (2000). Oxygen Processing in Photosynthesis: Regulation and Signaling. *New Phytologist, 146*, 359–388.

Foyer, C. H., & Noctor, G. (2003). Redox sensing and signalling associated with reactive oxygen in chloroplasts, peroxisomes and mitochondria. *Physiologia Plantarum, 119*, 355–364.

Foyer, C. H., & Noctor, G. (2009). Redox regulation in photosynthetic organisms: Signaling, acclimation, and practical implications. *Antioxidants & Redox Signaling, 11*, 861–905.

Foyer, C. H., & Shigeoka, S. (2011). Understanding oxidative stress and antioxidant functions to enhance photosynthesis. *Plant Physiology, 155*, 93–100.

Foyer, C. H., & Noctor, G. (2011). Ascorbate and glutathione: the heart of the redox hub. *Plant physiology, 155*(1), 2–18.

Foyer, C. H., Bloom, A. J., Queval, G., & Noctor, G. (2009). Photorespiratory metabolism: genes, mutants, energetics, and redox signaling. *Annual review of plant biology, 60*, 455–484.

Foyer, C. H., Baker, A., Wright, M., Sparkes, I. A., Mhamdi, A., Schippers, J. H. M., et al. (2021). On the move: Redox-dependent protein relocation in plants. *Journal of Experimental Botany, 71*, 620–631.

Furuichi, M., Suzuki, N., Dhakshnamoorhty, B., Minagawa, H., Yamagishi, R., Watanabe, Y., et al. (2008). X-ray structures of Aerococcus viridians lactate oxidase and its complex with d-lactate at pH 4.5 show an α-hydroxy acid oxidation mechanism. *Journal of Molecular Biology, 378*, 436–446.

Gemes, K., Kim, Y. J., Park, K. Y., Moschou, P. N., Andronis, E., Valassaki, C., et al. (2016). An NADPH-oxidase/polyamine oxidase feedback loop controls oxidative burst under salinity. *Plant Physiology, 172*, 1418–1431.

Graeve, K., von Schaewen, A., & Scheibe, R. (1994). Purification, characterization, and cDNA sequence of glucose-6-phosphate dehydrogenase from potato (Solanum tuberosum L.). *The Plant Journal, 5*, 353–361.

Halliwell, B, & Gutteridge, J. M. C (2015). *Free Radicals in Biology and Medicine* (5th edition). New York.: Oxford University Press.

Han, Y., Mhamdi, A., Chaouch, S., & Noctor, G. (2013). Regulation of basal and oxidative stress-triggered jasmonic acid-related gene expression by glutathione. *Plant, Cell & Environment, 36*, 1135–1146.

Han, Y., Chaouch, S., Mhamdi, A., Queval, G., Zechmann, B., & Noctor, G. (2013). Functional analysis of Arabidopsis mutants points to novel roles for glutathione in coupling H_2O_2 to activation of salicylic acid accumulation and signaling. *Antioxidants & Redox Signaling, 18*, 2106–2121.

Hansch, R., Lang, C., Riebeseel, E., Lindigkeit, R., Gessler, A., Rennenberg, H., et al. (2006). Plant sulfite oxidase as novel producer of H_2O_2: Combination of enzyme catalysis with a subsequent non-enzymatic reaction step. *Journal of Biological Chemistry, 281*, 6884–6888.

He, H., Van Breusegem, F., & Mhamdi, A. (2018). Redox-dependent control of nuclear transcription in plants. *Journal of Experimental Botany, 69*, 3359–3372.

Herbette, S., Lenne, C., Leblanc, N., Julien, J., Drevet, J. R., & Roeckel-Drevet, P. (2002). Two GPX-like proteins from Lycopersicon esculentum and Helianthus annuus are antioxidant enzymes with phospholipid hydroperoxide glutathione peroxidase and thioredoxin peroxidase activities. *European Journal of Biochemistry, 269*, 2414–2420.

Hicks, L. M., Cahoon, R. E., Bonner, E. R., Rivard, R. S., Sheffield, J., & Jez, J. M. (2007). Thiol-based regulation of redox-active glutamate-cysteine ligase from Arabidopsis thaliana. *The Plant Cell, 19*, 2653–2661.

Hirt, H. (2016). Aquaporins link ROS signaling to plant immunity. *Plant Physiology, 171*, 1540.

Hothorn, M., Wachter, A., Gromes, R., Stuwe, T., Rausch, T., & Scheffzek, K. (2006). Structural basis for the redox control of plant glutamate cysteine ligase. *Journal of Biological Chemistry, 281*, 27557–27565.

Huang, J., Willems, P., Wei, B., Tian, C., Ferreira, R. B., Bodra, N., et al. (2019). Mining for protein S-sulfenylation in Arabidopsis uncovers redox-sensitive sites. *Proceedings of the National Academy of Sciences of the United States of America, 116*, 21256–21261.

Iqbal, A., Yabuta, Y., Takeda, T., Nakano, Y., & Shigeoka, S. (2006). Hydroperoxide reduction by thioredoxin-specific glutathione peroxidase isoenzymes of Arabidopsis thaliana. *FEBS Journal, 273*, 5589–5597.

Jacques, S., Ghesquière, B., De Bock, P. J., Demol, H., Wahni, K., Willems, P., et al. (2015). Protein methionine sulfoxide dynamics in *Arabidopsis thaliana* under oxidative stress. *Molecular Cell Proteomics, 14*, 1217–1229.

Jiménez-Quesada, M. J., Traverso, J.Á., & Alché, J. D. (2016). NADPH oxidase-dependent superoxide production in plant reproductive tissues. *Frontiers in Plant Sciences, 7*, 359.

Kangasjarvi, S., Neukermans, J., Li, S. C., Aro, E. M., & Noctor, G. (2012). Photosynthesis, photorespiration, and light signalling in defence responses. *Journal of Experimental Botany, 63*, 1619–1636.

Karpinski, S., Escobar, C., Karpinska, B., Creissen, G., & Mullineaux, P. M. (1997). Photosynthetic electron transport regulates the expression of cytosolic ascorbate peroxidase genes in Arabidopsis during excess light stress. *The Plant Cell, 9*, 627–640.

Kelly, G. J., & Gibbs, M. (1973). Nonreversible D-glyceraldehyde 3-phosphate dehydrogenase of plant tissues. *Plant Physiol, 52*, 111–118.

Kendall, A. C., Keys, A. J., Turner, J. C., Lea, P. J., & Miflin, B. J. (1983). The isolation and characterisation of a catalase-deficient mutant of barley (Hordeum vulgare L.). *Planta, 159*, 505–511.

Kerchev, P., Waszczak, C., Lewandowska, A., Willems, P., Shapiguzov, A., Li, Z., et al. (2016). Lack of GLYCOLATE OXIDASE1, but not GLYCOLATE OXIDASE2, attenuates the photorespiratory phenotype of CATALASE2-deficient Arabidopsis. *Plant Physiol, 171*, 1704–1719.

Kirchsteiger, K., Ferrandez, J., Pascual, M. B., Gonzalez, M., & Cejudo, F. J. (2012). NADPH thioredoxin reductase C is localized in plastids of photosynthetic and non-photosynthetic tissues and is involved in lateral root formation in Arabidopsis. *The Plant Cell, 24*, 1534–1548.

Kliebenstein, D. J., Monde, R. A., & Last, R. L. (1998). Superoxide dismutase in Arabidopsis: An eclectic enzyme family with disparate regulation and protein localization. *Plant Physiology, 118*, 637–650.

Koussevitzky, S., Suzuki, N., Huntington, S., Armijo, L., Sha, W., Cortes, D., et al. (2008). Ascorbate peroxidase 1 plays a key role in the response of Arabidopsis thaliana to stress combination. *The Journal of Biological Chemistry, 283*, 34197–34203.

Lai, L. B., Tausta, L. S., & Nelson, T. M. (2002). Differential regulation of transcripts encoding cytosolic NADPmalic enzyme in C3 and C4 Flaveria species. *Plant Physiology, 128*, 140–149.

Laohavisit, A., Wakatake, T., Ishihama, N., Mulvey, H., Takizawa, K., Suzuki, T., et al. (2020). Quinone perception in plants via leucine-rich-repeat receptor-like kinases. *Nature, 587*, 92–97.

Lemaire, S. D. (2004). The glutaredoxin family in oxygenic photosynthetic organisms. *Photosynthesis Research, 79*, 305–318.

Leterrier, M., Del Rio, L. A., & Corpas, F. J. (2007). Cytosolic NADP–isocitrate dehydrogenase of pea plants: genomic clone characterization and functional analysis under abiotic stress conditions. *Free Radical Research, 41*, 191–199.

Li, S., Mhamdi, A., Clement, C., Jolivet, Y., & Noctor, G. (2013). Analysis of knockout mutants suggests that Arabidopsis NADP-MALIC ENZYME2 does not play an essential role in responses to oxidative stress of intracellular or extracellular origin. *Plant, Cell & Environment, 64*, 3605–3614.

Li, S., Mhamdi, A., Trotta, A., Kangasjärvi, S., & Noctor, G. (2014). The protein phosphatase subunit PP2A-B′γ is required to suppress day length-dependent pathogenesis responses triggered by intracellular oxidative stress. *New Phytologist, 202*, 145–160.

Liedschulte, V., Wachter, A., Zhigang, A., & Rausch, T. (2010). Exploiting plants for glutathione (GSH) production: Uncoupling GSH synthesis from cellular controls results in unprecedented GSH accumulation. *Journal of Plant Biotechnology, 8*, 807–820.

Macheroux, P., Kieweg, V., Massey, V., Söderlind, E., Stenberg, K., & Lindqvist, Y. (1993). Role of tyrosine 129 in the active site of spinach glycolate oxidase. *European Journal of Biochemistry, 213*, 1047–1054.

Marino, D., Gonzalez, E. M., Frendo, P., Puppo, A., & Arrese-Igor, C. (2007). NADPH recycling systems in oxidative stressed pea nodules: A key role for the NADP +-dependent isocitrate dehydrogenase. *Planta, 225*, 413–421.

Marty, L., Siala, W., Schwarzländer, M., Fricker, M. D., Wirtz, M., Sweetlove, L. J., et al. (2009). The NADPH-dependent thioredoxin system constitutes a functional backup for cytosolic glutathione reductase in Arabidopsis. *Proceedings of the National Academy of Sciences of the United States of America, 106*, 9109–9114.

Maurino, V. G., Wheeler, M. C. G., Andreo, C. S., & Drincovich, M. F. (2009). Redundancy is sometimes seen only by the uncritical: Does Arabidopsis need six malic enzyme isoforms? *Plant Sci, 176*, 715–721.

Melicher, P., Dvořák, P., Krasylenko, Y., Shapiguzov, A., Kangasjärvi, J., Šamaj, J., et al. (2022). Arabidopsis iron superoxide dismutase FSD1 protects against methyl viologen-induced oxidative stress in a copper-dependent manner. *Frontiers in Plant Sciences, 13*, 823561.

Meyer, Y., Siala, W., Bashandy, T., Riondet, C., Vignols, F., & Reichheld, J. P. (2008). Glutaredoxins and thioredoxins in plants. *Biochimica Biophysica Acta, 1783*, 589–600.

Mhamdi, A., & Noctor, G. (2015). Analysis of the roles of the Arabidopsis peroxisomal isocitrate dehydrogenase in leaf metabolism and oxidative stress. *Environmental and Experimental Botany, 114*, 22–29.

Mhamdi, A., & Noctor, G. (2016). High CO_2 primes plant biotic stress defences through redox-linked pathways. *Plant Physiology, 172*, 929–942.

Mhamdi, A., & Van Breusegem, F. (2018). Reactive oxygen species in plant development. *Development, 9*, 145.

Mhamdi, A., Noctor, G., & Baker, A. (2012). Plant catalases: Peroxisomal redox guardians. *Archives of Biochemistry and Biophysics, 525*, 181–194.

Mhamdi, A., Han, Y., & Noctor, G. (2013). Glutathione-dependent phytohormone responses: teasing apart signaling and antioxidant functions. *Plant signaling & behavior, 8*(5), e24181.

Mhamdi, A., Mauve, C., Gouia, H., Saindrenan, P., Hodges, M., & Noctor, G. (2010). Cytosolic NADP-dependent isocitrate dehydrogenase contributes to redox homeostasis and the regulation of pathogen responses in Arabidopsis leaves. *Plant, Cell & Environment, 33*, 1112–1123.

Mhamdi, A., Queval, G., Chaouch, S., Vanderauwera, S., Van Breusegem, F., & Noctor, G. (2010). Catalase function in plants: A focus on Arabidopsis mutants as stress-mimic models. *Journal of Experimental Botany, 61*, 4197–4220.

Mhamdi, A., Hager, J., Chaouch, S., Queval, G., Han, Y., Taconnat, L., et al. (2010). Arabidopsis GLUTATHIONE REDUCTASE 1 plays a crucial role in leaf responses to intracellular hydrogen peroxide and in ensuring appropriate gene expression through both salicylic acid and jasmonic acid signaling pathways. *Plant Physiology, 153*, 1144–1160.

Miao, Y., Lv, D., Wang, P., Wang, X. C., Chen, J., Miao, C., & Song, C. P. (2006). An Arabidopsis glutathione peroxidase functions as both a redox transducer and a scavenger in abscisic acid and drought stress responses. *The Plant cell, 18*(10), 2749–2766. https://doi.org/10.1105/tpc.106.044230

Mignolet, L., Xu, E., Idänheimo, N., Hoeberichts, F., Mühlenbock, P., Brosché, M., et al. (2016). Spreading the news: subcellular and organellar reactive oxygen species production and signalling. *Journal of Experimental Botany, 67*, 3831–3844.

Mittler, R. (2002). Oxidative stress, antioxidants and stress tolerance. *Trends in plant science*, 405–410.

Mittler, R. (2017). ROS are good. *Trends in Plant Sciences, 22*, 11–19.

Mittler, R., Zandalinas, S. I., Fichman, Y., & Van Breusegem, F. (2022). Reactive oxygen species signalling in plant stress responses. *Nature Reviews Molecular Cell Biology*. https://doi.org/10.1038/s41580-022-00499-2

Morgan, M. J., Lehmann, M., Schwarzländer, M., Baxter, C. J., Sienkiewicz-Porzucek, A., Williams, T. C. R., et al. (2008). Decrease in manganese superoxide dismutase leads to reduced root growth and affects tricarboxylic acid cycle flux and mitochondrial redox homeostasis. *Plant Physiology, 147*, 101–114.

Mubarakshina, M. M., Ivanov, B. N., Naydov, I. A., Hillier, W., Badger, M. R., & Krieger-Liszkay, A. (2010). Production and diffusion of chloroplastic H_2O_2 and its implication to signalling. *Journal of Experimental Botany, 61*, 3577–3587.

Myouga, F., Hosoda, C., Umezawa, T., Iizumi, H., Kuromori, T., Motohashi, R., et al. (2008). A heterocomplex of iron superoxide dismutases defends chloroplast nucleoids against oxidative stress and is essential for chloroplast development in Arabidopsis. *The Plant Cell, 20*, 3148–3162.

Noctor, G., & Foyer, C. H. (1998). Ascorbate and glutathione, keeping active oxygen under control. *Annual Review of Plant Physiology and Plant Molecular Biology, 49*, 249–279.

Noctor, G., & Foyer, C. H. (2017). Update on redox compartmentation intracellular redox compartmentation and ROS-related communication in regulation and signaling. *Plant Physiology, 171*, 1581–1592.

Noctor, G., Lelarge-Trouverie, C., & Mhamdi, A. (2015). The metabolomics of oxidative stress. *Phytochemistry, 112*, 33–53.

Noctor, G., Mhamdi, A., & Foyer, C. H. (2016). Oxidative stress and antioxidative systems: recipes for successful data collection and interpretation. *Plant, cell & environment, 39*(5), 1140–1160.

Noctor, G., Reichheld, J. P., & Foyer, C. H. (2018). ROS-related redox regulation and signaling in plants. *Seminars in Cell & Developmental Biology, 80*, 3–12.

Noctor, G., Strohm, S., Jouanin, L., Kunert, K. J., Foyer, C. H., & Rennenberg, H. (1996). Synthesis of glutathione in leaves of transgenic poplar overexpressing γ-glutamylcysteine synthetase. *Plant Physiology, 112*, 1071–1078.

Noctor, G., Mhamdi, A., Chaouch, S., Han, Y., Neukermans, J., Marquez-Garcia, B., et al. (2012). Glutathione in plants: An integrated overview. *Plant, Cell & Environment, 35*, 454–484.

Parisy, V., Poinssot, B., Owsianowski, L., Buchala, A., Glazebrook, J., & Mauch, F. (2007). Identification of PAD2 as a γ-glutamylcysteine synthetase highlights the importance of glutathione in disease resistance in Arabidopsis. *The Plant Journal, 49*, 159–172.

Pasternak, M., Lim, B., Wirtz, M., Hell, R., Cobbett, C. S., & Meyer, A. J. (2008). Restricting glutathione biosynthesis to the cytosol is sufficient for normal plant development. *The Plant Journal, 53*, 999–1012.

Pérez-Ruiz, J. M., & Cejudo, F. J. (2009). A proposed reaction mechanism for rice NADPH thioredoxin reductase C, an enzyme with protein disulfide reductase activity. *FEBS letters, 583*(9), 1399–1402. https://doi.org/10.1016/j.febslet.2009.03.067

Phua, S. Y., De Smet, B., Remacle, C., Chan, K. X., & Van Breusegem, F. (2021). Reactive oxygen species and organellar signaling. *Journal of Experimental Botany, 72*, 5807–5824.

Pilon, M., Ravet, K., & Tapken, W. (2011). The biogenesis and physiological function of chloroplast superoxide dismutases. *Biochimca Biophysica Acta (BBA) – Bioenergetics, 1807*, 989–998.

Pospisil, P. (2009). Production of reactive oxygen species by photosystem II. *Biochimca Biophysica Acta (BBA) – Bioenergetics, 1787*, 1151–1160.

Queval, G., Neukermans, J., Vanderauwera, S., Van Breusegem, F., & Noctor, G. (2012). Day length is a key regulator of transcriptomic responses to both CO_2 and H_2O_2 in Arabidopsis. *Plant, Cell & Environment, 35*, 374–387.

Queval, G., Issakidis-Bourguet, E., Hoeberichts, F. A., Vandorpe, M., Gakière, B., Vanacker, H., et al. (2007). Conditional oxidative stress responses in the Arabidopsis photorespiratory mutant cat2 demonstrate that redox state is a key modulator of day-length-dependent gene expression, and define photoperiod as a crucial factor in the regulation of H_2O_2-induced cell death. *The Plant Journal, 52*, 640–657.
Rahantaniaina, M. S., Tuzet, A., Mhamdi, A., & Noctor, G. (2013). Missing links in understanding redox signalling via thiol/disulfide modulation: How is glutathione oxidized in plants? *Frontiers in Plant Science, 25*, 4–477.
Rahantaniaina, M. S., Li, S., Chatel-Innocenti, G., Tuzet, A., Issakidis-Bourguet, E., Mhamdi, A., et al. (2017). Cytosolic and chloroplastic DHARs cooperate in oxidative stress-driven activation of the salicylic acid pathway. *Plant Physiology, 174*, 956–971.
Reichheld, J. P., Khafif, M., Riondet, C., Droux, M., Bonnard, G., & Meyer, Y. (2007). Inactivation of thioredoxin reductases reveals a complex interplay between thioredoxin and glutathione pathways in Arabidopsis development. *The Plant Cell, 19*, 1851–1865.
Reumann, S., Babujee, L., Ma, C., Wienkoop, S., Siemsen, T., Antonicelli, G. E., et al. (2007). Proteome analysis of Arabidopsis leaf peroxisomes reveals novel targeting peptides, metabolic pathways, and defense mechanisms. *The Plant Cell, 19*, 3170–3193.
Reumann, S., Quan, S., Aung, K., Yang, P., Manandhar-Shrestha, K., Holbrook, D., ... Hu, J. (2009). In-depth proteome analysis of Arabidopsis leaf peroxisomes combined with in vivo subcellular targeting verification indicates novel metabolic and regulatory functions of peroxisomes. *Plant physiology, 150*(1), 125–143. https://doi.org/10.1104/pp.109.137703
Rius, S. P., Casati, P., Iglesias-Diego, A., & Gomez-Casati, F. (2006). Characterization of an Arabidopsis thaliana mutant lacking a cytosolic non-phosphorylating glyceraldehyde-3-phosphate dehydrogenase. *Plant Molecular Biology, 61*, 945–957.
Roos, G., & Messens, J. (2011). Protein sulfenic acid formation: From cellular damage to redox regulation. *Free Radicals Biology and Medecine, 51*, 314–326.
Rouhier, N., & Jacquot, J. P. (2005). The plant multigenic family of thiol peroxidases. *Free Radicals Biology and Medecine, 38*, 1413–1421.
Sarry, J. E., Kuhn, L., Ducruix, C., Lafaye, A., Junot, C., Hugouvieux, V., et al. (2006). The early responses of Arabidopsis thaliana cells to cadmium exposure explored by protein and metabolite profiling analyses. *Proteomics, 6*, 2180–2198.
Scheible, W. R., Krapp, A., & Stitt, M. (2000). Reciprical diurnal changes of phosphoenolpyruvate carboxylase expression and cytosolic pyruvate kinase, citrate synthase and NADP-isocitrate dehydrogenase expression regulate organic acid metabolism during nitrate assimilation in tobacco leaves. *Plant, Cell & Environment, 23*, 1155–1167.
Scheible, W. R., Gonzalez-Fontes, A., Lauerer, M., Müller-Röber, B., Caboche, M., & Stitt, M. (1997). Nitrate acts as a signal to induce organic acid metabolism and repress starch metabolism in tobacco. *The Plant Cell, 9*, 783–798.
Schilmiller, A. L., Koo, A. J., & Howe, G. A. (2007). Functional diversification of acyl-coenzyme A oxidases in jasmonic acid biosynthesis and action. *Plant Physiology, 143*, 812–824.
Serrato, A., Pérez-Ruiz, J. M., Spínola, M. C., & Cejudo, F. J. (2004). A novel NADPH thioredoxin reductase, localized in the chloroplast, which deficiency causes hypersensitivity to abiotic stress in Arabidopsis thaliana. *Journal Biological Chemistry, 279*, 43821–43827.
Shigeoka, S., Ishikawa, T., Tamoi, M., Miyagawa, Y., Takeda, T., Yabuta, Y., et al. (2002). Regulation and function of ascorbate peroxidase isoenzymes. *Journal of Experimental Botany, 53*, 1305–1319.
Smirnoff, N., & Arnaud, D. (2019). Hydrogen peroxide metabolism and functions in plants. *New Phytologist, 221*, 1197–1214.

Smirnoff, N., Conklin, P. L., & Loewus, F. A. (2001). Biosynthesis of ascorbic acid in plants: A renaissance. *Annual Review of Plant Physiology and Plant Molecular Biology, 52*, 437.

Smith, P. M. C., & Atkins, C. A. (2002). Urine biosynthesis. Big in cell division, even bigger in nitrogen assimilation. *Plant Physiology, 128*, 793–802.

Suzuki, N., Miller, G., Salazar, C., Mondal, H. A., Shulaev, E., Cortes, D. F., et al. (2013). Temporal-spatial interaction between reactive oxygen species and abscisic acid regulates rapid systemic acclimation in plants. *The Plant Cell, 25*, 3553–3569.

Tamaoki, M., Freeman, J. L., & Pilon-Smits, E. A. H. (2008). Cooperative ethylene and jasmonic acid signaling regulates selenite resistance in Arabidopsis. *Plant Physiology, 146*, 1219–1230.

Terai, Y., Ueno, H., Ogawa, T., Sawa, Y., Miyagi, A., Kawai-Yamada, M., et al. (2020). Dehydroascorbate reductases and glutathione set a threshold for high-light–induced ascorbate accumulation. *Plant Physiology, 183*, 112–122.

Timm, S., & Bauwe, H. (2013). The variety of photorespiratory phenotypes employing the current status for future research directions on photorespiration. *Plant Biology, 15*, 737–747.

Ugalde, J. M., Fuchs, P., Nietzel, T., Cutolo, E. A., Homagk, M., Vothknecht, U. C., et al. (2021). Chloroplast-derived photooxidative stress causes changes in H_2O_2 and EGSH in other subcellular compartments. *Plant Physiology, 186*, 125–141.

Valderrama, R., Corpas, F. J., Carrera, s A., Gómez-Rodriguez, M. V., Chaki, M., Pedrajas, J. R., et al. (2006). The dehydrogenase-mediated recycling of NADPH is a key antioxidant system against salt-induced oxidative stress in olive plants. *Plant, Cell & Environment, 29*, 1449–1459.

Van Breusegem, F., Slooten, L., Stassart, J. M., Moens, T., Botterman, J., Van Montagu, M., et al. (1999). Overproduction of Arabidopsis thaliana FeSOD confers oxidative stress tolerance to transgenic maize. *Plant Cell Physiology, 40*, 515–523.

Van Camp, W., Capiau, K., Van Montagu, M., Inzé, D., & Slooten, L. (1996). Enhancement of oxidative stress tolerance in transgenic tobacco plants overproducing Fe-superoxide dismutase in chloroplasts. *Plant Physiology, 112*, 1703–1714.

Vandenabeele, S., Vanderauwera, S., Vuylstecke, M., Rombauts, S., Langebartels, C., Seidlitz, H. K., et al. (2004). Catalase deficiency drastically affects gene expression induced by high light in Arabidopsis thaliana. *The Plant Journal, 39*, 45–58.

Vanderauwera, S., Zimmermann, P., Rombauts, S., Vandenabeele, S., Langebartels, C., Gruissem, W., et al. (2005). Genome-wide analysis of hydrogen peroxide-regulated gene expression in Arabidopsis reveals a high light-induced transcriptional cluster involved in anthocyanin biosynthesis. *Plant Physiology, 139*, 806–821.

Vanderauwera, S., Suzuki, N., Miller, G., van de Cotte, B., Morsa, S., Ravanat, J.-L., et al. (2011). Extranuclear protection of chromosomal DNA from oxidative stress. *Proceedings of the National Academy of Sciences of the United States of America, 108*, 1711–1716.

Vernoux, T., Wilson, R. C., Seeley, K. A., Reichheld, J. P., Muroy, S., Brown, S., et al. (2000). The ROOT MERISTEMLESS1/CADMIUM SENSITIVE2 gene defines a glutathione-dependent pathway involved in initiation and maintenance of cell division during postembryonic root development. *The Plant Cell, 12*, 97–110.

Wachter, A., Wolf, S., Steiniger, H., Bogs, J., & Rausch, T. (2005). Differential targeting of GSH1 and GSH2 is achieved by multiple transcription initiation: Implications for the compartmentation of glutathione biosynthesis in the Brassicaceae. *The Plant Journal, 41*, 15–30.

Wakao, S., & Benning, C. (2005). Genome-wide analysis of glucose-6-phosphate dehydrogenases in Arabidopsis. *The Plant journal : for cell and molecular biology, 41*(2), 243–256. https://doi.org/10.1111/j.1365-313X.2004.02293.x

Waszczak, C., Carmody, M., & Kangasjärvi, J. (2018). Reactive oxygen species in plant signaling. *Annual Review of Plant Biology, 69*, 209–236.

Waszczak, C., Akter, S., Eeckhout, D., Persiau, G., Wahni, K., & Bodra, N. (2014). Sulfenome mining in Arabidopsis thaliana. *Proceedings of the National Academy of Sciences of the United States of America, 111*, 11545–11550.

Wei, B., Willems, P., Huang, J., Tian, C., Yang, J., Messens, J., et al. (2020). Identification of sulfenylated cysteines in Arabidopsis thaliana proteins using a disulfide-linked peptide reporter. *Frontiers in Plant Science, 11*, 777.

Wendt, U. K., Wenderoth, I., Tegeler, A., & von Schaewen, A. (2000). Molecular characterization of a novel glucose-6-phosphate dehydrogenase from potato (Solanum tuberosum L.). *The Plant Journal, 23*, 723–733.

Willekens, H, Chamnongpol, S., Davey, M., Schraudner, M, Langebartels, C, Van Montagu, M, ... Van Camp, W. (1997). Catalase is a sink for H_2O_2 and is indispensable for stress defense in C3 plants. *EMBO J, 16*, 4806–4816.

Willekens, H., Inzé, D., Van Montagu, M., & Van Camp, W. (1995). Catalases in plants. *Molecular Breeding, 1*, 207–228.

Wu, B., Li, P., Hong, X., Xu, C., Wang, R., & Liang, Y. (2022). The receptor-like cytosolic kinase RIPK activates NADP-malic enzyme 2 to generate NADPH for fueling ROS production. *Molecular Plant, 15*, 887–903.

Wu, F., Chi, Y., Jiang, Z., Xu, Y., Xie, L., Huang, F., et al. (2020). Hydrogen peroxide sensor HPCA1 is an LRR receptor kinase in Arabidopsis. *Nature, 578*, 577–581.

Xiang, C., Werner, B. L., Christensen, E. M., & Oliver, D. J. (2001). The biological functions of glutathione revisited in Arabidopsis transgenic plants with altered glutathione levels. *Plant Physiology, 126*, 564–574.

Yang, Z., Mhamdi, A., & Noctor, G. (2019). Analysis of catalase mutants underscores the essential role of CATALASE2 for plant growth and day length-dependent oxidative signalling. *Plant, Cell & Environment, 42*, 688–700.

Yin, L., Wang, S., Eltayeb, A. E., Uddin, M. I., Yamamoto, Y., Tsuji, W., et al. (2010). Overexpression of dehydroascorbate reductase, but not monodehydroascorbate reductase, confers tolerance to aluminium stress in transgenic tobacco. *Planta, 231*, 609–621.

Yoshida, S., Tamaoki, M., Shikano, T., Nakajima, N., Ogawa, D., Ioki, M., et al. (2006). Cytosolic dehydroascorbate reductase is important for ozone tolerance in Arabidopsis thaliana. *Plant Cell Physiology, 47*, 304–308.

Zamocky, M., Furtmüller, P. G., & Obinger, C. (2008). Evolution of catalases from bacteria to humans. *Antioxidants & Redox Signaling, 10*, 1527–1547.

Zandalinas, S. I., Fichman, Y., & Mittler, R. (2020). Vascular bundles mediate systemic reactive oxygen ignaling during light stress. *The Plant Cell, 32*, 3425–3435.

Zandalinas, S. I., Sengupta, S., Burks, D., Azad, R. K., & Mittler, R. (2019). Identification and characterization of a core set of ROS wave-associated transcripts involved in the systemic acquired acclimation response of Arabidopsis to excess light. *The Plant Journal, 98*, 126–141.

Zandalinas, S. I., Sengupta, S., Fritschi, F. B., Azad, R. K., Nechushtai, R., & Mittler, R. (2021). The impact of multifactorial stress combination on plant growth and survival. *New Phytologist, 230*, 1034–1048.

Zarepour, M., Kaspari, K., Stagge, S., Rethmeier, R., Mendel, R. R., & Bittner, F. (2010). Xanthine dehydrogenase AtXDH1 from Arabidopsis thaliana is a potent producer of superoxide anions via its NADH oxidase activity. *Plant molecular biology, 72*(3), 301–310. https://doi.org/10.1007/s11103-009-9570-2

Zechmann, B. (2011). Subcellular distribution of ascorbate in plants. *Plant Signaling & Behavior, 6*, 360–363.

CHAPTER THREE

The ascorbate/glutathione cycle

Karl Josef Kunert[a,1,*], **and Christine Helen Foyer**[b,2,*]
[a]Department of Plant and Soil Sciences, FABI, University of Pretoria, Pretoria, South Africa
[b]School of Biosciences, College of Life and Environmental Sciences, University of Birmingham, Edgbaston, United Kingdom
*Corresponding authors. e-mail address: karl.kunert@up.ac.za; C.H.Foyer@bham.ac.uk

Contents

Abstract

Reactive oxygen species (ROS) are essential signals that regulate plant growth, development and defense. Low molecular weight antioxidants, such as ascorbate and glutathione, and antioxidant enzyme are present in most, if not all, plant cell compartments, where they serve either to propagate redox signals or limit ROS signalling. The ascorbate/glutathione cycle, which involves ascorbate peroxidase (APX), monodehydroascorbate reductase (MDHAR), dehydroascorbate reductase (DHAR) and glutathione reductase (GR) was first described in chloroplasts, but has now been extensively characterized in all cellular compartments. In this system, NADPH/NADH is considered to provide the reducing power to regenerate reduced glutathione (GSH) from its oxidized form, glutathione disulphide (GSSG). Here, we review the functions of the ascorbate/glutathione cycle in terms of how this pathway serves to maintain the cellular pools of reduced ascorbate and glutathione, as well metabolising ROS. We consider the synthesis and functions of ascorbate and glutathione, the evidence

[1] ORCID: 0000-0002-7740-3508
[2] ORCID: 0000-0001-5989-6989

Advances in Botanical Research, Volume 105
ISSN 0065-2296, https://doi.org/10.1016/bs.abr.2022.11.004

regarding the interactions between these antioxidants in the ascorbate/glutathione cycle, and the position of this cycle within the network of cellular redox systems.

Nomenclature

ABA abscisic acid
ABI4 aba insensitive4
AMR1 ascorbic acid mannose pathway regulator 1
APX ascorbate peroxidase
CSN5B COP9 signalosome subunit 5B
DHAR dehydroascorbate reductase
EIN3 ethylene-insensitive3
ERF ethylene response factor
Fd ferredoxin
L-GalDH L-galactose dehydrogenase
GGP GDP-L-galactose phosphorylase
GMP GDP-D-mannose pyrophosphorylase
GME GDP-D-mannose epimerase
GGP GDP-L-galactose phosphorylase
GSH reduced glutathione
GR glutathione reductase
GRX glutaredoxins
GSSG glutathione disulphide
HMW high molecular weight
LMW low molecular weight
MDHA monodehydroascorbate
MDHAR monodehydroascorbate reductase
PCD programmed cell death
PETC photosynthetic electron transport chain
PTM post translational modifications
PRX peroxiredoxin
RBOH respiratory burst homolog proteins
ROS reactive oxygen species
SAR systemic acquired resistance
SOD superoxide dismutase
TET ten−eleven translocation
TRX thioredoxin
VTC vitamin C

1. Introduction

Since the ascorbate/glutathione cycle (sometimes called the Asada–Halliwell-Foyer cycle or Foyer–Halliwell–Asada pathway; Fig. 1) was first described in chloroplasts (Foyer & Halliwell, 1976), followed by the identification of the chloroplast ascorbate-specific ascorbate peroxidases

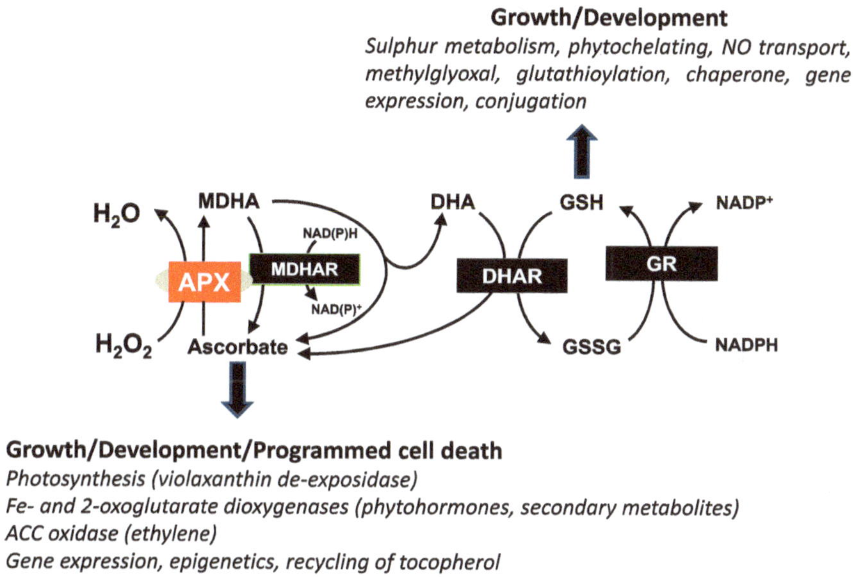

Fig. 1 The role of the ascorbate glutathione cycle (Asada–Halliwell-Foyer cycle) in regenerating the reduced forms of ascorbate and glutathione to maintain a wide range of biological functions. *APX*, Ascorbate peroxidase; *DHAR*, dehydroascorbate reductase; *GSH*, reduced glutathione; *GR*, glutathione reductase; *GSSG*, glutathione disulphide; *MDHA*, monodehydroascorbate; *MDHAR*, monodehydroascorbate reductase.

(APXs) (Groden & Beck, 1979; Kelly & Latzko, 1979), ascorbate, glutathione and other components of this cycle have been found in every compartment of the plant cell (Noctor & Foyer, 1998). The main function of this cycle is to preserve the redox state of cellular proteins by preventing uncontrolled accumulation of hydrogen peroxide (H_2O_2) and other reactive oxygen species (ROS) in the different subcellular compartments. However, as discussed below the functions of ascorbate/glutathione cycle extend far beyond the policing of ROS signals because this pathway serves to maintain the cellular pools of reduced ascorbate and glutathione. These multifunctional metabolites have important and diverse roles in plant growth, development as well as defence.

1.1 The central importance of ROS signalling

ROS are essential markers or signals of living cells (Van Breusegem, Foyer, & Mann, 2018). It is often stated that ROS are kept at low levels in plant cells to prevent oxidative damage, allowing them to be effective signaling molecules upon increased synthesis. However, such concepts are misleading in plants because there is little evidence that oxidative damage accumulates in plant cells to such an extent that it limits cellular functions. In many cases, ROS and oxidised lipids and proteins have been shown to function as signals that regulate gene expression to ensure approprate acclimation or cell death responses. We, therefore, propose that all ROS molecules are potentially signals and periods of high ROS accumulation in specific cellular compartments are required to effectvely transmit signals. Together with peroxiredoxins (PRX) and thioredoxins (TRX), the ascorbate/glutathione cycle polices ROS acculmulation in each compartment, except the apolplast/cell wall compartment, in which ascorbate degardation occurs (Foyer, Kyndt, & Hancock, 2020). Superoxide and H_2O_2 act as either electron donors (**red**uctants) or acceptors (**ox**idants) and they thus engage in electron transfer (redox) processes with cellular metabolites and proteins.

ROS signals fulfil important roles in the regulation of numerous development processes from root development (Eljebbawi, del Carmen Rondón Guerrero, Dunand, & Estevez, 2021; Mase & Tsukagoshi, 2021), the transition to flowering (Huang et al., 2021) to leaf senescence (Zentgraf, Andrade-Galan, & Bieker, 2022). They contribute to the elicitation of genetic and epigenetic responses that allow acclimation and adaptation to metabolic, developmental and environmental triggers (Ramakrishnan et al., 2022). ROS and the ascorbate/glutathione cycle function synchonously to regulate plant growth, development, as well as defence. For example, ROS generation is the driver and first requirement for many developmental processes such as the cell cycle, pollen viability, microspore reprogramming towards sporophytic development, the regulation of female gametophyte patterning and the maintenance of embryo sac polarity, as well as the prevention of self-pollonation (de Simone, Hubbard, Viñegra de la Torre, Velappan, & Wilson, 2017; Sankaranarayanan, Ju, & Kessler, 2020; Zhang et al., 2021; Zur et al., 2021). In such systems, the ascorbate/glutathione cycle together with TRX, PRX, glutaredoxins (GRX) and antioxidant enzymes, such as superoxide dismutase (SOD) and catalase (CAT), ensures appropriate redox-mediated regulation, so that ROS, ascorbate and glutathione can accumulate in the required compartment-specific manner.

Succesive waves of ROS accumulation and removal are important in cell-to-cell communication (Fichman & Mittler, 2021a, 2021b; Zandalinas et al., 2020a, 2020b, 2021). The activation of the respiratory burst homolog

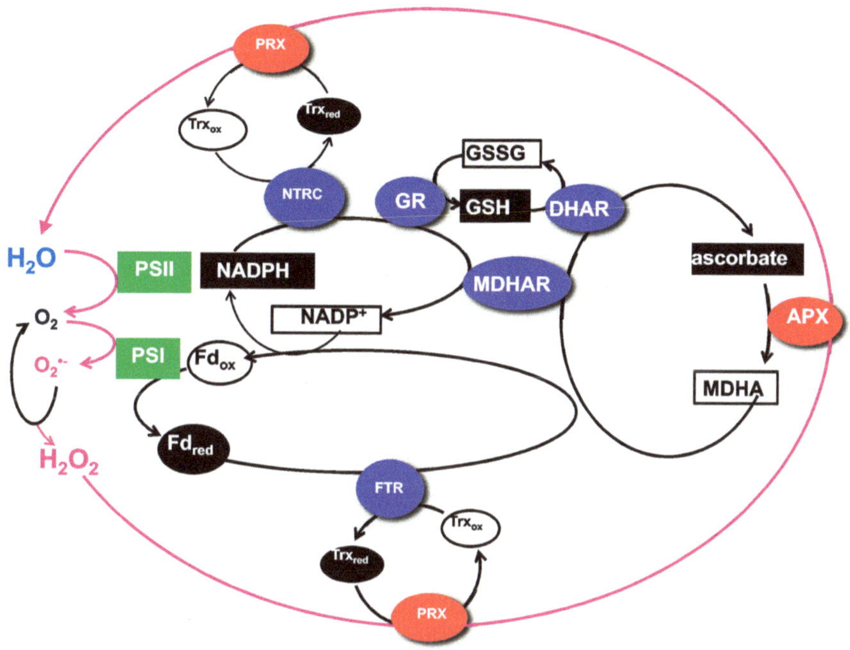

Fig. 2 The role of antioxidants in modulating the ROS wave pathway of systemic signalling.

(RBOH) proteins on the plasma membrane generates superoxide radicals in the apoplast, which are converted to H_2O_2 either through spontaneous dismutation of the action of SOD. Cell-to-cell ROS signalling plays a pivotal role in local and systemic responses, leading to stress acclimation and survival (Fichman & Mittler, 2021a, 2021b; Waszczak, Carmody, & Kangasjärvi, 2018; Zandalinas et al., 2020a, 2020b). Auto-propagating waves of ROS, calcium and electric signals function together to generate rapid systemic cell-to-cell communication (Wang et al., 2019).

The process of cell-to-cell ROS signalling, which is called the 'ROS wave' (Fig. 2), is linked to cell-to-cell calcium and membrane potential signalling and is essential for systemic stress signalling and systemic acquired acclimation (Fichman & Mittler, 2020, 2021a, 2021b). In this process, ROS production by the RBOHs, RBOHD and RBOHF, is triggered in the cells that are directly subjected to stress, resulting in a state of 'activated ROS production'. The leucine-rich-repeat receptor-like kinase HPCA1 (H_2O_2-induced Ca^{2+} increases 1) is required for coordination of ROS and calcium signals during the cell-to-cell propagation of ROS signals (Fichman, Zandalinas, Peck, Luan, & Mittler, 2022). Once the state of activated ROS production reaches cells and tissues other than those initiating the signal it triggers acclimation mechanisms and enhances overall stress resilience (Fichman & Mittler, 2021a). Little attention has as yet been paid to how the ascorbate/gluathione cycle regulates the lifetime of ROS signals in any given cellular compartment, but ROS processing and removal in activated cells is an essential feature of the progression of the ROS wave.

1.2 The ascorbate/glutathione-dependent water/water cycle

The ascorbate-glutathione cycle is comprised of metabolites (ascorbate, glutathione and NADPH) and enzymes which essentially serve to regenerate the reduced forms of ascorbate and glutathione (Fig. 1). In a first step of the pathway, H_2O_2 is reduced to water by ascorbate peroxidase (APX), which uses ascorbate acting as the electron donor. Oxidized ascorbate (monodehydroascorbate, MDA) can spontaneously disproportionate to ascorbate and dehydroascorbate (DHA). However, ascorbate is also regenerated by the enzyme monodehydroascorbate reductase (MDHAR), using the reducing power of NAD(P)H. Direct reduction of MDHA by the photosynthetic electron transport chain has also been reported (Miyake & Asada, 1994). DHA is reduced back to ascorbate by the enzyme DHA reductase (DHAR) using reduced glutathione (GSH) as the reductant. The enzyme glutathione reductase (GR) reduces the oxidised form of glutathione, glutathione disulphide (GSSG) to GSH using NADPH as the reductant.

In the water/water cycle the activity of the ascorbate/glutathione cycle is coupled to the production of hydrogen peroxide by the photosynthetic electron transport chain in the chloroplasts (Foyer & Hanke, 2022; Fig. 3). In photosynthesis, the splitting of water at photosystem II (PSII) results in the production of molecular oxygen and the transfer of elections to the photosynthetic electron transport chain for the reduction of ferredoxin and NADP. However, molecular oxygen can also act as an electron acceptor at various points in the photosynthetic electron transport chain (Foyer & Hanke, 2022). The reduction of molecular oxygen is called the Mehler reaction, or pseudocyclic electron flow. The single electron reduction of oxygen produces superoxide ($O_2^{\cdot-}$) radicals at the stromal surface of the thylakoid membranes. Superoxide is rapidly converted to H_2O_2 by the action of the thylakoid superoxide dismutases (SODs). Thereafter, H_2O_2 is reduced to water by chloroplast APXs and the ascorbate/glutathione cycle and also by the action of 2- Cys peroxiredoxins (PRXs) that are

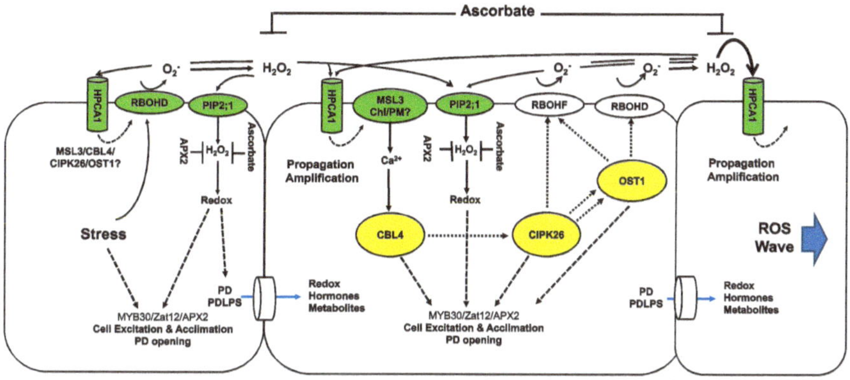

Fig. 3 The water-water cycle allowing the dissipation of excess excitation energy and electrons providing an alternative electron sink for protection of PSII from inhibition and supporting ATP production. *APX,* Ascorbate peroxidase; *PRX,* 2-Cys peroxiredoxins; *GSH,* reduced glutathione; $O_2^{\cdot -}$, superoxide anion radicalsL; *PSII,* photosystem II; *PRX,* peroxiredoxins; *TRX,* thioredoxin.

regenerated by the chloroplast TRX system (Foyer & Hanke, 2022). Taken together, these reactions are known as the water-water cycle because two electrons are used to produce H_2O_2 and two more are required to metabolize H_2O_2 to water, allowing the dissipation of excess excitation energy and electrons (Fig. 3). The water-water cycle provides an alternative electron sink that protects PSII from inhibition and still supports ATP production (Neubauer & Yamamoto, 1992). The water–water cycle may also be important in determining the oxidation state of the chloroplast-targeted 2-Cys PRX, which together with specific atypical TRXs such as ACHT1–4 and TRXL2, are involved in the transfer of oxidative equivalents from H_2O_2 to target chloroplast proteins, such as those of the reductive pentose pathway (Ojeda, Pérez-Ruiz, & Cejudo, 2018; Vaseghi et al., 2018; Yokochi, Fukushi, Wakabayashi, Yoshida, & Hisabori, 2021).

APXs are heme-containing enzymes that belong to class I of the peroxidase-catalase superfamily. They are encoded by small gene families, with different isoforms targeted to the cytosol, plastids, mitochondria, and peroxisomes. Some APX forms are associated with membranes, such as the plasmalemma, the peroxisomal membranes and the thylakoid membranes, often together with MDHARs, while other APX forms are soluble phase. The APX forms differ in substrate affinities, dimer formation, and in the presence of transmembrane domains. Moreover, the cytosolic APX of *Oncidium* orchid (*Og*cytAPX1) can use GSH as a substrate as well as ascorbate but using different active sites (Chin et al., 2019). The Pro63, Asp75 and Tyr97 residues are required for GSH oxidation by *Og*cytAPX1, whereas the corresponding site in *At*APX1 is composed of Asp63, His75 and His97 and has no GSH-binding activity. In addition to *Og*cytAPX1, the recombinant cytosolic APX forms from maize, rice and soybean also possess GSH oxidation activity (Chin et al., 2019). GSH consumption by *Og*CytAPX1 may provide a new biochemical step in the ascorbate glutathione cycle. Moreover, some APX forms, such as the Arabidopsis *At*APX1 (Kaur et al., 2021) and the rice *Os*APX2 (Hong et al., 2018), have chaperone functions. These proteins show a structural and functional transition from high molecular weight (HMW) to low molecular weight (LMW) forms. Only the HMW complexes of the *At*APX1 and *Os*APX2 display chaperone activity, whereas the LMW forms exhibit predominantly peroxidase activity (Hong et al., 2018).

Like other enzymes of the ascorbate glutathione cycle, APXs are subjected to posttranslational modifications (PTMs). For example, the peroxidase activity of APX1 is regulated by S-nitrosylation, tyrosine nitration, and S-sulfhydration either negatively or positively, depending on the plant species (Aroca, Gotor, & Romero, 2018; Begara-Morales et al., 2016). The Arabidopsis APX1 protein has 5 cysteine residues, of which 2 cysteine residues (Cys-32 and Cys-49) are S-nitrosylated (Yang et al., 2015). The Cys-32 residue is also the target for S-sulfhydration (Aroca, Serna, Gotor, & Romero, 2015), which could regulate the binding affinity of APX1 for ascorbate, resulting in increased peroxidase activity. Tyrosine nitration has been implicated in the loss of APX1 activity in pea and tobacco (Begara-Morales et al., 2013; Clark, Durner, Navarre, & Klessig, 2000). The APX enzymes can also be regulated by protein phosphorylation. For example, APX2 was found to bind to calcium-dependent protein kinases (CPK 28), a process that results in the phosphorylation of the APX2 protein at Thr-59 and Thr-164 and improves thermotolerance (Hu et al., 2021).

The chloroplast MDHAR forms are activated by TRXs. For example, the plastidial MDHAR form is activated by TRX y2 and the activity of a recombinant plastid Arabidopsis MDHAR isoform (MDHAR6) increases in the presence of reduced TRX y, and not by other plastidial TRXs (Vanacker, Guichard, Bohrer, & Issakidis-Bourguet, 2018). In addition to MDHA, MDHARs can use organic radicals as substrates (Hossain, Nakano, & Asada, 1984) and might be capable of recycling the oxidation products of other powerful antioxidants, such as phenolics (Sakihama, Mano, Sano, Asada, & Yamasaki, 2000). Mutants lacking MDHAR do not always show changes in ascorbate accumulation. For example, the peroxisomal membrane–associated ascorbate-dependent electron transfer system involves APX as well as MDHAR. While the Arabidopsis peroxisomal membrane APX isoform (APX3) is dispensable for growth and development (Narendra et al., 2006), the seedling-lethal *sugar-dependent2* mutant is deficient in the peroxisomal membrane MDHAR isoform (MDHAR4). MDHAR4 mutants have lower ascorbate to DHA ratios but have similar total ascorbate (ascorbate + DHA) levels to the wild type (Eastmond, 2007). Overexpression of the MDHAR increased the ascorbate contents only in some studies (Eltayeb et al., 2007; Eltelib, Fujikawa, & Esaka, 2012; Kavitha, George, Venkataraman, & Parida, 2010; Li et al., 2010; Yin et al., 2010). Hence, MDHAR may not be a rate-limiting step in ascorbate recycling.

The DHAR enzymes belong to the glutathione S-transferase (GST) superfamily. The DHAR reaction is likely to proceed via a "ping-pong" mechanism, where DHA binds to the free reduced form of DHAR followed by binding of GSH (Ding, Wang, Han, & Li, 2020). In addition, some peroxiredoxins PRXs can oxidise GSH via the action of GRX (Rahantaniaina, Tuzet, Mhamdi, & Noctor, 2013). While DHAR was found to contribute to GSH oxidation under stress conditions (Rahantaniaina, Li, Chatel-Innocenti, Tuzet, Issakidis-Bourguet, et al., 2017), such observations led to doubts about the role of DHARs in ascorbate regeneration (Rahantaniaina, Li, Chatel-Innocenti, Tuzet, Mhamdi, et al., 2017). However, a recent study has also demonstrated that GSH is involved in recycling ascorbate under high light conditions both by nonenzymic reduction or via DHAR activity (Terai et al., 2020). Moreover, the increased expression of DHARs has consistently generated increased concentrations of reduced ascorbate in a wide range of species (see for example Table 2 in Broad, Bonneau, Hellens, & Johnson, 2020). For example, the overexpression of DHAR in multi-vitamin white corn enhanced kernel ascorbate levels by 6-fold (Naqvi et al., 2009). The increased expression of the DHAR genes has also been consistently associated with enhanced abiotic stress tolerance (Broad et al., 2020). Moreover, loss-of-function mutations in the Arabidopsis DHAR3 gene reduced the ratio of ascorbate to DHA but did not affect the pool size of total ascorbate, supporting the role of DHAR in ascorbate recycling (Yoshida et al., 2006).

1.3 Ascorbate functions in plants

Ascorbate is an efficient broad spectrum, nonenzymatic antioxidant and ubiquitous ROS scavenger (Foyer et al., 2020). Although similar to thiols in its ability to react with ROS, ascorbate is much better (> 100× faster) than GSH at scavenging superoxide and singlet oxygen. While ascorbate is a major plant metabolite that accumulates to very high concentrations, its synthesis is tightly regulated by environmental factors, particularly light and feedback controls. The reasons for the accumulation of millimolar amounts of ascorbate in plant cells are not completely understood but may be linked to the wide range of ascorbate functions and the lack of toxicity. The concentrations of ascorbate, as determined in the different compartments of Arabidopsis cells are as follows: mitochondria (10.4 mM), chloroplasts (10.8 mM), peroxisomes (22.8 mM), nuclei (16.3 mM), vacuole (2.3 mM) and cytosol (21.7 mM) (Zechmann, Stumpe, & Mauch, 2011). In contrast, the concentrations of ascorbate (0.002 mM) and DHA (0.36 mM) in the apoplast are relatively low (Booker, Burkey, & Jones, 2012).

The synthesis and accumulation of ascorbate are determined on the level and quality of light level to which the leaves are exposed, being highest under high light (Bartoli, Tambussi, Fanello, & Foyer, 2009; Foyer et al., 2020). With the exception of the peroxisomes, the ascorbate content in most cell compartments increases under high light (Zechmann, 2011; Zechmann et al., 2011). While ascorbate is perhaps best known as an antioxidant and universal ROS scavenger, it also has multiple roles in plant growth and development, from seed maturation and germination to flowering, fruit ripening, and senescence (Bulley & Laing, 2016; Foyer et al., 2020). The importance of ascorbate in plant biology is certainly linked to the complex interactions between ascorbate, ROS and hormone signalling, which remain poorly characterised.

Ascorbate is an essential metabolite in plants because the Arabidopsis *vtc2/vtc5* double mutants, which are unable to synthesise ascorbate, are not viable (Dowdle, Ishikawa, Gatzek, Rolinski, & Smirnoff, 2007). Ascorbate is a specific co-factor for a large family of enzymes known as the Fe- and 2-oxoglutarate-dependent dioxygenases that catalyse the addition of a hydroxyl group to various substrates (Wei, Zhang, Fu, & Zhang, 2021). Ascorbate is required for the maintenance of activity of Fe(II)/2-oxoglutarate-dependent dioxygenases via reduction of Fe(III). As such, ascorbate is involved in the synthesis of phytohormones and secondary metabolites. For example, ascorbate is required for opening the ring structure of 1-aminocyclopropane-1-carboxylic acid (ACC) by supplying the electron to the active site of ACC oxidase, which catalyses the last step of ethylene biosynthesis (Smirnoff, 2018). Ascorbate is likely to be involved in the synthesis of abscisic acid and gibberellins through its functions with different dioxygenases, as well as auxin catabolism. Ascorbate is the natural substrate for many types of peroxidases in plants in addition to APX (Mehlhorn, Lelandais, Korth, & Foyer, 1996). Thus, ascorbate is able to influence the accumulation of a wide range of phenolic compounds, particularly in the cell wall/apoplastic compartment of plant cells. Ascorbate regulates the expression of genes involved in flavonol and anthocyanin precursor synthesis (Page, Sultana, Paszkiewicz, Florance, & Smirnoff, 2012) such as PHENYLALANINE AMMONIA-LYASE1 (PAL1), 4-COUMARATE:COENZYME A LIGASE3, CHALCONE SYNTHASE (CHS), as well as the MYB transcription factor PAP1 and an ELONGATED HYPOCOTYL5 (HY5) homologue HYH (Munné-Bosch, Queval, & Foyer, 2013). The low levels of leaf ascorbate in ascorbate deficient mutants (*vtc2–1* and *vtc2–4*) causes a significant decrease in leaf anthocyanin contents (Plumb et al., 2018).

Mutants that have low ascorbate contents show significant reprogramming of gene expression, including genes involved in hormone synthesis and signalling, as well as photosynthesis and defence (Kiddle et al., 2003; Pastori et al., 2003). These changes are accompanied by increases in the levels of salicylic acid (SA), pathogenesis-related proteins, and camalexin that demonstrate the activation of the ROS signalling branch of plant innate immunity (Mukherjee et al., 2010; Pavet et al., 2005). In this way, ascorbate can exert a key role in plant immunity, as well as defence responses to abiotic environmental stresses (Akram, Shafiq, & Ashraf, 2017; Pastori et al., 2003; Pavet et al., 2005; Venkatesh & Park, 2014) such as salt tolerance (Shalata, Mittova, Volokita, Guy, & Tal, 2001). Ascorbate accumulation is, therefore, important in the regulation of plant defences against biotrophic pathogens that rely on SA signalling such as *Pseudomonas syringae* and *Peronospora parasitica* (Mukherjee et al., 2010; Pavet et al., 2005), and also phloem-feeding insects (Kerchev et al., 2013). In contrast, ascorbate deficiency enhances susceptibility to the necrotrophic pathogen *Alternaria brassicicola*, in which defence is mediated by jasmonic acid and ethylene signalling (Botanga, et al., 2012). The application of exogenous ascorbate acts as an inducer of disease resistance in plant interactions with different types of pathogens including viruses (Fujiwara, Shimura, Masuta, Sano, & Inukai, 2013). The mechanisms involved in such strategies are complex because reduced ascorbate is highly susceptible to oxidation in aqueous solution and, moreover, it is likely to be oxidised by the ascorbate oxidase activities in the apoplast/cell wall compartments of the plant cell before it enters the cytoplasm. The role of ascorbate in programmed cell death (PCD) is related to its role in the control of the activation of the ROS signalling branch of innate immune responses

(Mukherjee et al., 2010; Pavet et al., 2005) Localized PCD, similar to that occurring during hypersensitive responses to plant pathogen, is observed in the leaves of the ascorbate-deficient mutants (Pavet et al., 2005). Increased ascorbate synthesis resulting from supplying L-galactono-1,4-lactone delays PCD during kernel maturation in durum wheat, with a consequent postponement of dehydration and improvement in kernel filling (Paradiso, de Pinto, Locato, & De Gara, 2012).

Ascorbate may play a key role in the epigenetic memory of plant cells because of its central role in cellular redox metabolism (Ramakrishnan et al., 2022). In human embryonic stem cells and differentiated cells, ascorbate acts as a cofactor for the ten−eleven translocation (TET1−3) family of proteins that remove cytosine methylation in DNA (Zhithovich, 2020). Cytosine methylation occurs at its C5 position in CpG dinucleotides and usually acts as a transcription-repressive mark. Ascorbate drives the active removal of DNA methylation by enhancing the activities of TET enzymes, which helps to erase DNA methylation and hence the epigenetic memory of specific cell types. Interestingly, TET enzymes are Fe (II)-dependent dioxygenases that catalyse a series of consecutive oxidations of 5-methylcytosine by first hydroxylating it to 5-hydroxymethylcytosine, which is further oxidized to 5-formylcytosine and 5-carboxycytosine. Although there is some evidence for the presence of 5-methylcytosine oxidation products, particularly 5-hydroxymethylcytosine (5hmC) in various plant species (Mahmood & Dunwell, 2019), no TET-like enzymes have as yet been identified in plants. However, this may prove difficult because the expression of TET-like genes may be limited to specific cell types, as occurs in humans.

Ascorbate fulfils a number of important roles in the regulation of photosynthesis, particularly in the acclimation of plants to high light (Karpinska et al., 2017; Müller-Moulé, Golan, & Niyogi, 2014). In addition to its participation in the water–water cycle, ascorbate is also required for the regeneration of lipid-soluble antioxidants, particularly tocopherols and tocotrienols (vitamin E), which protect the polyunsaturated fatty acids in the thylakoid membranes from oxidation to chromanoxyl radicals by singlet oxygen. These radicals are converted back to vitamin E by the reducing power of ascorbate, or by reaction with carotenoids. Ascorbate is also required for the conversion of violaxanthin to zeaxanthin in the light-dependent xanthophyll cycle, which is a key component of the thermal energy dissipation mechanisms measured by the non-photoenergy quenching component of chlorophyll a fluorescence (Müller-Moulé, Conklin, & Niyogi, 2002). Knockout mutants of the chloroplast envelope ascorbate transporter AtPHT4;4 are compromised in thermal energy dissipation (Miyaji et al., 2015). Moreover, ascorbate is a potent specific inhibitor of the expression of 2-Cys PRX A and other chloroplast antioxidant enzymes (Baier, Ströher, & Dietz, 2004; Horling et al., 2003), influencing chloroplast to nucleus signalling pathways via the redox-sensitive transcription factor Rap2.4a (Shaikhali et al., 2008). Conversely, the expression of chloroplast APX and MDHAR is induced in lines defective in 2-Cys peroxiredoxins (Baier, Noctor, Foyer, & Dietz, 2000). Ascorbate is also able to donate as well as accept electrons from the photosynthetic electron transport chain acting as an alternative electron donor for PSII (Mano, Hideg, & Asada, 2004; Tóth, Puthur, Nagy, & Garab, 2009).

While ascorbate has been largely discounted as a significant factor in NO metabolism (Wang & Hargrove, 2013), the ascorbate-mediated regulation of flowering in plants, such as *Oncidium*, acts through the NO-mediated flowering-repression pathway (Kumar et al., 2016). Arabidopsis low ascorbate mutants have long been known to show early flowering (Barth, De Tullio, & Conklin, 2006), a trait that is linked to the altered expression of genes, such as flowering locus T (*FT*) and CONSTANS (*CO*) that regulate flowering (Kotchoni, Larrimore, Mukherjee, Kempinski, & Barth, 2009). Similar effects on flowering have been reported in plants with altered APX or ascorbate oxidase (AO) activities (Pignocchi et al., 2006; Pnueli, Liang, Rozenberg, & Mittler, 2003). Moreover, the exogenous application of ascorbate or its precursor L-galactono-1,4-lactone (GAL) delays flowering (Shen, Krishnamurthy, & Yeh, 2009).

1.4 Ascorbate synthesis and regulation

Ascorbate is an essential vitamin in the human diet and there have has been considerable emphasis to significantly increase the levels of ascorbate in fruits and vegetables (Fenech, Amaya, Valpuesta, & Botella, 2019). Several strategies have been used to increase the ascorbate levels including increased biosynthesis, enhanced recycling, and modulation of regulatory factors (Broad et al., 2020). A large number of transcription factors and other regulators of ascorbate synthesis have not been identified. Ascorbate synthesis and recycling are tightly regulated at the transcriptional, translational, and post-translational levels.

The concept that the L-Gal pathway (Smirnoff–Wheeler pathway; Smirnoff & Wheeler, 2000) is the main pathway of ascorbate synthesis in plants is now firmly embedded in the literature (Bulley & Laing, 2016; Fenech et al., 2021; Hancock & Viola, 2005). In this pathway, D-Glucose 6-P is sequentially transformed into D-fructose 6-P, D-mannose 6-P, and

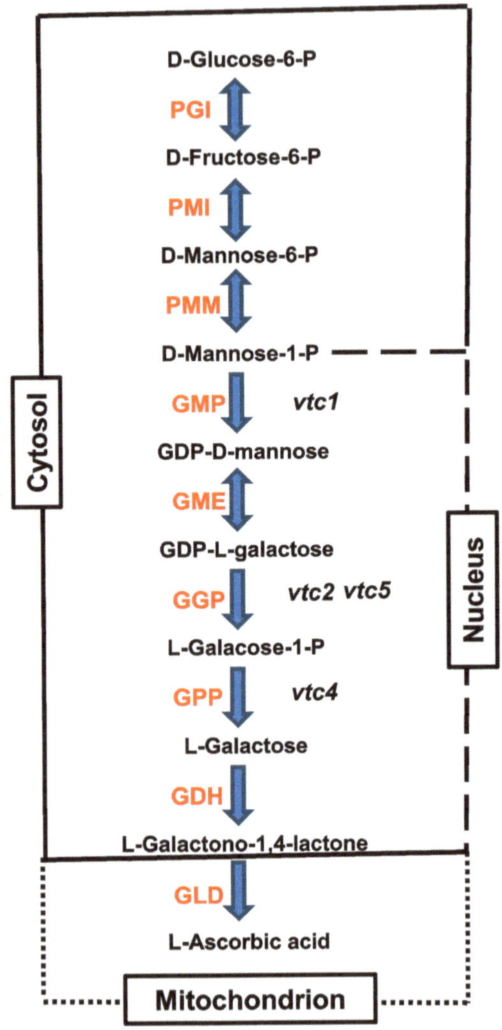

Fig. 4 Ascorbate biosynthesis in higher plants with various mutants (*vtc*) in the biosynthesis steps indicated. *GDH*, L-Galactose dehydrogenase; *GLDH*, L-galactono-1,4-lactone dehydrogenase; *GGP*, GDP-L-galactose phosphorylase; *GPP*, L-galactose-1-P phosphatase; *PGI*, phosphoglucose isomerase; *GMP*, GDP-D-mannose pyrophosphorylase; *GME*, GDP-D-mannose 3′,5′-epimerase; *PMI*, phosphomannose isomerase; *PMM*, phosphomannomutase.

D-mannose 1-P by the enzymes phosphoglucose isomerase, phosphomannose isomerase, and phosphomannomutase (Fig. 4). Next, the conversion of D-mannose 1-P into GDP-D-mannose is catalysed by GDP-D-mannose pyrophosphorylase (GMP), which is then epimerized by

GDPD-mannose epimerase (GME) into GDP-L-galactose. Thereafter, GDP-L-galactose is converted to L-galactose 1-P by GDP-L galactose phosphorylase (GGP), producing L-galactose 1-P (Fig. 4). This enzyme, which is encoded by two paralogous genes in Arabidopsis, *Vitamin C* (*VTC2* and *VTC5*), is a main control point in the pathway (Dowdle et al., 2007; Laing et al., 2015; Linster et al., 2007). Next, L-galactose 1-phosphate phosphatase (GPP) and other unidentified phosphatases (Laing et al., 2004; Torabinejad, Donahue, Gunesekera, Allen-Daniels, & Gillaspy, 2009), transform L-galactose 1-phosphate into L-galactose. L-Galactose is then oxidized to L-galactono1,4-lactone by L-galactose dehydrogenase (L-GalDH). Most of the enzymes of this pathway are localized in the cytosol, where they may form part of a large complex (Fenech et al., 2021). However, all the enzymes from GMP up to L-GalDH have a dual cytosolic/nuclear localization and L-GalDH is localised in the mitochondria, where it is an integral component of respiratory electron transport Complex 1 (Pineau, Layoune, Danon, & De Paepe, 2008; Schimmeyer, Bock, & Meyer, 2016). This pathway is connected to cell wall biosynthesis through the requirement of this process for GDP-D-mannose and GDP-L-galactose (Fenech et al., 2021).

Ascorbate synthesis is regulated through a highly conserved, cis-acting upstream open reading frame (uORF) in the long 5′ leader sequence of the GGP mRNA, which encodes a peptide that functions in the ascorbate inhibition of translation (Laing et al., 2015). Genome editing of the uORF of the GDP-L-galactose phosphorylase (GGP) gene in lettuce increased the ascorbate content by 1.5-fold (Zhang, Si, Ji, Fan, Liu, & Chen, 2018; Zhang, Ye, Singh, Townsend, & Tew, 2018). A kinetic modelling analysis based on the known properties of the pathway enzymes and the proposed posttranscriptional repression of GGP via the 50-uORF (Laing et al., 2015), showed that GGP is the only step in the biosynthetic pathway that controls pathway flux, and hence ascorbate accumulation (Fenech et al., 2021). *VTC3* has been suggested to play a role in the GGP uORF sensing mechanism of ascorbate levels (Bulley & Laing, 2016).

In addition to the above mechanism of post-transcriptional regulation, many transcriptional regulators of ascorbate synthesis have now been identified. This includes the ascorbic acid mannose pathway regulator 1 (AMR1) gene, which was identified in an activation-tagged Arabidopsis mutant that contained 40% of wild-type ascorbate levels (Zhang, Lorence, Gruszewski, Chevone, & Nessler, 2009), the tomato basic helix-loop-helix 59 (bHLH59) gene, which underlies the ascorbate quantitative trait locus

TFA9 in tomato (Ye, Li, et al., 2019; Ye, Liu, et al., 2019), the Arabidopsis calcium sensor, calmodulin-like 10 (CML10; Cho et al., 2016), the Arabidopsis COP9 signalosome subunit 5B (CSN5B) and a component of the photomorphogenic COP9 signalosome (Wang et al., 2013). The COP9 complex subunit CSN5B interacts with GMP *to* promote ubiquitination and degradation of the enzyme through the 26S proteasome pathway. Plants, with a point mutation in CSN5B, had lower GMP degradation rates and higher ascorbate levels (Wang et al., 2013). The nucleotide sugar phosphorylase-like proteins, KONJAC 1 and 2 (KJC1 and KJC2), which were identified based on their phylogenetic relationship to GMP, were suggested to post-translationally regulate GMP and act in opposition to the COP9 signalosome in the regulation of GDP-D-mannose (Sawake et al., 2015). The tomato C2H2–type zinc finger 3 (ZF3) protein also interacts with CSN5B and inhibits the interaction between CSN5B and GMP (Ouyang et al., 2007).

The DNA-binding with one finger 22 (Dof22) transcription factor is also a negative transcriptional regulator of ascorbate accumulation targeting genes in the L- galactose pathway, ascorbate recycling and the ascorbate-glutathione cycle (Cai et al., 2016). Similarly, the NBS-LRR 33 (NL33) resistance protein negatively regulates the transcription of ascorbate biosynthesis, recycling, and ascorbate-glutathione cycle genes (Ye, Li, et al., 2019; Ye, Liu, et al., 2019). In contrast, the ethylene response factor (ERF) 98 positively regulates the transcription of ascorbate biosynthesis, ascorbate recycling, and ascorbate-glutathione cycle genes (Zhang, Wang, Zhang, & Huang, 2012). Ethylene and abscisic acid (ABA) also regulate ascorbate synthesis via the ETHYLENE-INSENSITIVE3 (EIN3) and ABA INSENSITIVE4 (ABI4) transcription factors (Yu et al., 2019). ABI4 is a key component of ABA signalling that directly interacts with RbohD and *VTC2* to modulate ascorbate synthesis during seed germination under conditions of salt stress (Luo et al., 2021). ABI4 represses the expression of *VTC2* by directly binding its promoter fragment and promoted ROS scavenging (Yu et al., 2019). Similarly, the expression of ABI4 is repressed by EIN3, suggesting that ethylene and ABA act antagonistically to regulate ascorbate synthesis (Yu et al., 2019). Abiotic stresses, such as high salt, induce the ABI4-NADPH oxidase (RbohD)/VTC2 module to regulate ROS accumulation and restrict seed germination in Arabidopsis (Luo et al., 2021). The HD-Zip I family transcription factor 24 (HZ24) binds to the promoter region of tomato GMP3 and also regulates the expression of L- galactose pathway genes in tomato (Hu et al., 2016). Another gene,

AIR12 (Auxin induced in root culture 12) that mediates cold tolerance by altering H_2O_2 in apoplast, regulates ascorbate homeostasis as well as the CBF mediated chilling response pathway (Wang et al., 2021). The WAX1 gene encodes a MYB transcription factor that positively regulates ascorbate synthesis genes, as well as wax related genes (Zhu, Guo, Zhu, & Zhou, 2014).

Increased expression of ascorbate synthesis genes has been widely used to increase ascorbate concentrations and enhance abiotic stress tolerance. However, increasing ascorbate concentrations is likely to impede ROS signalling pathways. The increased expression of ascorbate synthesis and recycling genes has been reported to reduce the responsiveness of guard cell (Chen & Gallie, 2004) and disturb embryo development (Chen & Gallie, 2012), as well as induce seedless fruit (Bulley et al., 2012) and pollen sterility in tomato (Deslous et al., 2021).

2. Glutathione

Reduced glutathione (γ-glutamyl-cysteinyl-glycine: GSH) is one of the most abundant low molecular weight non-protein thiols in plants. It is an essential metabolite with different functions beyond its well studied antioxidative role (Noctor et al., 2012). Recent reviews have focused on current knowledge concerning the pathways of GSH biosynthesis, metabolism and degradation, and the role of glutathione in abiotic stress tolerance, particularly phyto-hormone crosstalk with jasmonate and calcium signalling (Aslam et al., 2021; Dorion, Ouellet, & Rivoal, 2021; Dumanović, Nepovimova, Natić, Kuča, & Jaćević, 2021; Hasanuzzaman, Nahar, Anee, & Fujita, 2017).

2.1 Glutathione functions

Mutants that are completely deficient in glutathione do not survive beyond the embryonic stage, but a range of mutants with low glutathione levels have been widely studied, particularly in relation to environmental stresses (Cairns, Pasternak, Wachter, Cobbett, & Meyer, 2006). Like glutathione synthesis, glutathione degradation is controlled by environmental conditions, with regulation occurring at the transcriptional and posttranslational levels. Similar to ascorbate, glutathione also accumulates in the different compartments of plant cells (Noctor et al., 2012). The concentrations of glutathione determined in the different compartments of Arabidopsis cells are as follows: mitochondria (15.0 mM), chloroplasts (1.2 mM), peroxisomes (4.4 mM), nuclei (6.4 mM), vacuole (0.08 mM) and cytosol (4.5 mM) (Zechmann, 2020). Very small amounts of glutathione are found in the apoplast (0.0003 mM), even under stress conditions (Hasanuzzaman et al., 2017; Khan et al., 2020; Zechmann, 2020). GSH is important for many detoxification mechanisms and protects plant cells against heavy metals and xenobiotics. It is the substrate for phytochelatins, oligomers of glutathione, produced by the enzyme phytochelatin synthase. The formation of glutathione conjugates with electrophilic compounds is mediated by a glutathione-S-transferases (Eaton and Bammler, 1999; Strange et al., 2000). This allows glutathione to function as a chaperone for a wide range of metabolites including malonyl-dialdehyde and 4-hydroxy-2-nonenal (Comporti, 1987). GSH also has a fundamental role in buffering nitric oxide (NO), which can exist in different redox forms with distinct and highly reactive properties, such as nitroxyl (HNO), oxides (NO_2, N_2O_4 and N_2O_3), peroxynitrite ($ONOO^{\cdot -}$) and S-nitrosothiols (RSNO). S-nitrosoglutathione (GSNO) is considered an endogenous NO reserve that contributes to the intracellular regulation of nitrosative stress, as well as the regulation of many different enzymes.

Glutathione is required to detoxify methylglyoxal, which is produced by the glycolytic pathway (Inagi et al., 2010). Methylglyoxal interacts with molecules containing free amino groups, such as amino acids and nucleotide bases, and also the cysteine residues in proteins (Li et al., 2008). Glutathione is a cofactor in methylglyoxal elimination, a process that involves two enzymes called glyoxalases (Inagi et al., 2010). GSH maintains the thiol state of protein cysteine residues preventing oxidation (Dickinson & Forman, 2002; Dorion et al., 2021; Hasanuzzaman et al., 2017). Glutathione acts as a reserve of cysteine and sulphur and is involved in the production of deoxyribonucleotides and iron–sulfur cluster in proteins. Changes in the intracellular GSH/GSSG ratio triggers signal transduction cascades that mediate plant stress responses. For example, it regulates ABA signal transduction in the control of plant development (Koramutla, Negi, & Ayele, 2021) and modulates the expression of transcription factors particularly through the action of glutaredoxins (GRXs), playing a central role in plant development as well as responses to the environment (Considine & Foyer, 2014, 2021). For example, the ROXY family of CC GRXs, which have adjacent cysteines, is important in the control of flower development. The Arabidopsis *roxy1* mutants have a reduced number of petals, with many of the remaining ones being folded or smaller than normal. There are

21 CC GRX family members in Arabidopsis, named ROXY1–21, depending on their degree of relatedness to ROXY1 (Li et al., 2009). Some ROXY family members interact with bZIP transcription factors of the TGA class, such as PERIANTHIA (PAN, TGA8), which is important in defining the floral blueprint of the Arabidopsis flower. Together, with TRXs, GSH fulfils essential roles in root development (Bashandy et al., 2010; Diaz-Vivancos, de Simone, Kiddle, & Foyer, 2015; Schnaubelt et al., 2015). GSH deficiency disrupts auxin homeostasis because of decreased expression of PIN-formed proteins (Bashandy et al., 2010). It also inhibits the auxin-related, ubiquitination-dependent degradation of auxin (IAA) proteins and the transcriptional activation of early auxin-responsive genes (Pasternak, Palme, & Paponov, 2020). GSH is also required for the conversion of indole butyric acid (IBA) to indole acetic acid (Trujillo-Hernandez et al., 2020). It has also been implicated in the regulation of translation (Cheng, et al., 2015).

The GSH-deficient *rootmeristemless1* (*rml1*) mutant is unable to develop roots because of impaired root apical meristem functions (Vernoux et al., 2000). This mutant is unable to establish and maintain root meristem development, but cell division occurs in the shoot apical meristem, which produces a small shoot (Vernoux et al., 2000). Mutants lacking glutathione peroxidases have modified root phenotypes (Schnaubelt et al., 2015). GRXS8 and GRXS17 are required for the biogenesis of iron-sulphur clusters and they involved in the regulation of root system architecture (Ehrary et al., 2020; Martins et al., 2020).

2.2 Glutathione biosynthesis

The synthesis of glutathione occurs in two ATP-dependent steps (Galant, Preuss, Cameron, & Jez, 2011; Hasanuzzaman et al., 2017). The first step of glutathione biosynthesis is catalysed by glutamate–cysteine ligase (GCL, also known as γ-glutamylcysteine synthetase). This enzyme catalyzes the formation of γ-glutamylcysteine from cysteine and glutamate. GCL is a heterodimer and dissociates under non-denaturing conditions into a modulatory, or light subunit (GCLM), and a catalytic or heavy subunit (Lu, 2013). The enzyme forms an unusual peptide bond between the γ-carboxyl of glutamate and the amino group of cysteine using the energy provided by the hydrolysis of ATP. The second step in the pathway is catalysed by glutathione synthetase (GS), which adds glycine to γ-glutamylcysteine to form glutathione. The overexpression of GCL, but not GS, leads to substantial increases in glutathione accumulation (Foyer et al., 1995; Noctor et al., 1996). While

GCL is a chloroplast-localized enzyme, GS is localized in both the chloroplasts and cytosol (Galant et al., 2011; Hell & Bergmann, 1990). While higher plant GCL is encoded by a single gene, the two genes encoding GS have alternate transcription start sites for targeting the protein into the plastids or cytosol (Zechmann, 2014). The redox state of the chloroplast regulates GCL activity and hence glutathione production, which is particularly important in plant responses to abiotic and biotic stresses (Watson, Sowden, & Jarvis, 2018). GCL is post-translationally activated by formation of an intramolecular disulfide bridge [Cys186-Cys406], the oxidized GCL protein forming homodimers (Yang et al., 2019). While homo-dimer formation of redox-activated GCL is likely to occur in vivo, this transition is not required for enzyme activation (Yang et al., 2019). However, mutants lacking the disulphide bond have a decreased half-life (Kumar, Kasturia, Sharma, Datt, & Bachhawat, 2013). GCL is the rate-limiting enzyme in glutathione biosynthesis (Foyer et al., 1995; Lu, 2013; Noctor et al., 1996). The expression of *GSH1* is induced in response to heavy metals, chilling, pathogens and H_2O_2 (Lu, Luo, Gong, & Bai, 2021; Queval et al., 2009; Zhang, Zhang, Chen, Lu, & An, 2022). The expression of GSH1 in response to drought stress is also associated with higher drought tolerance (Lu et al., 2021).

Glutathione degradation also plays a role in the regulation of cellular glutathione accumulation (Bachhawat & Kaur, 2017). Glutathione is degraded in two enzymatic steps catalysed by γ-glutamyl cyclotransferase (GGCT; acting on γ-glutamyl amino acids) and the Cys-Gly dipeptidase. GGCTs are able to degrade glutathione to 5-oxoproline and Cys-Gly, but they do not show significant activity with regard to γ-glutamyl cysteine (Kumar, Kaur, Chattopadhyay, & Bachhawat, 2015).

2.3 Glutathione and cellular redox homeostasis

The concept of cellular redox homeostasis, a process that ensures an optimal balance between reducing and oxidizing reactions within cells, is firmly embedded in the literature (Le Gal, Schmidt, & Sayin, 2021). The production of ROS, particularly H_2O_2, in discrete cellular compartments shifts the redox balance for signalling to occur in response to changing environmental conditions (Considine & Foyer, 2021). The glutathione couple GSH/GSSG, together with other redox-active couples, such as NADPH/NADP+, TRX-SH/TRX-SS, not only function in maintaining cellular redox homeostasis but also in the propagation of redox signals (Foyer & Noctor, 2011). In particular, the redox-active cysteine residues of

GSH and TRX interact with protein thiol groups both to preserve redox state or transmit signals.

The enzyme glutathione reductase (GR) is a flavoprotein oxidoreductase which reduces glutathione disulphide to GSH using NADPH. The enzyme ensures that the glutathione pools of the major cellular compartments are largely reduced. GR is an important component of the ascorbate/glutathione cycle and plant responses to environmental stress (Foyer et al., 1995; Noctor et al., 1998). GR is encoded in higher plants by two genes, *GR1* and *GR2*. While *GR1* encodes a cytosolic, or peroxisomal, form of the enzyme, *GR2*, which contains a long N-terminal sequence, encodes mitochondrial and chloroplastic forms. The chloroplast form represents about 80% of the total GR activity. GR has also two Cys residues that form a redox-active disulfide bridge at the active site. Glutathione disulphide binds to the active site to form a disulfide bond separately with a Cys residue and a histidine (His) residue at the active site allowing reduction to GSH (Couto, Wood, & Barber, 2016; Kataya & Reumann, 2010; Yousuf, Hakeem, Chandna, & Ahmad, 2012). Overexpression of the chloroplast form of GR significantly increases the glutathione (GSH) content of plants and increases tolerance to a range of abiotic stresses (Foyer et al., 1995; Gill et al., 2013; Li, Dhankher, Carreira, Balish, & Meagher, 2005). Since increased GSH will decrease GCL activity thereby slowing down glutathione biosynthesis (Galant et al., 2011), this problem could be overcome by expression of bacterial forms of the enzyme that lack the regulatory feedback loop.

2.4 Glutathione as a cofactor for proteins

Superoxide and H_2O_2 react with reduced glutathione, but this reaction is relatively slow (Winterbourn, 2016). Glutathione peroxidases (GPX) may be considered as part of the glutathione/ascorbate cycle. Plant GPX protein sequences show high sequence similarities with mammalian phospholipid hydroperoxide GPX4 (Faltin et al., 2010), containing three conserved non-selenium Cys residues at the active sites. However, they catalyse the reduction of H_2O_2 using TRX and GRX as the electron donor rather than GSH and so are more correctly called thiol peroxidases than GPX (Bela et al., 2015). The plant GPX protein family consists of multiple isoenzymes located in different subcellular compartments that have distinct expression patterns with respect to tissues and developmental stages (Gao et al., 2014). These enzymes play an important role in protection against environmental stress (Zhang et al., 2019). For example, transgenic plants overexpressing

GPX genes have better stress tolerance (Diao et al., 2014; Zhang et al., 2019). Some GSTs also have GPX activity. These enzymes can detoxify lipid hydroperoxides and thus participate in antioxidative defence (Ding et al., 2020; Dixon, Skipsey, Grundy, & Edwards, 2005). Plant GSTs are mostly cytosolic enzymes, and they can represent up to 2% of soluble proteins (Pascal & Scalla, 1999).

Glutaredoxins (also known as thioltransferase) are small (12 kD) redox enzymes that fulfil important functions in plants (Meyer et al., 2008; Meyer, Belin, Delorme-Hinoux, Reichheld, & Riondet, 2012). GRXs not only catalyze the reduction of disulfides, but also the reduction of mixed disulfides, in a process called deglutathionylation. Glutaredoxins are defined on the basis of their active site sequences. While two types are present in Chlamydomonas, *E. coli* and yeast, a third type of GRX with a Cys-Cys-X-Cys or Cys-Cys-X-Ser active site is specific to higher plants (Ströher & Millar, 2012; Zaffagnini, Michelet, Massot, Trost, & Lemaire, 2008). GRXs require two distinct interaction sites for efficient redox catalysis (Begas, Liedgens, Moseler, Meyer, & Deponte, 2017). The first site interacts with the glutathione moiety of glutathionylated disulfide substrates. The second site activates glutathione as the reducing agent (Begas et al., 2017).

PRXs also catalyse H_2O_2 reduction. Due to their thiol-based catalytic mechanisms, these enzymes can reduce alkyl hydroperoxide and peroxynitrite as well as H_2O_2 (Liebthal, Maynard, & Dietz, 2018). Peroxiredoxin 2 (Prx2) is an effective antioxidant, regulating the concentration of cellular peroxides, and also acting as a sensor of redox signals. The active site Cys residues of Prx2 can form stable mixed disulfides with reduced GSH (Peskin et al., 2016). This glutathionylation process is reversed by glutaredoxin 1 (Grx1).

2.5 Glutathione and post translational protein modification

Protein glutathionylation is an important redox post transcriptional modification (PTM) that participates in redox sensing and signal transduction pathways (Sevilla et al., 2015). Redox signalling pathways involve redox-sensitive protein cysteines that undergo redox PTMs (Waszczak et al., 2015). Redox PTMs, such as protein S-nitrosylation and glutathionylation, modify protein structure and function (Hartl & Finkemeier, 2012). Studies of S-glutathionylation of specific target proteins have revealed that such PTMs are crucial to plant stress responses (Dixon et al., 2005; Waszczak et al., 2015). Glutathionylation is also important to generate thiolating agents

(Dalle-Donne et al., 2007). Glutathionylation is mediated by GSTs which add glutathione to the thiolate form of protein cysteines and are highly induced by stress conditions (Gullner, Komives, Király, & Schröder, 2018). Oxidative stress generally promotes S-glutathionylation, which protects protein cysteine residues from irreversible oxidation (Roos & Messens, 2011) and allows the transduction of redox signals due to changes in the structure/function of target proteins. S-glutathionylation also occurs spontaneously by thiol-disulphide exchange (Zhang, Si, et al., 2018; Zhang, Ye, et al., 2018). H_2O_2 can be directly involved in the glutathionylation process through the direct oxidation of protein Cys, which produces a thiyl intermediate, and also by reaction with reduced GSH to form mixed disulfides (Grek et al., 2013). Reversible protein glutathionylation is an important reversible covalent PTM that functions in the redox signalling pathways of animal and plant cells (Mieyal et al., 2008; Townsend, Lushchak, & Cooper, 2014). The reverse reaction, deglutathionylation, is catalysed by GRXs, in a pathway that involves GSH and GR (Waszczak et al., 2015; Zhang, Si, et al., 2018; Zhang, Ye, et al., 2018).

3. Future Perspectives

The above discussion has highlighted the importance of the ascorbate/glutathione cycle in policing ROS levels in the different sub-cellular compartments. It is likely that the concetrations of ascorbate and gluathione that occur in different intracellular compartments allow appropriate regulation of the cell-to-cell ROS signalling wave, as well key development processes including the cell cycle (Pellny et al., 2009) and embryogenesis induction in isolated microspore cultures (Zur et al., 2021). However, as highlighted above, the functions of ascorbate/glutathione cycle extend far beyond the policing of ROS signals. This pathway also serves to prevent decreases in the levels of ascorbate and glutathione in the different cellular compartments and preserves these metabolic pools in the reduced state. Ascorbate and glutathione are versatile metabolites that are central to plant growth and development as well as defence. A key question concerns how ascorabte and glutathione function in the nucleus to modulate and protect nuclear processes and participate in genetic and epigenetic regulation, particularly under stress conditions (Diaz-Vivancos et al., 2015; García-Giménez, Romá-Mateo, Pérez-Machado, Peiró-Chova, & Pallardó, 2017). Simlarly, little is known about how glutathione interacts with the nuclear thioredoxin system and nitric oxide to regulate nuclear processes. The nucleus is rapidly oxidised in response to stresses, such as high temperatures (Babbar, Karpinska, Grover, & Foyer, 2021), and this will have a significant impact on the redox state of nuclear protein thiols. The direct impacts of stress-induced oxidation of nuclei have significant implications for current concepts of redox sensing and regulation, as well as associated signal transduction pathways (Sevilla et al., 2015). The control of nuclear thiol-disulfide redox states by nucleoredoxins and TRX1 remains, however, laregely uncharacterised (Kneeshaw et al., 2017). Similarly, how the nuclear glutathione and ascorbate pools influence the regulation of cell cycle proteins (de Simone et al., 2017; Diaz-Vivancos et al., 2015). The in vivo imaging and mesaurement of glutathione and H_2O_2 using genetically encoded redox biosensors will be, therefore, essential to investigate the dynamic nature of these redox controls (Kostyuk et al., 2020) together with the involvement and regulation of the TRX/PRX systems, particularly under abiotic stress conditions.

Acknowledgments

KJK was funded by NRF SA incentive funding.

References

Akram, N. A., Shafiq, F., & Ashraf, M. (2017). Ascorbic acid-a potential oxidant scavenger and its role in plant development and abiotic stress tolerance. *Frontiers in Plant Science, 8*, 613.

Aroca, A., Gotor, C., & Romero, L. C. (2018). Hydrogen sulfide signaling in plants: Emerging roles of protein persulfidation. *Frontiers in Plant Science, 9*, 1369. https://doi.org/10.3389/fpls.2018.01369

Aroca, Á., Serna, A., Gotor, C., & Romero, L. C. (2015). S-sulfhydration: A cysteine posttranslational modification in plant systems. *Plant Physiology, 168*, 334–342.

Aslam, S., Gul, N., Mir, M. A., Asgher, M., Al-Sulami, N., Abulfaraj, A. A., & Qari, S. (2021). Role of jasmonates, calcium, and glutathione in plants to combat abiotic stresses through precise signaling cascade. *Frontiers in Plant Science, 12*, 668029. https://doi.org/10.3389/fpls.2021.668029

Babbar, R., Karpinska, B., Grover, A., & Foyer, C. H. (2021). Heat-induced oxidation of the nuclei and cytosol. *Frontiers in Plant Science, 11*, 617779.

Bachhawat, A. K., & Kaur, A. (2017). Glutathione degradation. *Antioxidants & Redox Signaling, 27*, 1200–1216. https://doi.org/10.1089/ars.2017.7136

Baier, M., Ströher, E., & Dietz, K. J. (2004). The acceptor availability at photosystem I and ABA control nuclear expression of 2-Cys peroxiredoxin-A in Arabidopsis thaliana. *Plant and Cell Physiology, 45*, 997–1006.

Baier, M., Noctor, G., Foyer, C. H., & Dietz, K. J. (2000). Antisense suppression of 2-cysteine peroxiredoxin in Arabidopsis specifically enhances the activities and expression of enzymes associated with ascorbate metabolism but not glutathione metabolism. *Plant Physiology, 124*, 823–832.

Barth, C., De Tullio, M., & Conklin, P. L. (2006). The role of ascorbic acid in the control of flowering time and the onset of senescence. *Journal of Experimental Botany, 57*, 1657–1665.

Bartoli, C. G., Tambussi, E. A., Fanello, D., & Foyer, C. H. (2009). Control of ascorbic acid synthesis and accumulation and glutathione by the incident light red/far red ratio in Phaseolus vulgaris leaves. *FEBS Letters, 583*, 118–122.

Bashandy, T., Guilleminot, J., Vernoux, T., Caparros-Ruiz, D., Ljung, K., Meyer, Y., & Reichheld, J. P. (2010). Interplay between the NADP-linked thioredoxin and glutathione systems in Arabidopsis auxin signaling. *The Plant Cell, 22*, 376–391.

Begara-Morales, J. C., Chaki, M., Sanchez-Calvo, B., Mata-Perez, C., Leterrier, M., Palma, J. M., et al. (2013). Protein tyrosine nitration in pea roots during development and senescence. *Journal of Experimental Botany, 64*, 1121–1134.

Begara-Morales, J. C., Sánchez-Calvo, B., Chaki, M., Valderrama, R., Mata-Pérez, C., Padilla, M. N., et al. (2016). Antioxidant systems are regulated by nitric oxide-mediated post-translational modifications (NO-PTMs). *Frontiers in Plant Science, 7*, 152.

Begas, P., Liedgens, L., Moseler, A., Meyer, A. J., & Deponte, M. (2017). Glutaredoxin catalysis requires two distinct glutathione interaction sites. *Nature Communications, 8*, 14835. https://doi.org/10.1038/ncomms14835

Bela, K., Horvath, E., Galle, A., Szabados, L., Tari, I., & Csiszar, J. (2015). Plant glutathione peroxidases: Emerging role of the antioxidant enzymes in plant development and stress responses. *Journal of Plant Physiology, 176*, 192–201.

Booker, F. L., Burkey, K. O., & Jones, A. M. (2012). Re-evaluating the role of ascorbic acid and phenolic glycosides in ozone scavenging in the leaf apoplast of Arabidopsis thaliana L. *Plant, Cell and Environment, 35*, 1456–1466. https://doi.org/10.1111/j.1365-3040.2012.02502.x

Botanga, C. J., Bethke, G., Chen, Z., Gallie, D. R., Fiehn, O., & Glazebrook, J. (2012). Metabolite profiling of Arabidopsis inoculated with Alternaria brassicicola reveals that ascorbate reduces disease severity. *Molecular Plant Microbe Interactions, 25*, 1628–1638.

Broad, R. C., Bonneau, J. P., Hellens, R. P., & Johnson, A. A. T. (2020). Manipulation of ascorbate biosynthetic, recycling, and regulatory pathways for improved abiotic stress tolerance in plants. *International Journal of Molecular Sciences, 21*, 1790. https://doi.org/10.3390/ijms21051790

Bulley, S., & Laing, W. (2016). The regulation of ascorbate biosynthesis. *Current Opinion in Plant Biology, 33*, 15–22.

Bulley, S., Wright, M., Rommens, C., Yan, H., Rassam, M., Lin-Wang, K., et al. (2012). Enhancing ascorbate in fruits and tubers through over-expression of the L- galactose pathway gene GDP-L-galactose phosphorylase. *Plant Biotechnology Journal, 10*, 390–397.

Cai, X., Zhang, C., Shu, W., Ye, Z., Li, H., & Zhang, Y. (2016). The transcription factor SlDof22 involved in ascorbate accumulation and salinity stress in tomato. *Biochemical and Biophysical Research Communications, 474*, 736–741.

Cairns, N. G., Pasternak, M., Wachter, A., Cobbett, C. S., & Meyer, A. J. (2006). Maturation of Arabidopsis seeds is dependent on glutathione biosynthesis within the embryo. *Plant Physiology, 141*, 446–455.

Chen, Z., & Gallie, D. R. (2004). The ascorbic acid redox state controls guard cell signaling and stomatal movement. *The Plant Cell, 16*, 1143–1162.

Chen, Z., & Gallie, D. R. (2012). Induction of monozygotic twinning by ascorbic acid in tobacco. *PLoS One, 7*, e39147.

Cheng, M.-C., Ko, K., Chnag, W.-L., Kuo, W.-C., Chen, G.-H., & Lin, T.-P. (2015). Increased glutathione contributes to stress tolerance and global translational changes in Arabidopsis. *Plant Journal, 83*, 926–939.

Chin, D.-C., Kumar, R. S., Suen, C.-S., Chien, C.-Y., Hwang, M.-J., Hsu, C.-H., et al. (2019). Plant cytosolic ascorbate peroxidase with dual catalytic activity modulates abiotic stress tolerances. *iScience, 16*, 31–49.

Cho, K. M., Nguyen, H. T. K., Kim, S. Y., Shin, J. S., Cho, D. H., Hong, S. B., et al. (2016). CML 10, a variant of calmodulin, modulates ascorbic acid synthesis. *New Phytologist, 209*, 664–678.

Clark, D., Durner, J., Navarre, D. A., & Klessig, D. F. (2000). Nitric oxide inhibition of tobacco catalase and ascorbate peroxidase. *Molecular Plant Microbe Interactions, 13*, 1380–1384.

Considine, M. J., & Foyer, C. H. (2014). Redox regulation of plant development. *Antioxidants and Redox Signaling, 21*, 1305–1326.

Considine, M. J., & Foyer, C. H. (2021). Oxygen and reactive oxygen species (ROS) dependent regulation of plant growth and development. *Plant Physiology, 186*, 79–92.

Couto, N., Wood, J., & Barber, J. (2016). The role of glutathione reductase and related enzymes on cellular redox homoeostasis network. *Free Radical Biology & Medicine, 95*, 27–42.

Dalle-Donne, I., Rossi, R., Giustarini, D., Colombo, R., & Milzani, A. (2007). S-glutathionylation in protein redox regulation. *Free Radical Biology and Medicine, 43*, 883–898. https://doi.org/10.1016/j.freeradbiomed.2007.06.014

Deslous, P., Bournonville, C., Decros, G., Okabe, Y., Mauxion, J. P., Jorly, J., et al. (2021). Overproduction of ascorbic acid impairs pollen fertility in tomato. *Journal of Experimental Botany, 72*, 3091–3107.

Diao, Y., Xu, H., Li, G., Yu, A., Yu, X., Hu, W., et al. (2014). Cloning a glutathione peroxidase gene from Nelumbo nucifera and enhanced salt tolerance by overexpressing in rice. *Molecular Biology Reporter, 41*, 4919–4927. https://doi.org/10.1007/s11033-014-3358-4

Diaz-Vivancos, P., de Simone, A., Kiddle, G., & Foyer, C. H. (2015). Glutathione – Linking cell proliferation to oxidative stress. *Free Radical Biology & Medicine, 89*, 1154–1164. https://doi.org/10.1016/j.freeradbiomed.2015.09.023

Dickinson, D. A., & Forman, H. J. (2002). Cellular glutathione and thiols metabolism. *Biochemical Pharmacology, 64*, 1019–1026. https://doi.org/10.1016/S0006-2952(02)01172-3

Ding, H., Wang, B., Han, Y., & Li, S. (2020). The pivotal function of dehydroascorbate reductase in glutathione homeostasis in plants. *Journal of Experimental Botany, 22*, 3405–3416. https://doi.org/10.1093/jxb/eraa107

Dixon, D. P., Skipsey, M., Grundy, N. M., & Edwards, R. (2005). Stress-induced protein S-glutathionylation in Arabidopsis. *Plant Physiology, 138*, 2233–2244. https://doi.org/10.1104/pp.104.058917

Dorion, S., Ouellet, J. C., & Rivoal, J. (2021). Glutathione metabolism in plants under stress: Beyond reactive oxygen species detoxification. *Metabolites, 11*, 641. https://doi.org/10.3390/metabo11090641

Dowdle, J., Ishikawa, T., Gatzek, S., Rolinski, S., & Smirnoff, N. (2007). Two genes in Arabidopsis thaliana encoding GDP-L-galactose phosphorylase are required for ascorbate biosynthesis and seedling viability. *Plant Journal, 52*, 673–689.

Dumanović, J., Nepovimova, E., Natić, M., Kuča, K., & Jaćević, V. (2021). The significance of reactive oxygen species and antioxidant defense system in plants: A Concise Overview. *Frontiers in Plant Science, 11*, 552969. https://doi.org/10.3389/fpls.2020.552969

Eastmond, P. J. (2007). Monodehyroascorbate reductase 4 is required for seed storage oil hydrolysis and postgerminative growth in Arabidopsis. *The Plant Cell, 19*, 1376–1387.

Ehrary, A., Rosas, M., Carpinelli, S., Davalos, O., Cowlong, C., Fernandez, F., & Escobar, M. (2020). Glutaredoxin AtGRXS8 represses transcriptional and developmental responses to nitrate in Arabidopsis thaliana roots. *Plant Direct, 4*, e00227.

Eljebbawi, H., del Carmen Rondón Guerrero, Y., Dunand, C., & Estevez, J. M. (2021). Highlighting reactive oxygen species (ROS) as multitaskers in root development. *iScience, 24*, 101978.

Eltayeb, A. E., Kawano, N., Badawi, G. H., Kaminaka, H., Sanekata, T., Shibahara, T., et al. (2007). Overexpression of monodehydroascorbate reductase in transgenic tobacco confers enhanced tolerance to ozone, salt and polyethylene glycol stresses. *Planta, 225*, 1255–1264.

Eltelib, H., Fujikawa, Y., & Esaka, M. (2012). Overexpression of the acerola (Malpighia glabra) monodehydroascorbate reductase gene in transgenic tobacco plants results in increased ascorbate levels and enhanced tolerance to salt stress. *South African Journal of Botany, 28*, 295–301.

Faltin, Z., Holland, D., Velcheva, M., Tsapovetsky, M., Roeckel-Drevet, P., Handa, A. K., et al. (2010). Glutathione peroxidase regulation of reactive oxygen species level is crucial for in vitro plant differentiation. *Plant and Cell Physiology, 51*, 1151–1162. https://doi.org/10.1093/pcp/pcq082

Fenech, M., Amaya, I., Valpuesta, V., & Botella, M. A. (2019). Vitamin C content in fruits: Biosynthesis and regulation. *Frontiers in Plant Science, 9*, 2006.

Fenech, M., Amorim-Silva, V., Esteban del Valle, A., Arnaud, D., Ruiz-Lopez, N., Castillo, A. G., et al. (2021). The role of GDP-L-galactose phosphorylase in the control of ascorbate biosynthesis. *Plant Physiology, 185*, 1574–1594.

Fichman, Y., & Mittler, R. (2020). Rapid systemic signaling during abiotic and biotic stresses: Is the ROS wave master of all trades? *Plant Journal, 102*, 887–896.

Fichman, Y., & Mittler, R. (2021a). Integration of electric, calcium, reactive oxygen species and hydraulic signals during rapid systemic signaling in plants. *Plant Journal, 107*, 7–20.

Fichman, Y., & Mittler, R. (2021b). A systemic whole-plant change in redox levels accompanies the rapid systemic response to wounding. *Plant Physiology, 186*, 4–8.

Fichman, Y., Zandalinas, S. I., Peck, S., Luan, S., & Mittler, R. (2022). HPCA1 is required for systemic reactive oxygen species and calcium cell-to-cell signaling and plant acclimation to stress. *The Plant Cell, 241*. https://doi.org/10.1093/plcell/koac241

Foyer, C. H., & Halliwell, B. (1976). The presence of glutathione and glutathione reductase in chloroplasts: A proposed role in ascorbic acid metabolism. *Planta, 133*, 21–25.

Foyer, C. H., & Noctor, G. (2011). Ascorbate and glutathione: The heart of the redox hub. *Plant Physiology, 155*, 2–18.

Foyer, C. H., & Hanke, G. (2022). ROS production and signalling in chloroplasts: Cornerstones and evolving concepts. *Plant Journal, 111*, 642–661. https://doi.org/10.1111/tpj.15856

Foyer, C. H., Kyndt, T., & Hancock, R. D. (2020). Vitamin C in Plants: Novel concepts, new perspectives and outstanding issues. *Antioxidants and Redox Signalling, 32*, 463–485.

Foyer, C. H., Souriau, N., Perret, S., Lelandais, M., Kunert, K. J., Pruvost, C., & Jouanin, L. (1995). Overexpression of glutathione reductase but not glutathione synthetase leads to increases in antioxidant capacity and resistance to photoinhibition in poplar trees. *Plant Physiology, 109*, 1047–1057. https://doi.org/10.1104/pp.109.3.1047

Fujiwara, A., Shimura, H., Masuta, C., Sano, S., & Inukai, T. (2013). Exogenous ascorbic acid derivatives and dehydroascorbic acid are effective antiviral agents against Turnip mosaic virus in Brassica rapa. *Journal of General Plant Pathology, 79*, 198–204.

Galant, A., Preuss, M. L., Cameron, J. C., & Jez, J. M. (2011). Plant glutathione biosynthesis: diversity in biochemical regulation and reaction products. *Frontiers in Plant Science, 2*, 45. https://doi.org/10.3389/fpls.2011.00045

Gao, F., Chen, J., Ma, T., Li, H., Wang, N., & Li, Z. (2014). The glutathione peroxidase gene family in Thellungiella salsuginea: genome-wide identification, classification, and gene and protein expression analysis under stress conditions. *International Journal of Molecular Sciences, 15*, 3319–3335. https://doi.org/10.3390/ijms15023319

García-Giménez, J. L., Romá-Mateo, C., Pérez-Machado, G., Peiró-Chova, L., & Pallardó, F. V. (2017). Role of glutathione in the regulation of epigenetic mechanisms in disease. *Free Radical Biology & Medicine, 112*, 36–48. https://doi.org/10.1016/j.freeradbiomed.2017.07.008

Gill, S. S., Anjum, N. A., Hasanuzzaman, M., Gill, R., Trivedi, D. K., Ahmad, I., et al. (2013). Glutathione and glutathione reductase: A boon in disguise for plant abiotic stress defense operations. *Plant Physiology and Biochemistry, 70*, 204–212. https://doi.org/10.1016/j.plaphy.2013.05.032

Grek, C. L., Zhang, J., Manevich, Y., Townsend, D. M., & Tew, K. D. (2013). Causes and consequences of cysteine S-glutathionylation. *Journal of Biological Chemistry, 288*, 26497–26504. https://doi.org/10.1074/jbc.R113.461368

Groden, D., & Beck, E. (1979). H_2O_2 destruction by ascorbate-dependent systems from chloroplasts. *Biochimica Biophysica Acta, 546*, 426–435.

Gullner, G., Komives, T., Király, L., & Schröder, P. (2018). Glutathione S-Transferase enzymes in plant-pathogen interactions. *Frontiers in Plant Science, 9*, 1836. https://doi.org/10.3389/fpls.2018.01836

Hancock, R. D., & Viola, R. (2005). Biosynthesis and catabolism of L-ascorbic acid in plants. *Critical Reviews in Plant Sciences, 24*, 167–188.

Hartl, M., & Finkemeier, I. (2012). Plant mitochondrial retrograde signaling: post-translational modifications enter the stage. *Frontiers in Plant Science, 3*, 253. https://doi.org/10.3389/fpls.2012.00253

Hasanuzzaman, M., Nahar, K., Anee, T. I., & Fujita, M. (2017). Glutathione in plants: Biosynthesis and physiological role in environmental stress tolerance. *Physiology and Molecular Biology of Plants, 23*, 249–268. https://doi.org/10.1007/s12298-017-0422-2

Hell, R., & Bergmann, L. (1990). Gamma-glutamylcysteine synthetase in higher plants: Catalytic properties and subcellular localization. *Planta, 180*, 603–612. https://doi.org/10.1007/BF02411460

Hong, S. H., Tripathi, B. N., Chung, M.-S., Cho, C., Lee, S., Kim, J.-H., et al. (2018). Functional switching of ascorbate peroxidase 2 of rice (OsAPX2) between peroxidase and molecular chaperone. *Scientific Reports, 8*, 9171.

Horling, F., Lamkemeyer, P., König, J., Finkemeier, I., Kandlbinder, A., Baier, M., & Dietz, K. J. (2003). Divergent light-, ascorbate-, and oxidative stress-dependent regulation of expression of the peroxiredoxin gene family in Arabidopsis. *Plant Physiology, 131*, 317–325.

Hossain, M. A., Nakano, Y., & Asada, K. (1984). Monodehydroascorbate reductase in spinach chloroplasts and its participation in regeneration of ascorbate for scavenging hydrogen peroxide. *Plant Cell Physiology, 25*, 385–395.

Hu, T., Ye, J., Tao, P., Li, H., Zhang, J., Zhang, Y., & Ye, Z. (2016). The tomato HD-Zip I transcription factor SlHZ24 modulates ascorbate accumulation through positive regulation of the d-mannose/l-galactose pathway. *Plant Journal, 85*, 16–29.

Hu, Z., Li, J., Ding, S., Cheng, F., Li, X., Jiang, Y., ... Shi, K. (2021). The protein kinase CPK28 phosphorylates ascorbate peroxidase and enhances thermotolerance in tomato. *Plant Physiology, 186*, 1302–1317.

Huang, X., Chen, S., Li, W., Tang, L., Zhang, Y., Yang, N., et al. (2021). ROS regulated reversible protein phase separation synchronizes plant flowering. *Nature Chemical Biology, 17*, 549–557.

Inagi, R., Kumagai, T., Fujita, T., & Nangaku, M. (2010). The role of glyoxalase system in renal hypoxia. *Advances in Experimental Medicine and Biology, 662*, 49–55. https://doi.org/10.1007/978-1-4419-1241-1_6

Karpinska, B., Zhang, K., Rasool, B., Pastok, D., Morris, J., Verrall, S. R., et al. (2017). The redox state of the apoplast influences the acclimation of photosynthesis and leaf metabolism to changing irradiance. *Plant, Cell & Environment, 41*, 1083–1097.

Kataya, A. R., & Reumann, S. (2010). Arabidopsis glutathione reductase 1 is dually targeted to peroxisomes and the cytosol. *Plant Signaling & Behavior Plant, 5*, 171–175.

Kaur, S., Prakash, P., Bak, D.-H., Hong, S. H., Cho, C., Chung, M.-S., et al. (2021). Regulation of dual activity of ascorbate peroxidase 1 from Arabidopsis thaliana by conformational changes and posttranslational modifications. *Frontiers in Plant Science, 12*, 678111.

Kavitha, K., George, S., Venkataraman, G., & Parida, A. (2010). A salt-inducible chloroplastic monodehydroascorbate reductase from halophyte Avicennia marina confers salt stress tolerance on transgenic plants. *Biochimie, 92*, 1321–1329.

Kelly, G. J., & Latzko, E. (1979). Soluble ascorbate peroxidase, detection in plants and use in vitamin C estimation. *Die Naturwissenschaften, 66*, 617–619.

Kerchev, P. I., Karpińska, B., Morris, J. A., Hussain, A., Verrall, S. R., Hedley, P. E., et al. (2013). Vitamin C and the abscisic acid-insensitive 4 (ABI4) transcription factor are important determinants of aphid resistance in Arabidopsis. *Antioxidants and Redox Signaling, 18*, 2091–2105.

Khan, M., Samrana, S., Zhang, Y., Malik, Z., Khan, M. D., & Zhu, S. (2020). Reduced glutathione protects subcellular compartments from Pb-induced ROS injury in leaves and roots of upland cotton (Gossypium hirsutum L.). *Frontiers in Plant Science, 11*, 412. https://doi.org/10.3389/fpls.2020.00412

Kiddle, G., Pastori, G. M., Bernard, B., Pignocchi, C., Antoniw, J., Verrier, P. J., & Foyer, C. H. (2003). Effects of leaf ascorbate content on defense and photosynthesis gene expression in Arabidopsis thaliana. *Antioxidants and Redox Signalling, 5*, 23–32.

Kneeshaw, S., Keyani, R., Delorme-Hinoux, V., Imrie, L., Loake, G. J., Le Bihan, T., et al. (2017). Nucleoredoxin guards against oxidative stress by protecting antioxidant enzymes. *Proceedings of the National Academy of Sciences U S A, 114*, 8414–8419. https://doi.org/10.1073/pnas.1703344114

Koramutla, M. K., Negi, M., & Ayele, B. T. (2021). Roles of glutathione in mediating abscisic acid signaling and its regulation of seed dormancy and drought tolerance. *Genes, 12*, 1620.

Kostyuk, A. I., Panova, A. S., Kokova, A. D., Kotova, D. A., Maltsev, D. I., Podgorny, O. V., et al. (2020). In vivo imaging with genetically encoded redox biosensors. *International Journal of Molecular Sciences, 21*, 8164.

Kotchoni, S. O., Larrimore, K. E., Mukherjee, M., Kempinski, C. F., & Barth, C. (2009). Alterations in the endogenous ascorbic acid content affect flowering time in Arabidopsis. *Plant Physiology, 149*, 803–815.

Kumar, R. S., Shen, C.-H., Wu, P.-Y., Kumar, S., Hua, M. S., & Yeh, K.-Y. (2016). Nitric oxide participates in plant flowering repression by ascorbate. *Science Reports, 6*, 35246.

Kumar, S., Kaur, A., Chattopadhyay, B., & Bachhawat, A. K. (2015). Defining the cytosolic pathway of glutathione degradation in Arabidopsis thaliana: Role of the ChaC/GCG family of γ-glutamyl cyclotransferases as glutathione-degrading enzymes and AtLAP1 as the Cys-Gly peptidase. *Biochemical Journal, 468*, 73–85. https://doi.org/10.1042/BJ20141154

Kumar, S., Kasturia, N., Sharma, A., Datt, M., & Bachhawat, A. K. (2013). Redox-dependent stability of the γ-glutamylcysteine synthetase enzyme of *Escherichia coli*: A novel means of redox regulation. *Biochemical Journal, 449*, 783–794. https://doi.org/10.1042/BJ20120204

Laing, W. A., Martınez-Sanchez, M., Wright, M. A., Bulley, S. M., Brewster, D., Dare, A. P., et al. (2015). An upstream open reading frame is essential for feedback regulation of ascorbate biosynthesis in Arabidopsis. *The Plant Cell, 27*, 772–786.

Laing, W. A., Bulley, S., Wright, M., Cooney, J., Jensen, D., Barraclough, D., & MacRae, E. (2004). A highly specific L-galactose-1-phosphate phosphatase on the path to ascorbate biosynthesis. *Proceedings of the National Academy of Sciences of the United States of America, 101*, 16976–16981.

Le Gal, K., Schmidt, E. E., & Sayin, V. I. (2021). Cellular redox homeostasis. *Antioxidants, 10*, 1377. https://doi.org/10.3390/antiox10091377

Li, F., Wu, Q. Y., Sun, Y. L., Wang, L. Y., Yang, X. H., & Meng, Q. W. (2010). Overexpression of chloroplastic monodehydroascorbate reductase enhanced tolerance to temperature and methyl viologen-mediated oxidative stresses. *Physiologia Plantarum, 139*, 421–434.

Li, Y., Cohenford, M. A., Dutta, U., & Dain, J. A. (2008). The structural modification of DNA nucleosides by nonenzymatic glycation: an in vitro study based on the reactions of glyoxal and methylglyoxal with 2'-deoxyguanosine. *Analytical and Bioanalytical Chemistry, 390*, 679–688. https://doi.org/10.1007/s00216-007-1682-4

Li, Y., Dhankher, O. P., Carreira, L., Balish, R. S., & Meagher, R. B. (2005). Arsenic and mercury tolerance and cadmium sensitivity in Arabidopsis plants expressing bacterial γ-glutamylcysteine synthetase. *Environmental Toxicology and Chemistry, 24*, 1376–1386.

Liebthal, M., Maynard, D., & Dietz, K. J. (2018). Peroxiredoxins and redox signaling in plants. *Antioxidants & Redox Signaling, 28*, 609–624. https://doi.org/10.1089/ars.2017.7164

Linster, C. L., Gomez, T. A., Christensen, K. C., Adler, L. N., Young, B. D., Brenner, C., & Clarke, S. G. (2007). Arabidopsis VTC2 encodes a GDP-L-galactose phosphorylase, the last unknown enzyme in the smirnoff-wheeler pathway to ascorbic acid in plants. *Journal of Biological Chemistry, 282*, 18879–18885.

Lu, B., Luo, X., Gong, C., & Bai, J. (2021). Overexpression of γ-glutamylcysteine synthetase gene from Caragana korshinskii decreases stomatal density and enhances drought tolerance. *BMC Plant Biology, 21*, 444. https://doi.org/10.1186/s12870-021-03226-9

Lu, S. C. (2013). Glutathione synthesis. *Biochimica et Biophysica Acta, 1830*, 3143–3153. https://doi.org/10.1016/j.bbagen.2012.09.008

Luo, X., Dai, Y., Zheng, C., Yang, Y., Chen, W., Wang, Q., et al. (2021). The ABI4–RbohD/VTC2 regulatory module promotes Reactive Oxygen Species (ROS) accumulation to decrease seed germination under salinity stress. *New Phytologist, 229*, 950–962.

Mahmood, A. M., & Dunwell, J. M. (2019). Evidence for novel epigenetic marks within plants. *AIMS Genetics, 6*, 70–87.

Mano, É., Hideg, K., & Asada, K. (2004). Ascorbate in thylakoid lumen functions as an alternative electron donor to photosystem II and to photosystem I. *Archives of Biochemistry and Biophysics, 429*, 71–80.

Martins, L., Knuesting, J., Bariat, L., Dard, A., Freibert, S. A., Marchand, C. H., et al. (2020). Redox modification of the iron-sulfur glutaredoxin GRXS17 activates holdase activity and protects plants from heat stress. *Plant Physiology, 184*, 676–692.

Mase, K., & Tsukagoshi, H. (2021). Reactive oxygen species link gene regulatory networks during Arabidopsis root development. *Frontiers in Plant Science, 12*, 660274.

Mehlhorn, H., Lelandais, M., Korth, H. G., & Foyer, C. H. (1996). Ascorbate is the natural substrate for plant peroxidases. *FEBS Letters, 378*, 203–206 1996.

Meyer, Y., Belin, C., Delorme-Hinoux, V., Reichheld, J. P., & Riondet, C. (2012). Thioredoxin and glutaredoxin systems in plants: Molecular mechanisms, crosstalks, and functional significance. *Antioxidants & Redox Signaling, 17*, 1124–1160. https://doi.org/10.1089/ars.2011.4327

Meyer, Y., Siala, W., Bashandy, T., Riondet, C., Vignols, F., & Reichheld, J. P. (2008). Glutaredoxins and thioredoxins in plants. *Biochimica et Biophysica Acta – Molecular Cell Research, 1783*, 589–600. https://doi.org/10.1016/j.bbamcr.2007.10.017

Mieyal, J. J., Gallogly, M. M., Qanungo, S., Sabens, E. A., & Shelton, M. D. (2008). Molecular mechanisms and clinical implications of reversible protein S-glutathionylation. *Antioxidants & Redox Signaling, 10*, 1941–1988. https://doi.org/10.1089/ars.2008.2089

Miyaji, T., Kuromori, T., Takeuchi, Y., Yamaji, N., Yokosho, K., Shimazawa, A., et al. (2015). AtPHT4;4 is a chloroplast-localized ascorbate transporter in Arabidopsis. *Nature Communications, 6*, 5928.

Miyake, C., & Asada, K. (1994). Ferredoxin-dependent photoreduction of the monodehydroascorbate radical in spinach thylakoids. *Plant and Cell Physiology, 35*, 539–549.

Mukherjee, M., Larrimore, K. E., Ahmed, N. J., Bedick, T. S., Barghouthi, N. T., Traw, M. B., & Barth, C. (2010). Ascorbic acid deficiency in Arabidopsis induces constitutive priming that is dependent on hydrogen peroxide, salicylic acid, and the NPR1 gene. *Molecular Plant-Microbe Interactions, 23*, 340–351.

Müller-Moulé, P., Conklin, P. L., & Niyogi, K. K. (2002). Ascorbate deficiency can limit violaxanthin de-epoxidase activity in vivo. *Plant Physiology, 128*, 970–977.

Müller-Moulé, P., Golan, T., & Niyogi, K. K. (2014). Ascorbate-deficient mutants of Arabidopsis grow in high light despite chronic photo-oxidative stress. *Plant Physiology, 133*, 748–760.

Munné-Bosch, S., Queval, G., & Foyer, C. H. (2013). The impact of global change factors on redox signaling underpinning stress tolerance. *Plant Physiology, 161*, 5–19.

Naqvi, S., Zhu, C., Farre, G., Ramessar, K., Bassie, L., Breitenbach, J., et al. (2009). Transgenic multivitamin corn through biofortification of endosperm with three vitamins representing three distinct metabolic pathways. *Proceedings of the National Academy of Sciences of the United States of America, 106*, 7762–7767.

Narendra, S., Venkataramani, S., Shen, G., Wang, J., Pasapula, V., Lin, Y., et al. (2006). The Arabidopsis ascorbate peroxidase 3 is a peroxisomal membrane-bound antioxidant enzyme and is dispensable for Arabidopsis growth and development. *Journal of Experimental Botany, 57*, 3033–3042.

Neubauer, C., & Yamamoto, H. Y. (1992). Mehler-peroxidase reaction mediates zeaxanthin formation and zeaxanthin-related fluorescence quenching in intact chloroplasts. *Plant Physiology, 99*, 1354–1361.

Noctor, G., & Foyer, C. H. (1998). Ascorbate and Glutathione: Keeping active oxygen under control. *Annual Review Plant Physiology Plant Molecular Biology, 49*, 249–279.

Noctor, G., Strohm, M., Jouanin, L., Kunert, K. J., Foyer, C. H., & Rennenberg, H. (1996). Synthesis of glutathione in leaves of transgenic poplar overexpressing [gamma]-glutamylcysteine synthetase. *Plant Physiology, 112*, 1071–1078.

Noctor, G., Arisi, A.-C. M., Jouanin, L., Kunert, K. J., Rennenberg, H., & Foyer, C. H. (1998). Glutathione: Biosynthesis, metabolism and relationship to stress tolerance explored in transformed plants. *Journal of Experimental Botany, 49*, 623–647.

Noctor, G., Mhamdi, A., Chaouch, S., Han, Y., Neukermans, J., Marquez-Garcia, B., et al. (2012). Glutathione in plants: An integrated overview. *Plant, Cell & Environment, 35*, 454–484.

Ojeda, V., Pérez-Ruiz, J. M., & Cejudo, F. J. (2018). 2-Cys peroxiredoxins participate in the oxidation of chloroplast enzymes in the dark. *Molecular Plant, 11*, 1377–1388.

Ouyang, B., Yang, T., Li, H., Zhang, L., Zhang, Y., Zhang, J., et al. (2007). Identification of early salt stress response genes in tomato root by suppression subtractive hybridization and microarray analysis. *Journal of Experimental Botany, 58*, 507–520.

Page, M., Sultana, N., Paszkiewicz, K., Florance, H., & Smirnoff, N. (2012). The influence of ascorbate on anthocyanin accumulation during high light acclimation in Arabidopsis thaliana: Further evidence for redox control of anthocyanin synthesis. *Plant, Cell & Environment, 35*, 388–404.

Paradiso, A., de Pinto, M. C., Locato, V., & De Gara, L. (2012). Galactone-γ-lactone-dependent ascorbate biosynthesis alters wheat kernel maturation. *Plant Biology, 14*, 652–658. https://doi.org/10.1111/j.1438-8677.2011.00543.x

Pascal, S., & Scalla, R. (1999). Purification and characterization of a safener-induced glutathione S-transferase from wheat (Triticum aestivum). *Physiologia Plantarum, 106*, 17–27. https://doi.org/10.1034/j.1399-3054.1999.106103

Pasternak, T., Palme, K., & Paponov, I. A. (2020). Glutathione enhances auxin sensitivity in Arabidopsis roots. *Biomolecules, 10*, 1550.

Pastori, G. M., Kiddle, G., Antoniw, J., Bernard, S., Veljovic-Jovanovic, S., Verrier, P. J., et al. (2003). Leaf vitamin C contents modulate plant defense transcripts and regulate genes that control development through hormone signaling. *The Plant Cell, 15*, 939–951.

Pavet, V., Olmos, E., Kiddle, G., Mowla, S., Kumar, S., Antoniw, J., et al. (2005). Ascorbic acid deficiency activates cell death and disease resistance responses in Arabidopsis. *Plant Physiology, 139*, 1291–1303.

Pellny, T. K., Locato, V., Diaz Vivancos, P., Markovic, J., De Gara, L., Pallardó, F. V., & Foyer, C. H. (2009). Pyridine nucleotide cycling and control of intracellular redox state in relation to poly (ADP-ribose) polymerase activity and nuclear localisation of glutathione during exponential growth of Arabidopsis cells in culture. *Molecular Plant, 2*, 442–456.

Peskin, A. V., Pace, P. E., Behring, J. B., Paton, L. N., Soethoudt, M., Bachschmid, M. M., & Winterbourn, C. C. (2016). Glutathionylation of the active site cysteines of peroxiredoxin 2 and recycling by glutaredoxin. *Journal of Biological Chemistry, 291*, 3053–3062. https://doi.org/10.1074/jbc.M115.692798

Pignocchi, C., Kiddle, G., Hernández, I., Foster, S. J., Asensi, A., Taybi, T., et al. (2006). Ascorbate oxidase-dependent changes in the redox state of the apoplast modulate gene transcript accumulation leading to modified hormone signaling and orchestration of defense processes in tobacco. *Plant Physiology, 141*, 423–435.

Pineau, B., Layoune, O., Danon, A., & De Paepe, R. (2008). L-galactono-1,4-lactone dehydrogenase is required for the accumulation of plant respiratory complex I. *Journal of Biological Chemistry, 283*, 32500–32505.

Plumb, W., Townsend, A. J., Rasool, B., Alomrani, S., Razak, N., Karpinska, B., et al. (2018). Ascorbate-mediated regulation of growth, photoprotection, and photoinhibition in Arabidopsis thaliana. *Journal of Experimental Botany, 69*, 2823–2835.

Pnueli, L., Liang, H., Rozenberg, M., & Mittler, R. (2003). Growth suppression, altered stomatal responses, and augmented induction of heat shock proteins in cytosolic ascorbate peroxidase (Apx1)-deficient Arabidopsis plants. *Plant Journal, 34*, 187–203 2003.

Queval, G., Thominet, D., Vanacker, H., Miginiac-Maslow, M., Gakière, B., & Noctor, G. (2009). H_2O_2-Activated up-regulation of glutathione in Arabidopsis involves induction of genes encoding enzymes involved in cysteine synthesis in the chloroplast. *Molecular Plant, 2*, 344–356. https://doi.org/10.1093/mp/ssp002

Rahantaniaina, M., Tuzet, A., Mhamdi, A., & Noctor, G. (2013). Missing links in understanding redox signaling via thioldisulfide modulation how is glutathione oxidized in plants. *Frontiers in Plant Science, 4*, 477.

Rahantaniaina, M. S., Li, S., Chatel-Innocenti, G., Tuzet, A., Mhamdi, A., Vanacker, H., & Noctor, G. (2017). Glutathione oxidation in response to intracellular H2O2: Key but overlapping roles for dehydroascorbate reductases. *Plant Signaling & Behavior, 12*, e1356531.

Rahantaniaina, M. S., Li, S., Chatel-Innocenti, G., Tuzet, A., Issakidis-Bourguet, E., Mhamdi, A., & Noctor, G. (2017). Cytosolic and chloroplastic DHARs cooperate in oxidative stress-driven activation of the salicylic acid pathway. *Plant Physiology, 174*, 956–971.

Ramakrishnan, M., Papolu, P. K., Satish, L., Vinod, K. K., Wei, Q., Sharma, A., et al. (2022). Redox status of the plant cell determines epigenetic modifications under abiotic stress conditions and during developmental processes. *Journal of Advanced Research, S2090–1232*(22), 00103–00105. https://doi.org/10.1016/j.jare.2022.04.007

Roos, G., & Messens, J. (2011). Protein sulfenic acid formation: From cellular damage to redox regulation. *Free Radical Biology and Medicine, 51*, 314–326. https://doi.org/10.1016/j.freeradbiomed.2011.04.031

Sakihama, Y., Mano, J., Sano, S., Asada, K., & Yamasaki, H. (2000). Reduction of phenoxyl radicals mediated by monodehydroascorbate reductase. *Biochemical and Biophysical Research Communications, 279*, 949–954.

Sankaranarayanan, S., Ju, Y., & Kessler, S. A. (2020). Reactive oxygen species as mediators of gametophyte development and double fertilization in flowering plants. *Frontiers in Plant Science, 11*, 1199.

Sawake, S., Tajima, N., Mortimer, J. C., Lao, J., Ishikawa, T., Yu, X., et al. (2015). KONJAC1 and 2 are key factors for GDP-mannose generation and affect L- ascorbic acid and glucomannan biosynthesis in Arabidopsis. *The Plant Cell, 27*, 3397–3409.

Schimmeyer, J., Bock, R., & Meyer, E. H. (2016). L-Galactono-1,4-lactone dehydrogenase is an assembly factor of the membrane arm of mitochondrial complex I in Arabidopsis. *Plant Molecular Biology, 90*, 117–126.

Schnaubelt, D., Queval, G., Dong, Y., Diaz-Vivancos, P., Makgoba, M. E., Howell, G., et al. (2015). Low glutathione regulates gene expression and the redox potentials of the nucleus and cytosol in Arabidopsis thaliana. *Plant, Cell & Environment, 38*, 266–279.

Sevilla, F., Camejo, D., Ortiz-Espín, A., Calderón, A., Lázaro, J. J., & Jiménez, A. (2015). The thioredoxin/peroxiredoxin/sulfiredoxin system: current overview on its redox function in plants and regulation by reactive oxygen and nitrogen species. *Journal of Experimental Botany, 66*, 2945–2955 doi: 0.1093/jxb/erv146.

Shaikhali, J., Heiber, I., Seidel, T., Ströher, E., Hiltscher, H., Birkmann, S., et al. (2008). The redox-sensitive transcription factor Rap2.4a controls nuclear expression of 2-Cys peroxiredoxin A and other chloroplast antioxidant enzymes. *BMC Plant Biology, 8*, 48.

Shalata, A., Mittova, V., Volokita, M., Guy, M., & Tal, M. (2001). Response of the cultivated tomato and its wild salt-tolerant relative Lycopersicon pennellii to salt-dependent oxidative stress: The root antioxidative system. *Physiologia Plantarum, 112*, 487–494.

Shen, C. H., Krishnamurthy, R., & Yeh, K. (2009). Decreased L-ascorbate content mediating bolting is mainly regulated by the galacturonate pathway in Oncidium. *Plant and Cell Physiology, 50*, 935–946.

de Simone, A., Hubbard, R., Viñegra de la Torre, N., Velappan, Y., Wilson, M., et al. (2017). Redox changes during the cell cycle in the embryonic root meristem of Arabidopsis thaliana. *Antioxidants and Redox Signaling, 27*, 1505–1519.

Smirnoff, N. (2018). Ascorbic acid metabolism and functions: A comparison of plants and mammals. *Free Radical Biology & Medicine, 122*, 116–129.

Smirnoff, N., & Wheeler, G. L. (2000). Ascorbic acid in plants: Biosynthesis and function. *Critical Reviews in Biochemistry and Molecular Biology, 35*, 291–314.

Ströher, E., & Millar, A. H. (2012). The biological roles of glutaredoxins. *Biochemical Journal, 446*, 333–348. https://doi.org/10.1042/BJ20112131

Terai, Y., Ueno, H., Ogawa, T., Sawa, Y., Miyagi, A., Kawai-Yamada, M., et al. (2020). Dehydroascorbate reductases and glutathione set a threshold for high-light-induced ascorbate accumulation. *Plant Physiology, 183*, 112–122. https://doi.org/10.1104/pp.19.01556

Torabinejad, J., Donahue, J. L., Gunesekera, B. N., Allen-Daniels, M. J., & Gillaspy, G. E. (2009). VTC4 is a bifunctional enzyme that affects myoinositol and ascorbate biosynthesis in plants. *Plant Physiology, 150*, 951–961.

Tóth, S. Z., Puthur, J. T., Nagy, V., & Garab, G. (2009). Experimental evidence for ascorbate-dependent electron transport in leaves with inactive oxygen evolving complexes. *Plant Physiology, 149*, 1568–1578.

Townsend, D. M., Lushchak, V. I., & Cooper, A. J. (2014). A comparison of reversible versus irreversible protein glutathionylation. *Advances in Cancer Research, 122*, 177–198. https://doi.org/10.1016/B978-0-12-420117-0.00005-0

Trujillo-Hernandez, J. A., Bariat, L., Enders, T. A., Strader, L. C., Reichheld, J. P., & Belin, C. A. (2020). Glutathione-dependent control of the indole butyric acid pathway supports Arabidopsis root system adaptation to phosphate deprivation. *Journal of Experimental Botany, 71*, 4843–4857. https://doi.org/10.1093/jxb/eraa195

Van Breusegem, F., Foyer, C. H., & Mann, G. E. (2018). Reactive oxygen species are crucial "pro-life" survival signals in plants. *Free Radical Biology and Medicine, 122*, 1–3.

Vanacker, V., Guichard, M., Bohrer, A.-S., & Issakidis-Bourguet, E. (2018). Redox regulation of monodehydroascorbate reductase by thioredoxin y in plastids revealed in the context of water stress. *Antioxidants, 7*, 183.

Vaseghi, M.-J., Chibani, K., Telman, W., Liebthal, M. F., Gerken, M., Schnitzer, H., ... Dietz, K.-J. (2018). The chloroplast 2-cysteine peroxiredoxin functions as thioredoxin oxidase in redox regulation of chloroplast metabolism. *Elife, 7*, e38194.

Venkatesh, J., & Park, S. W. (2014). Role of L-ascorbate in alleviating abiotic stresses in crop plants. *Botanical Studies, 55*, 38.

Vernoux, T., Wilson, R. C., Seeley, K. A., Reichheld, J. P., Muroy, S., Brown, S., et al. (2000). The ROOT MERISTEMLESS1/CADMIUM SENSITIVE2 gene defines a glutathione-dependent pathway involved in initiation and maintenance of cell division during postembryonic root development. *The Plant Cell, 12*, 97–110.

Wang, G., Hu, C., Zhou, J., Liu, Y., Cai, J., Pan, C., et al. (2019). Systemic root-shoot signaling drives jasmonate-based root defense against nematodes. *Current Biology, 29*, 3430–3438.

Wang, J., Yu, Y., Zhang, Z., Quan, R., Zhang, H., Ma, L., et al. (2013). Arabidopsis CSN5B interacts with VTC1 and modulates ascorbic acid synthesis. *The Plant Cell, 25*, 625–663.

Wang, Q., Shi, H., Huang, R., Ye, R., Luo, Y., Guo, Z., & Lu, S. (2021). AIR12 confers cold tolerance through regulation of the CBF cold response pathway and ascorbate homeostasis. *Plant, Cell & Environment, 44*, 1522–1533.

Wang, X., & Hargrove, M. S. (2013). Nitric oxide in plants: The roles of ascorbate and hemoglobin. *PLoS One, 8*, e82611.

Waszczak, C., Carmody, M., & Kangasjärvi, J. (2018). Reactive oxygen species in plant signaling. *Annual Review of Plant Biology, 69*, 209–236.

Waszczak, C., Akter, S., Jacques, S., Huang, J., Messens, J., & Van Breusegem, F. (2015). Oxidative post-translational modifications of cysteine residues in plant signal transduction. *Journal of Experimental Botany, 66*, 2923–2934. https://doi.org/10.1093/jxb/erv084

Watson, S. J., Sowden, R. G., & Jarvis, P. (2018). Abiotic stress-induced chloroplast proteome remodelling: A mechanistic overview. *Journal of Experimental Botany, 69*, 2773–2781.

Wei, S., Zhang, W., Fu, R., & Zhang, Y. (2021). Genome-wide characterization of 2-oxoglutarate and Fe(II)-dependent dioxygenase family genes in tomato during growth cycle and their roles in metabolism. *BMC Genomics, 22*, 126.

Winterbourn, C. C. (2016). Revisiting the reactions of superoxide with glutathione and other thiols. *Archives of Biochemistry and Biophysics, 595*, 68–71. https://doi.org/10.1016/j.abb.2015.11.028

Yang, H., Mu, J., Chen, L., Feng, J., Hu, J., Li, L., et al. (2015). S-nitrosylation positively regulates ascorbate peroxidase activity during plant stress responses. *Plant Physiology, 167*, 1604–1615.

Yang, Y., Lenherr, E. D., Gromes, R., Wang, S., Wirtz, M., Hell, R., et al. (2019). Plant glutathione biosynthesis revisited: Redox-mediated activation of glutamylcysteine ligase does not require homo-dimerization. *Biochemical Journal, 476*, 1191–1203.

Ye, J., Li, W., Ai, G., Li, C., Liu, G., Chen, W., et al. (2019). Genome-wide association analysis identifies a natural variation in basic helix-loop-helix transcription factor regulating ascorbate biosynthesis via D-mannose/L-galactose pathway in tomato. *PLoS Genetics, 15*, e1008149.

Ye, J., Liu, G., Chen, W., Zhang, F., Li, H., Ye, Z., & Zhang, Y. (2019). Knockdown of SlNL33 accumulates ascorbate, enhances disease and oxidative stress tolerance in tomato (Solanum lycopersicum). *Plant Growth Regulation, 89*, 49–58.

Yin, L., Wang, S., Eltayeb, A. E., Uddin, M. I., Yamamoto, Y., Tsuji, W., et al. (2010). Overexpression of dehydroascorbate reductase, but not monodehydroascorbate reductase, confers tolerance to aluminum stress in transgenic tobacco. *Planta, 231*, 609–621.

Yokochi, Y., Fukushi, Y., Wakabayashi, K., Yoshida, K., & Hisabori, T. (2021). Oxidative regulation of chloroplast enzymes by thioredoxin and thioredoxin-like proteins in Arabidopsis thaliana. *Proceedings of the National Academy of Sciences (USA), 118*, 51.

Yoshida, S., Tamaoki, M., Shikano, T., Nakajima, N., Ogawa, D., Ioki, M., et al. (2006). Cytosolic dehydroascorbate reductase is important for ozone tolerance in Arabidopsis thaliana. *Plant and Cell Physiology, 47*, 304–308.

Yousuf, P. Y., Hakeem, K. U. R., Chandna, R., & Ahmad, P. (2012). Role of glutathione reductase in plant abiotic stress. In P. Ahmad, & M. N. V. Prasad (Vol. Eds.), *Abiotic stress responses in plants: 2012*, (pp. 149–158). New York: Springer.

Yu, Y. W., Wang, J., Li, S. H., Kakan, X., Zhou, Y., Miao, Y., et al. (2019). Ascorbic acid integrates the antagonistic modulation of ethylene and abscisic acid in the accumulation of reactive oxygen species. *Plant Physiology, 179*, 1861–1875.

Zaffagnini, M., Michelet, L., Massot, V., Trost, P., & Lemaire, S. D. (2008). Biochemical characterization of glutaredoxins from Chlamydomonas reinhardtii reveals the unique properties of a chloroplastic CGFS-type glutaredoxin. *Journal of Biological Chemistry, 283*, 8868–8876. https://doi.org/10.1074/jbc.M709567200

Zandalinas, S. I., Fichman, Y., Devireddy, A. R., Sengupta, S., Azad, R. K., & Mittler, R. (2020a). Systemic signaling during abiotic stress combination in plants. *Proceedings of the National Academy of Sciences of the United States of America, 117*, 13810–13820.

Zandalinas, S. I., Fichman, Y., Devireddy, A. R., Sengupta, S., Azad, R. K., & Mittler, R. (2020b). Systemic signaling during abiotic stress combination in plants. *Proceedings of the National Academy of Sciences of the United States of America, 117*, 13810–13820.

Zandalinas, S. I., Sengupta, S., Fritschi, F. B., Azad, R. K., Nechushtai, R., & Mittler, R. (2021). The impact of multifactorial stress combination on plant growth and survival. *New Phytologist, 230*, 1034–1048.

Zechmann, B. (2011). Subcellular distribution of ascorbate in plants. *Plant Signaling & Behavior, 6*, 360–363. https://doi.org/10.4161/psb.6.3.14342

Zechmann, B. (2014). Compartment-specific importance of glutathione during abiotic and biotic stress. *Frontiers in Plant Science, 5*, 566. https://doi.org/10.3389/fpls.2014.00566

Zechmann, B. (2020). Subcellular roles of glutathione in mediating plant defense during biotic stress. *Plants, 9*, 1067.

Zechmann, B., Stumpe, M., & Mauch, F. (2011). Immunocytochemical determination of the subcellular distribution of ascorbate in plants. *Planta, 233*, 1–12.

Zentgraf, U., Andrade-Galan, A. G., & Bieker, S. (2022). Specificity of H_2O_2 signaling in leaf senescence: Is the ratio of H_2O_2 contents in different cellular compartments sensed in Arabidopsis plants? *Cellular & Molecular Biology Letters, 27*, 4.

Zhang, H., Si, X., Ji, X., Fan, R., Liu, J., Chen, K., et al. (2018). Genome editing of upstream open reading frames enables translational control in plants. *Nature Biotechnology, 36*, 894–898.

Zhang, J., Ye, Z. W., Singh, S., Townsend, D. M., & Tew, K. D. (2018). An evolving understanding of the S-glutathionylation cycle in pathways of redox regulation. *Free Radical Biology & Medicine, 120*, 204–216. https://doi.org/10.1016/j.freeradbiomed.2018.03.038

Zhang, L., Wu, M., Teng, Y., Jia, S., Yu, D., Wei, T., et al. (2019). Overexpression of the glutathione peroxidase 5 (RcGPX5) gene from Rhodiola crenulata increases drought tolerance in Salvia miltiorrhiza. *Frontiers in Plant Science, 9*, 1950. https://doi.org/10.3389/fpls.2018.01950

Zhang, L., Huang, J., Su, S., Wei, X., Yang, L., Zhao, H., et al. (2021). FERONIA receptor kinase-regulated reactive oxygen species mediate self-incompatibility in Brassica rapa. *Current Biology, 31*, 3004–3016.

Zhang, W., Lorence, A., Gruszewski, H. A., Chevone, B. I., & Nessler, C. L. (2009). AMR1, an Arabidopsis gene that coordinately and negatively regulates the mannose/l-galactose ascorbic acid biosynthetic pathway. *Plant Physiology, 150*, 942–950.

Zhang, X., Zhang, L., Chen, L., Lu, Y., & An, Y. (2022). Ectopic expression γ-glutamylcysteine synthetase of Vicia sativa increased cadmium tolerance in Arabidopsis. *Gene, 823*, 146358. https://doi.org/10.1016/j.gene.2022.146358

Zhang, Z., Wang, J., Zhang, R., & Huang, R. (2012). Ethylene response factor AtERF98 enhances tolerance to salt through the transcriptional activation of ascorbic acid synthesis in Arabidopsis. *Plant Journal, 71*, 273–287.

Zhithovich, A. (2020). Nuclear and cytoplasmic functions of vitamin C. *Chemical Research in Toxicology, 33*, 2515–2526.

Zhu, L., Guo, J., Zhu, J., & Zhou, C. (2014). Enhanced expression of EsWAX1 improves drought tolerance with increased accumulation of cuticular wax and ascorbic acid in transgenic Arabidopsis. *Plant Physiology and Biochemistry, 75*, 24–35.

Zur, I., Kopeć, P., Surówka, E., Dubas, E., Krzewska, M., Nowicka, A., et al. (2021). Impact of Ascorbate—Glutathione cycle components on the effectiveness of embryogenesis induction in isolated microspore cultures of barley and triticale. *Antioxidants, 10*, 1254. https://doi.org/10.3390/antiox10081254

CHAPTER FOUR

Regulation of leaf development through the modulation of ROS homeostasis

Jos H.M. Schippers
Department of Molecular Genetics, Leibniz Institute of Plant Genetics and Crop Plant Research (IPK), Gatersleben, Germany

Contents

Abstract

Life evolved in the presence of reactive oxygen species (ROS) which resulted in an early integration of ROS signals and properties for cellular physiology. Originally ROS were viewed merely as metabolic by-products, but now they are viewed as conserved regulators of physiological and developmental processes throughout all kingdoms of life. Plants monitor constantly the fluctuating environmental conditions in which they grow, for which ROS signaling is essential. However, in recent years it has become clear that plants also use ROS homeostasis to guide their development. Here, an overview is presented for the role of ROS gradients and homeostasis in the development of leaves from the shoot apical meristem based mainly on the model plant species *Arabidopsis thaliana*. Furthermore, this chapter summarizes current knowledge on the involvement of different ROS species and molecular components in the regulation of leaf development. Although our current understanding of how plants

Advances in Botanical Research, Volume 105
ISSN 0065-2296, https://doi.org/10.1016/bs.abr.2022.10.001

employ ROS to guide their development is far from complete, it is clear that ROS is utilized during many developmental phases of leaf growth.

1. Introduction

Terrestrialization by plants occurred around 470 million years ago as indicated by the fossil records of spores or spore masses (Harrison & Morris, 2018; Wellman, Osterloff, & Mohiuddin, 2003). These initial land plants lacked leaves, and mainly entered reproductive development straight away or by sporangium formation, which is still found among extant bryophyte relatives of the earliest land plants (Edwards, Morris, Richardson, & Kenrick, 2014). Subsequently, around 430 million years ago land plants started to develop leaves (Harrison & Morris, 2018), a process that evolved independently in several plant lineages (Vasco et al., 2016; Zumajo-Cardona, Vasco, & Ambrose, 2019). Over time, leaves evolved as the main photosynthetic organs in plants, whereby leaf morphology greatly varies among species but also within a single plant (Tsukaya, 2014).

Leaves are usually flat lateral organs that show a typical bilateral symmetry. During leaf establishment at the shoot apical meristem (SAM), two anatomically distinct surfaces are formed, the upper (adaxial) and lower (abaxial) leaf sides. The upper side is specialized on light capture, while the lower side enables gas exchange and the regulation of transpiration. On both sides of the leaf, reactive oxygen species (ROS) plays an essential roles as signaling molecules to monitor photosynthesis but also facilitate gas exchange. Photosynthesis is a complex redox metabolic process that needs to be instantly adapted to conditions of fluctuating light (Dietz, Turkan, & Krieger-Liszkay, 2016). Inevitably, photosynthesis cause ROS formation, including singlet oxygen ($^1O_2$), superoxide anion radicals ($O_2^{\bullet -}$), and hydrogen peroxide (H_2O_2). The chloroplast-derived ROS signals are implicated in retrograde signaling to tune photosynthetic processes through transcriptional regulation in the nucleus. Several major ROS signaling pathways are linked to the chloroplast, including the singlet oxygen signaling through the EXECUTER pathway (Zhang, Apel, & Kim, 2014) or the OXI1 kinase pathway (Shumbe et al., 2016). Next to that, gas exchange through stomata opening and closing is largely depending on the transient formation, sensing, and scavenging of ROS (Medeiros, Barros, Fernie, & Araújo, 2020; Wu et al., 2020). Recently, the plasma membrane localized apoplastic H_2O_2 sensor HYDROGEN--PEROXIDE-INDUCED CA^{2+} INCREASES 1 (HPCA1) was identified to be required for stomata closing (Wu et al., 2020). Furthermore, stomata opening requires managing ROS homeostasis, which includes the elimination of oxidized guard cell peroxisomes through autophagy (Yamauchi et al., 2019). That ROS is implicated in the physiological functioning of leaves is clearly established, however, their role in the development of the organ is less well understood but has made enormous progress in recent years.

This chapter summarizes progress in the study of leaf development and growth through the modulation of ROS homeostasis, mainly in the model plant *Arabidopsis thaliana*. The following sections provide an overview of leaf initiation from the SAM, development and differentiation of cell types and expansion growth to reach the final leaf size and shape. Especially the underlying molecular networks and their control on ROS homeostasis are discussed.

2. ROS homeostasis and leaf initiation at the SAM

Multicellular organisms contain pluripotent stem cells from which new organs arise, lost cells are replenished and organs can be regenerated upon damage (Aichinger, Kornet, Friedrich, & Laux, 2012). In plants, stem cells are located in stem cell niches, which provide signals to prevent premature differentiation. The best studied stem cells niches in plants are found in the root apical meristem (RAM) and the shoot apical meristem (SAM), but also other niches exist in plants. The SAM gives rise to all above-ground organs of a plant, which relies on the one hand on maintaining an undifferentiated population of cells, while on the other on promoting differentiation of stem cell daughter cells into specific cell types that are required for building the lateral organ. In the case of leaves, a leaf primordium is formed whose meristematic activity is limited to enable determinate growth of the organ (Alvarez, Furumizu, Efroni, Eshed, & Bowman, 2016). Here the regulation of SAM activity through the establishment of a ROS gradient and the molecular regulation of leaf meristems is discussed.

2.1 ROS gradients underlie the functioning of the SAM

The establishment and maintenance of pluripotent stem cells in the shoot apical meristem (SAM) are key developmental processes in land plants. Still, the morphology of the SAM largely varies between plants, whereby bryophytes contain a single pluripotent stem cell, called a shoot apical cell

(Hata & Kyozuka, 2021). In contrast, in Arabidopsis the SAM contains around 35 stem cells (Reddy & Meyerowitz, 2005) while in *Oryza sativa* around 200 stem cells are present (Nosaka-Takahashi, Kato, Kumamaru, & Sato, 2022). Despite the morphological differences in the size and shape of the SAM, the principal architecture, whereby centrally located stem cell(s) are surrounded by regularly differentiating tissue is conserved. In addition, the role of the plant hormone auxin in promoting organ differentiation from the SAM occurs throughout all land plants (Aichinger et al., 2012; Liu, Shi, Chang, Ding, & Ding, 2022; Viaene et al., 2014). Furthermore, the role of cytokinin in promoting the undifferentiated stem cell pool in the SAM is preserved amongst land plants (Coudert et al., 2015; Gordon, Chickarmane, Ohno, & Meyerowitz, 2009; Tsuda, Ito, Sato, & Kurata, 2011). Despite the conserved function of these phytohormones, the downstream molecular pathways are not fully conserved between all plant lineages, suggesting the existence of additional mechanisms promoting stem cell identity and differentiation in plants.

In Arabidopsis the architecture of the SAM creates distinct functional domains: the central zone (CZ) containing the stem cells, the peripheral zone (PZ) containing the differentiating stem cells that give rise to the lateral organs or secondary meristems, the organizing center (OC) which maintains the stem cell population in the CZ, and the rib zone (RZ) from which cells differentiate to form the stem (Fig. 1A). Transcriptional and signaling networks underlying the maintenance of stem cells within the SAM have been largely uncovered and studied in detail in Arabidopsis. The role of the class I KNOTTED-LIKE HOMEOBOX (KNOXI) transcription factors in promoting stem cell proliferation is in Arabidopsis exemplified by the loss of *SHOOT MERISTEMLESS* (*STM*) which causes premature differentiation of stem cells and SAM disruption at the seedling stage (Long, Moan, Medford, & Barton, 1996). The promotion of stem cell proliferation by STM relies on the activation of cytokinin biosynthesis in Arabidopsis (Yanai et al., 2005). Of note, in mosses, cytokinin is required for SAM maintenance, however, this is not controlled by KNOXI (Sakakibara, Nishiyama, Deguchi, & Hasebe, 2008). In the OC, the mobile transcription factor WUSCHEL (WUS) is expressed and moves to the CZ to promote stemness (Chickarmane, Gordon, Tarr, Heisler, & Meyerowitz, 2012; Yadev et al., 2011). Within the CZ, WUS activates the most prominent member of the *CLAVATA3/EMBRYO SURROUNDING REGION-RELATED* (*CLE*) genes that encode extracellular signaling peptides, namely *CLAVATA3* (*CLV3*) (Brand, Grünewald, Hobe,

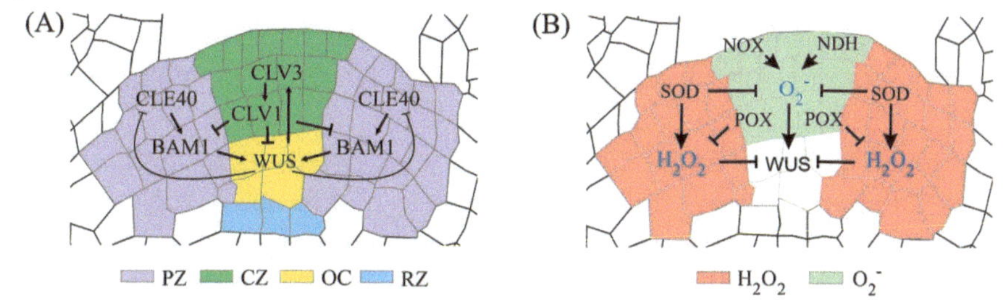

Fig. 1 The zones in the shoot apical meristem are defined by a ROS gradient. (A) Schematic representation of the different tissue zones presents within the SAM. Regulation of the SAM through the modulation of WUS activity by the CLV3-CLV1 and the CLE40-BAM1 pathways is shown. (B) The SAM zones are characterized by different levels of ROS. The central zone (CZ) contains elevated levels of superoxide (O_2^-) while the peripheral zone (PZ) is characterized by increased levels of H_2O_2 (Zeng et al., 2017). Accumulation of O_2^- within the CZ promotes the expression of WUS, thereby stimulating stemness. The build-up of O_2^- in the CZ is on the one hand attributed to the activity of NADPH oxidases (NOX) and NADH dehydrogenases (NDH), while at the same time SODs are not expressed in the CZ to prevent the dismutation of O_2^- and the formation of H_2O_2. In addition, peroxidases (POXs) are highly expressed in the CZ to avoid accumulation of H_2O_2. Once cells move out of the CZ into the PZ they are confronted with increased levels of H_2O_2 due to the expression of SODs within this zone. The elevated H_2O_2 levels promote differentiation of the stem cells into the founder cells for the different lateral organs.

& Simon, 2002). CLV3 is detected by its receptor CLV1, which is expressed in the OC, activating a phosphorelay that suppresses the expression of *WUS* and thereby creating a homeostatic feedback mechanism to control the size of the SAM (Betsuyaku et al., 2011). Although the CLV-WUS feedback loop ensures that a stem cell pool is maintained, is does not account for how stem cells are being primed for differentiation (Zhang, Jiao, Jiao, Zhao, & Zhu, 2017). In addition, the CLV-WUS system is not conserved in all land plants, suggesting that other mechanisms in stem cell maintenance exists. The cells in the PZ will give rise to new aerial organs, which need to be replenished with cells from the CZ, for which WUS activity is essential. Recently, the signal peptide CLE40 was discovered to operate in the PZ (Schlegel et al., 2021), to stimulate replenishment of cells by activating WUS expression in the OC through the activation of the receptor kinase BARELY ANY MERISTEM1 (BAM1) (Fig. 1A). WUS blocks CLE40 expression from the OC, thereby establishing a second negative feedback regulation. Thus, in the SAM two intertwined pathways modulate WUS activity in the OC, the CLV3-CLV1 module incorporates information on the actual size of the stem cell domain, while the CLE40-BAM1 module ensures that the growth requirements of the PZ are accommodated through timely cell replenishment.

Animal stem cells require low levels of H_2O_2 signaling for proliferation, while differentiation of stem cells is mediated by increased levels of H_2O_2 (Ludin et al., 2014; Zhou, Meng, Li, Ghebre, & Cooke, 2016). In analogy to animal stem cells, it was found that plant stem cells posses low levels of H_2O_2 (Zeng, Dong, Wu, Tian, & Zhao, 2017). The low level of H_2O_2 in the CZ of the SAM is the result of elevated expression of peroxidases, that efficiently scavenge H_2O_2 (Fig. 1B). In addition, plant stem cells show elevated levels of superoxide, which is required to maintain their stemness. The elevated levels of superoxide in stem cells are the results of the activity of mitochondrial NADH-ubiquinone oxidoreductase (NDUFSs) and plasma-membrane localized NADPH oxidases (NOXs). Of note, plant literature on NOX proteins mainly refers to these as respiratory burst oxidase homologs (RBOHs).

Moreover, genes encoding for superoxide dismutases (SODs) are strongly downregulated in the CZ. Either loss of NDUFSs, NOXs or expression of SODs in the CZ, resulted in a reduction or complete abolishment of the stem cell pool (Zeng et al., 2017). Thus, proliferating stem cells are characterized by enhanced levels of superoxide. One potential reason why superoxide could promote stemness is the fact that several chromatin remodeling factors contain an iron-sulfur (Fe-S) cluster (Buzas, Nakamura, & Kinoshita, 2014; Wang et al., 2016). Superoxide is well-known for its ability to inactivate enzymes containing Fe-S clusters (Py, Moreau, & Barras, 2011), which in the CZ might help to maintain stem cell identity. As soon as daughter stem cells from the CZ move into the PZ they will start to differentiate as they encounter increased levels of H_2O_2 (Zeng et al., 2017). The increased H_2O_2 levels are in part caused by the active dismutation of superoxide in the PZ by SODs. Thus, a ROS gradient in the SAM between superoxide and H_2O_2 is required to maintain stem cells and modulate their differentiation. Still, which other factors contribute to the increased H_2O_2 levels in the PZ is so far not know. In addition, to which extend there is an interplay with the CLE40 and CLV3 pathway and the ROS gradient in the SAM is not resolved, although they appear to be highly complementary. From the PZ leaf primordia are initiated during vegetative growth, representing a specialized meristem that enables determinate growth of the leaf.

2.2 ROS homeostasis and leaf meristems

In Arabidopsis, leaves emerge as a small population of founder cells flanking the SAM (Irish & Sussex, 1992). The initiation of leaf primordia is preceded by the formation of an auxin response maxima (Benková et al., 2003). The auxin maximum promotes leaf initiation, probably through the activation of AUXIN RESPONSE FACTOR (ARF) proteins (Hardtke et al., 2004; Krogan, Marcos, Weiner, & Berleth, 2016). Simultaneous loss of MONOPTEROS (MP/ARF5) and NONPHOTOTROPIC HYPOCOTYL 4 (NPH4/ARF7) results in a club-shaped seedling devoid of lateral organs (Hardtke et al., 2004). The impact of auxin on plant growth and development is partially linked to its effect on ROS production (Xia et al., 2015). For instance, tip-growth of root hairs underlies auxin-induced expression of ROS-producing enzymes (Mangano et al., 2017). Potentially, the outgrowth of leaf primordia underlies a similar mechanism. Leaf meristems show a stronger proliferative activity as cells in the SAM, causing a rapid outgrowth of the organ. The leaf primordium contains two meristematic zones, the plate meristem and the marginal meristem (Tsukaya, 2021). The plate meristem is localized to the basal (proximal) part of the leaf primordia, while the marginal meristem can be found at the leaf margin (distal). To date, it has not been shown if the leaf meristematic zones, like the CZ in the SAM, accumulate superoxide.

The marginal meristem is a meristematic zone at the margin of the leaf primordia that drives lateral expansion of the leaf lamina through polar growth. In Arabidopsis, WUSCHEL-RELATED HOMEOBOX1 (WOX1) and WOX3 are positive regulators of the marginal meristem whose activity and position are regulated through a spatial auxin signal (Guan et al., 2017). The determinate growth of the leaf margin is invoked by the action of by CINCINNATA-like TEOSINTE BRANCHED1--CYCLOIDEA-PCF (TCP-C) transcription factors (Alvarez et al., 2016; Palatnik et al., 2003). Repression of TCP-C activity results in marginal expansion of the leaf lamina due to excessive proliferation and delayed differentiation. Still, TCP-C activity alone does not account for the determinate nature of the marginal meristem. B3-type transcription factors of the NGATHA (NGA) subfamily were shown to act like the TCP-C genes to limit cell proliferation in the developing leaf (Lee et al., 2015). Moreover, simultaneous loss of TCP-C and NGA activity results in unrestricted formation of tissue at the leaf margin (Alvarez et al., 2016), indicating the activation of indeterminate growth. Interestingly, NGA factors provide drought stress tolerance through the activation of ABA biosynthesis and restriction of growth (Sato et al., 2018). During drought stress, NGA factors are stabilized at the protein level, while under control conditions they show a rapid turn-over. While phosphorylation was shown to only affect their transcriptional activity, it is possible that their stability is controlled by the redox state of the cell. Of note, a conserved cysteine residue is present in most NGA factors next to a TOPLESS-interaction domain. Interestingly, also TCP-C genes appear to become more active under drought stress as the miRNA that targets their transcript is down-regulated under these conditions (Barciszewska-Pacak et al., 2015). That said, restriction of proliferation by TCP-C and NGA factors also affect the plate meristem, therefore it was suggested that the short-acting marginal meristem might act similarly to the OC of the SAM, while the plate meristem acts like the CZ (Alvarez et al., 2016; Tsukaya, 2021).

The plate meristem consists of parallel layers of cells dividing anticlinal to give rise to the leaf blade and the petiole (Donnelly, Bonetta, Tsukaya, Dengler, & Dengler, 1999; Ichihashi et al., 2011). Proliferative activity in the plate meristem is largely regulated by the mobile transcriptional coactivator ANGUSTIFOLIA3 (AN3) through its interaction with GROWTH REGULATING FACTOR (GRF) transcription factors (Kawade, Horiguchi, Usami, Hirai, & Tsukaya, 2013; Lee et al., 2009). Interestingly, AN3 was shown to associate with the promoter of *RbohD* and *RbohF* (Vercruyssen et al., 2014), two NADPH oxidases that are required for maintenance of the stem cell pool in the SAM (Zeng et al., 2017). GRFs are positive regulators of cell proliferation but have also been implicated in the regulation of stress adaptation (Lantzouni, Alkofer, Falter-Braun, & Schwechheimer, 2020; Piya, Liu, Burch-Smith, Baum, & Hewezi, 2020), suggesting that stress signals modulate the activity of the AN3-GRF complex to control growth. Next to GRF factors, class I TCP transcription factors (TCP-P) have been implicated as positive regulators of cell proliferation during leaf development (Aguilar-Martínez & Sinha, 2013; Davière et al., 2014). TCP-P members act highly redundantly in the regulation of cell proliferation. Interestingly, TCP-P proteins contain a redox-sensitive cysteine residue in their DNA binding domain (Fig. 2A), which when oxidized prevents TCP-P proteins from binding DNA (Viola, Güttlein, & Gonzalez, 2013). Therefore, these transcription factors form a link between the cellular redox signaling hub and the regulation of the cell cycle (Schippers, Foyer, & van Dongen, 2016). TCP-C proteins also have

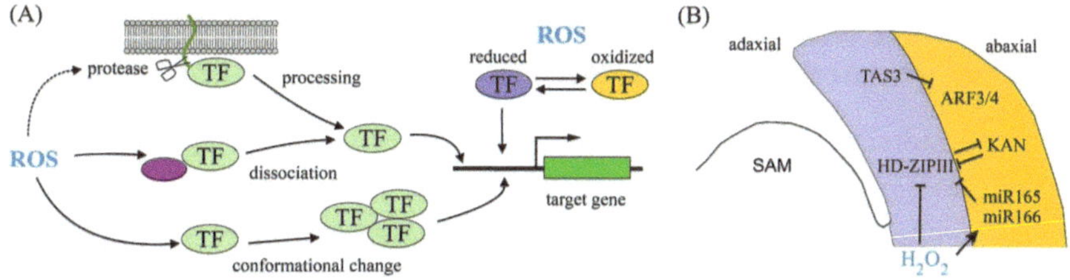

Fig. 2 Adaxial identity establishment is sensitive to oxidative stress. (A) Schematic representation of how ROS can modulate transcription factor activity. This includes the stimulation of proteolytic processing (as known for several NAC transcription factors), dissociation from a partner protein, conformational change of the protein or the formation of a protein complex (as known for several HSF transcription factors), and finally the regulation of DNA binding activity through the oxidation status of the transcription factors (known for TCP-P proteins). (B) Early on during the establishment of a leaf primordium, the upper and lower side of the leaf are set. ARF3/ARF4, KAN and miR165/166 accumulate at the abaxial side to promote the identity of the lower leaf surface, while HD-ZIP III transcription factors and TAS3 accumulate at the adaxial side. The adaxial and abaxial factors show mutual repression, whereby ARF3/ARF4 are prevented from the adaxial side through the activity of TAS3. In addition, KAN and HD-ZIP III proteins also oppose each other's activities, whereby the abaxial expressed *miR165/166* prevent accumulation of HD-ZIP III genes. Interestingly, the adaxial HD-ZIP III proteins are redox-sensitive and H_2O_2 inhibits their DNA binding activity. Moreover, oxidative stress promotes the expression of *miR165/166* which also impair HD-ZIP III activity.

several cysteine residues; however, their DNA binding is not impaired in an oxidized environment. The switch from proliferation to differentiation in leaves is paralleled by a switch in activity from TCP-P to TCP-C proteins which might be linked to a shift in ROS homeostasis during leaf development.

The impact of ROS on transcription factor action and activity is summarized in Fig. 2A, and has been extensively discussed previously (He, Van Breusegem, & Mhamdi, 2018). In short, ROS can promote the nuclear translocation of transcription factors by enabling their processing, their dissociation of interaction partners or by causing a conformational change. The nuclear accumulation enables the transcription factors to modulate transcriptional programs. Furthermore, as exemplified by TCP-P proteins, oxidation prevents DNA binding while under reducing conditions they are able to associate with target promoter (Viola et al., 2013). Especially in the light of ROS gradients in developing tissues, it will be of interest to determine how the different transcriptional regulators are affected.

3. Leaf polarities are established early during primordia formation

In eudicots, the primordium initially forms as a peg-like structure, whereas in cereals like maize, it is broader and appears as a half-ring shape. The peg-like primordium is portioned in a dorsal (adaxial) and ventral (abaxial) domain, which is essential to form a leaf blade. Once the dorsiventral asymmetry has been established, the primordium undergoes elongation and mediolateral (ML) and proximodistal (PD) polarity is established.

3.1 Adaxial-abaxial polarity, is there a ROS gradient involved?

The establishment of the upper and lower leaf surface is one of the most intensely studied processes in plant development biology (Kuhlemeier & Timmermans, 2016). Pioneering experiments by Ian Sussex (1951) suggested that an instructive signal from the SAM is required, since surgically separated leaf primordia from the SAM become radialized. The idea was taken up first skeptical but is now widely accepted. Still, other experimenters suggested that wounding might be the cause for the formation of radial centric organs and not a SAM-derived signal (Sachs, 1969). In 2005, laser ablation of the L1 layer of the meristem located to a primordia anlage of tomato was shown to be sufficient to cause radialization of leaves (Reinhardt, Frenz, Mandel, & Kuhlemeier, 2005). The proposed SAM-derived signal that initiates dorsoventrality has been coined the Sussex signal, but to date has remained elusive. One interesting observation of the centric organs is that in response to the surgical manipulation they become completely abaxialized (Steeves & Sussex, 1989).

The advent of molecular genetic studies on leaf development has resulted in the identification of transcription factors that control the development of the adaxial and abaxial surface of the leaf. Here, I first would like to the discuss the role of HD-ZIP III transcription factors in the regulation of adaxial identity and propose a potential involvement of wounding-induced ROS formation as cause for the centric organs observed by Sussex. The three redundantly acting HD-ZIP III transcription factors REVOLUTA (REV), PHABULOSA (PHB), and PHAVOLUTA (PHV) are required for adaxial surface cell identity (McConnell et al., 2001; Prigge et al., 2005). Loss-of-function mutants result in abaxialized leaves while gain-of-function mutants have the opposite effect. The expression of *HD-ZIP III* genes at the abaxial side of the leaf is prevented by the action of the antagonizing *miR165* and *miR166* (Juarez, Kui, Thomas, Heller, & Timmermans, 2004; Yao et al., 2009). Interestingly, ROS homeostasis could act as a regulator of the molecular components controlling adaxial leaf identity. First of all, the DNA binding of HD-ZIP III proteins, including PHV and REV, was shown to be redox-dependent (Comelli & Gonzalez, 2007; Xie et al., 2014). Under oxidizing conditions, HD-ZIP III proteins are no longer able to bind to DNA. Secondly, HD-ZIP III proteins contain a C-terminal PAS (per-ARNT-sim) like domain, which could act as a redox sensing domain (Magnani & Barton, 2011). The PAS domain, of around 100 residues, is a sensory module that can be found in all kingdoms of life. Although the protein structure is highly conserved, the amino-acid sequences are only partially conserved (Mei & Dvornyk, 2014). That said, several cysteine residues appear to be conserved between different PAS domain containing proteins and were shown to be important for the sensing function (Vogt & Schippers, 2015). Thirdly, the expression of *miRNA165* and *miRNA166* is induced upon oxidative stress (Barciszewska-Pacak et al., 2015; Iyer, Jia, Sunkar, Tang, & Mahalingam, 2012). Taken together, oxidative signals counteract the activity of HD-ZIP III proteins, suggesting that a ROS gradient might exists in the differentiating leaf primordia to establish dorsoventrality (Fig. 2B). Considering that wounding causes immediate formation of ROS (Orozco-Cárdenas, Narváez-Vásquez, & Ryan, 2001), which also occurs by surgical cutting and laser-ablation, it is possible that the HD-ZIP III function in the young meristem will be strongly inhibited, resulting in the observed centric organs. The wounding-induced hydrogen peroxide will limit HD-ZIP III DNA-binding and activate the expression of the two *miRNA* genes. As this wounding needs to be local, as surgical cutting at other positions of the meristem does not cause abaxilization of the leaf primordia, it suggests that systemic ROS signals do not impact on leaf polarity in the proposed model.

Repression of the HD-ZIP III genes enables expression of KANADI (KAN), YABBY (YAB) and ARF transcription factor genes, promoting abaxial cell identity (Eshed, Baum, Perea, & Bowman, 2001; Siegfried et al., 1999). *ARF3* and *ARF4* were shown to be abaxially expressed in lateral organs and shown to promote KAN activity by forming a transcriptional complex (Kelley, Arreola, Gallagher, & Gasser, 2012; Pekker, Alvarez, & Eshed, 2005). Moreover, ARF and YABBY transcription factors were shown to form a complex that represses *STM* (Chung et al., 2019), thereby restricting stemness. The expression of *ARF*, *KAN* and *YAB* genes is confined to the abaxial leaf side, which is established through the mutual repressive interaction with adaxial factors (Fig. 2B). The activity of ARF3 and ARF4 in the adaxial domain is repressed by the trans-acting small interfering RNA3 (Chitwood et al., 2009). Interestingly, ARF3 was shown to repress cytokinin biosynthesis genes, but activates genes involved in stress responses, including wounding stress, oxidative stress, salt stress, and drought stress (Zhang et al., 2018). Amongst the upregulated genes are classical ROS-related transcription factors like *ZAT10* and *ZAT12* (Mittler et al., 2006). Moreover, plants lacking *ARF3* showed an impaired expression of *ZAT12*, clearly indicating that ARF3 is involved in the modulation of ROS homeostasis (Zheng, Zhang, Guo, Liu, & Zhang, 2018). Reduced leaf growth upon osmotic stress in Arabidopsis was shown to be linked to auxin responses and sensitizing the leaf towards auxin perception (Kalve et al., 2020). In Arabidopsis, the activation-tagged mutant *yuc7–1D*, which shows increased expression of an auxin biosynthesis gene, exhibits curled leaf growth and increased drought tolerance (Li, Li, Shen, Huang, & Dong, 2012). Under control conditions the development alongside the adaxial–abaxial axes is essential to form a flat leaf lamina, however, upon stress there appears to be a dominance of one side over the other. Taken together, although establishment of adaxial-abaxial polarity of the leaf might be closely related to ROS homeostasis, more research efforts are needed to uncover the underlying mechanisms.

3.2 Proximal-distal polarity and redox-sensitive regulators

As the primordium elongates, proximal-distal polarity is established, whereby the side closer to the SAM is the proximal part and the side further away from the SAM is the distal part. The proximal domain gives rise to the petiole, while the distal domain becomes the leaf blade. In a sense, the plate meristem acts like the RZ, giving rise to the petiole and the CZ in connection with the marginal meristem to establish the leaf blade. BLADE-ON-PETIOLE1 (BOP1) and BOP2, belong to the family of NONEXPRESSOR OF PR GENES1 (NPR1) proteins, but lack the NIMIN-interaction domain (Backer, Naidoo, & van den Berg, 2019; Hepworth, Zhang, McKim, Li, & Haughn, 2005). Leaves of the *bop1 bop2* double mutant are elongated and grow rather indeterminate along the proximal-distal axis. In addition, blade development occurs along the petiole, resulting in the loss of distinct proximal and distal zones. Like NPR1, also BOP1 and BOP2 form a complex with TGA transcription factors to modulate gene expression (Hepworth et al., 2005; Wang et al., 2019). In contrast, NPR1 interacts with TGA1 under reducing conditions, however, BOP1 and BOP2 appear to interact only with the oxidizable form of TGA1 and TGA4 (Després et al., 2003; Wang et al., 2019). Furthermore, it is suggested that BOP1 and BOP2 promote the expression of boundary genes for which they recruit the clade I TGA factors.

CC-type glutaredoxins (named ROXYs in Arabidopsis) are major redox regulators and known to interact with TGA transcription factors (Gutsche, Thurow, Zachgo, & Gatz, 2015; Gutsche & Zachgo, 2016). It is speculated that the redox state of the TGA transcription factors affect the cofactors they recruit and thereby their effect on the target genes they control (Wang et al., 2019). Thus, the activity of the proximal-distal axis regulators is clearly redox-modulated, suggesting that a ROS gradient along this axis might modulate leaf development. Loss of function mutants of TGA1 and TGA4, and overexpression of ROXY8 or ROXY9, greatly impairs hyponastic leaf growth (Li et al., 2019). As hyponastic growth involves elongation in the proximal-distal direction at the abaxial side, this further supports a role for redox-dependent growth regulation. Furthermore, BOP1 and BOP2 negatively regulate the expression of class 1 *KNOX* genes in complex with the adaxial regulator ASYMMETRIC LEAVES2 (Jun, Ha, & Fletcher, 2010). This suggest that to some extent factors controlling adaxial-abaxial polarity or proximal-distal polarity also cooperate to fine-tune leaf growth.

4. Cell expansion during leaf growth

The cessation of cell proliferation during leaf growth occurs first at the tip of the leaf and moves towards the base of the organ. As soon as the cells stop to divide, they begin to expand and thereby determine the final size and shape of the leaf (Tsukaya, 2006). Cell expansion in plants is largely driven through the modulation of apoplastic ROS homeostasis. While the modulation of cell wall properties by different ROS types is well documented (Müller et al., 2009; Schmidt, Kunkowska, & Schippers, 2016; Schopfer, 1996), it still remains poorly understood how apoplastic ROS homeostasis is regulated.

4.1 ROS homeostasis and cell wall remodeling

In contrast to animal cells, plant cells are enclosed by a cell wall which links individual cells together and acts as an exoskeleton to provide strength to the plant body. Cell walls are dynamic structures that are adjusted to the cell type that they surround. Primary cell walls are initiated during cell division and develop along with expanding cells (Li, Fernie, & Persson, 2016). Secondary cell walls are made after cells have stopped expanding. The primary cell wall is highly hydrated and typically consists of a cellulose matrix that is cross-linked by the branched polysaccharides hemicelluloses and pectin (Lampugnani, Khan, Somssich, & Persson, 2018). The primary cell wall provides mechanical strength but at the same time needs to be extensible to enable cell expansion. Polar cell growth, as observed in root hairs and pollen tubes, is characterized by alternating periods of rapid and slow expansion (Monshausen, Messerli, & Gilroy, 2008). Similarly, cell expansion in leaves occurs in a diurnal fashion, whereby the largest growth rates are observed at dawn (Wiese, Christ, Virnich, Schurr, & Walter, 2007). Several factors are required to enable cell expansion, including turgor pressure to drive sliding and separation of cellulose microfibrils within the growing wall (Cosgrove, 2018). In addition, auxin induces cell wall acidification which promotes the activity of expansin proteins, which mediate acid-induced extension of plant cell walls without mechanically weakening the cell walls. Furthermore, xyloglucan endotransglucosylases/hydrolases are able to cut and re-join hemicellulose chains to modulate cell wall rigidity (Sharples, Nguyen-Phan, & Fry, 2017). Next to that, cell wall polymers can be non-enzymatically cleaved or linked through the action of different types of ROS to promote or restrict cell expansion (Schmidt et al., 2016), which will be discussed below.

Cell expansion involves switching between wall relaxing and wall stiffening processes (Wolf, Hématy, & Höfte, 2012). At the level of ROS this means switching between the accumulation of the hydroxyl radical ($^{\cdot}OH$) and H_2O_2, respectively. Cell wall relaxation by $^{\cdot}OH$ is achieved by cleavage of polymers like xyloglucan and pectin (Schmidt et al., 2016; Schweikert, Liszkay, & Schopfer, 2000). The $^{\cdot}OH$ is formed from H_2O_2 by either the Haber-Weiss reaction or in a reaction catalyzed by peroxidase in the presence of NADH and superoxide (Chen & Schopfer, 1999). As soon as the apoplastic-formed H_2O_2 is not used for the formation of $^{\cdot}OH$, the build-up of H_2O_2 will promote cell wall stiffening and restrict cell elongation (Schopfer, 2006). The accumulation of ROS as an initial response to stress has been linked to the inhibition of growth and the onset of adaptive responses (Schmidt et al., 2016). Thus, apoplastic ROS homeostasis during cell growth is strictly regulated and highly adaptable. As apoplastic ROS formation is largely enzymatically controlled, directed growth of the cell is possible by controlling the position of the required enzymes and their activity.

ROS formation in the apoplast occurs both enzymatically and non-enzymatically (see also Chapter 1 and 2 of this book), and in the case of the latter, the genes encoding the enzymes come in large gene families. In the apoplast, superoxide is formed by NADPH oxidases (NOXs, mainly known as respiratory burst oxidase homologs; RBOHs), H_2O_2 is formed by SODs, peroxidases, amine oxidases and oxalate oxidases, while $^{\cdot}OH$ are formed by peroxidases. NOX enzymes regulate a wide array of developmental processes in plants, including pollen germination and growth (Potocký, Jones, Bezvoda, Smirnoff, & Žárský, 2007), root and root hair development (Foreman et al., 2003; Yamauchi et al., 2017), seed germination (Ishibashi et al., 2015), SAM maintenance (Zeng et al., 2017), and leaf growth (Neuser et al., 2019). Similarly, apoplastic peroxidases have been implicated in numerous developmental processes in plants, amongst them seed germination (Jemmat et al., 2020), root development (Tsukagoshi, Busch, & Benfey, 2010), SAM maintenance (Zeng et al., 2017), leaf growth (Lu, Wang, Persson, Mueller-Roeber, & Schippers, 2014), and casparian strip formation (Rojas-Murcia et al., 2020). Next to that, SODs have been implicated in the regulation of the shoot apical meristem (Zeng et al., 2017), casparian strip formation (Lee, Rubio, Alassimone, & Geldner, 2013), and potentially plant reproduction (Zhou & Cai, 2018). Apoplastic amine oxidases have also been shown to play a role in plant development in multiple species (Tavladoraki, Cona, & Angelini, 2016). Taken together, regulation of ROS homeostasis in the apoplast is done through a dynamic and complex network of different proteins. One open question here is if the apoplastic proteins themselves are also targets of ROS to modulate their activity. The detection of ROS in the apoplast depends on cysteine-rich receptor-like kinases (CRKs; Castro et al., 2021), but also cysteine-rich peptides (CRPs) and extensins might be directly influenced by the apoplastic ROS homeostasis. Next to that the activity of the enzymes is post-transcirptionally regulated, their spatial and developmental phase specific occurrence is modulated at the transcriptional level.

4.2 Transcriptional regulation of ROS homeostasis during cell expansion

Under regulated environmental conditions, plant growth is highly reproducible, and the final size of the leaves is determined by genetic factors controlling cell division, expansion and differentiation. The gene

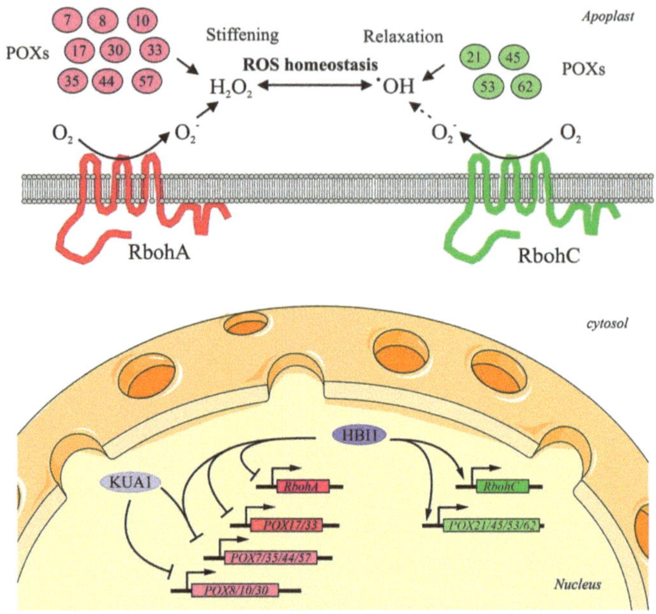

Fig. 3 Transcriptional regulation of apoplastic ROS homeostasis during cell expansion. Within the nucleus, KUA1 and HBI1 target genes encoding for peroxidases (POXs) or NADPH oxidases (NOXs). Whereas KUA1 only represses a subset of *POX* genes, HBI1 not only suppresses POX genes but also activates the NOX gene *RbohC* and several other *POX* genes. It is speculated that POXs repressed by KUA1 and HBI1 promote the formation of apoplastic H_2O_2, for which also the NOX protein RbohA appears to be required. On the other hand, RbohC and the upregulated POXs are proposed to be required for the controlled formation of hydroxyl radicals ($^{\cdot}OH$) to promote cell wall relaxation. The balance between the different sets of ROS-forming proteins in the apoplast determines if the cell wall will be relaxed or will be stiffened.

regulatory networks underlying cell proliferation are to some extent uncovered in Arabidopsis (Vercruysse, Baekelandt, Gonzalez, & Inzé, 2020). Still, how transcriptional programs establish and modulate ROS gradients during this phase is far from understood. In contrast, not many transcriptional regulators of cell expansion are known, however, for several it was found that they modulate ROS homeostasis to control cell expansion in leaves (Fig. 3).

Loss of KUODA1 (KUA1) results in plants with smaller leaves due to a decrease in cell expansion (Lu et al., 2014). Interestingly, these plants suffered from increased H_2O_2 levels due to the upregulation of several apoplastic class III peroxidases during leaf development. KUA1 directly binds to the promoters of POX genes to repress their expression during leaf growth (Lu et al., 2014). Thus, by preventing excessive accumulation of apoplastic H_2O_2, KUA1 prevents cell wall stiffening to enable cell expansion. Similarly, homolog of BEE2 interacting with IBH 1 (HBI1) was found to limit POX activity during leaf growth (Neuser et al., 2019). In addition, HBI1 was shown to promote the expression of *RbohC*, which is a positive regulator of leaf cell expansion, but inhibit the expression of *RbohA*, which was shown to act as a negative regulator of cell expansion. Interestingly, overexpression of KUA1 has no effect on cell proliferation, but plants overexpressing *HBI1* show next to an increase in cell size, a severely reduced cell number. This difference between KUA1 and HBI1 might be related to the activation of *NOX* expression by HBI1. Potentially, altered *NOX* expression might cause an early exit from cell proliferation during leaf growth. It can be expected that additional transcription factors modulate *NOX* and *POX* expression during cell proliferation and expansion in leaves. It would be of interest to understand how the different factors act together on the same target genes during leaf growth.

The Arabidopsis HB12 transcription factor, a homeodomain-leucine zipper class I (HD-Zip I) protein, was found to act as a positive regulator of leaf expansion (Hur et al., 2015). Genes related to endoreduplication, cell wall growth and gibberellin are upregulated in plants overexpressing *HB12*. Still, as direct target genes have not yet been determined, it remains to be uncovered what are the primary and secondary effects of the transcription factor on gene expression. With the advent of large data resources, like those generated using DNA affinity purification sequencing (DAP-seq; O'Malley et al., 2016), it is relatively easy to find transcription factors that potentially control ROS homeostasis during plant development. Therefore, it can be expected that in the coming years a better understanding on the gene regulatory networks underlying ROS homeostasis during leaf growth in plants will be obtained.

5. Concluding remarks

In this chapter, an overview has been given of the emerging fundamental role that ROS homeostasis plays in shaping plant tissues and the regulation of cell fate during leaf development and growth. It can be concluded that modulation of ROS homeostasis appears to be much more intimately connected with the development of plants as previously thought. This includes the regulation of stemness in the SAM through the formation of a ROS gradient, and the major impact of ROS homeostasis on cell expansion. However, new data implicate ROS in many other processes during leaf development, including cell proliferation and the determination of leaf polarity. Here, I hypothesize that the formation of radial organs by cutting or ablating SAM tissues is due to wound-induced ROS formation which inactivates the adaxial regulators. Although ablation of only the L1 layer of the SAM was shown to be sufficient to cause radialization of leaves (Reinhardt et al., 2005), it should be kept in mind that plants having a defect in the epidermal L1 layer still form normal leaves (Ogawa et al., 2015). It will be relatively easy to determine if the proposed model here holds up.

Since proteins controlling apoplastic ROS homeostasis appear to be encoded by large gene families with high redundancy between the different members, it is difficult to study the effect of these genes on ROS homeostasis and development. Nowadays, it can be expected that using CRISPR-CAS9 to target multiple members in one go will help to overcome these limitations. Finally, the role ROS have on stemness, cell proliferation and differentiation appears to be conserved amongst eukaryotes. This is in contrast to the hormones that guide the developmental processes in plants and animals. It is therefore likely that redox targets for these processes are conserved between plants and animals, and it would be helpful to generate a compendium of such components. Finding the molecular components that act as entry points for ROS homeostasis in cellular regulation is of the utmost importance to understand how ROS can influence so many cellular processes.

Acknowledgments

I thank the Leibniz Institute of Plant Genetics and Crop Plant Research (IPK) for their support. Furthermore, parts of the work were carried out in the framework of MAdLand (https://madland.science, DFG priority programme 2237), the author is grateful for funding by the DFG (SCHI 1130/9-1).

References

Aguilar-Martínez, J. A., & Sinha, N. (2013). Analysis of the role of Arabidopsis class I TCP genes AtTCP7, AtTCP8, AtTCP22, and AtTCP23 in leaf development. *Frontiers in Plant Science, 4*, 406.

Aichinger, E., Kornet, N., Friedrich, T., & Laux, T. (2012). Plant stem cell niches. *Annual Review of Plant Biology, 63*, 615–636.

Alvarez, J. P., Furumizu, C., Efroni, I., Eshed, Y., & Bowman, J. L. (2016). Active suppression of a leaf meristem orchestrates determinate leaf growth. *eLife, 5*, e15023.

Backer, R., Naidoo, S., & van den Berg, N. (2019). The nonexpressor of pathogenesis-related genes 1 (NPR1) and related family: Mechanistic insights in plant disease resistance. *Frontiers in Plant Science, 10*, 102.

Barciszewska-Pacak, M., Milanowska, K., Knop, K., Bielewicz, D., Nuc, P., Plewka, P., ... Szweykowska-Kulinska, Z. (2015). Arabidopsis microRNA expression regulation in a wide range of abiotic stress responses. *Frontiers in Plant Science, 6*, 410.
Benková, E., Michniewicz, M., Sauer, M., Teichmann, T., Seifertová, D., Jürgens, G., & Friml, J. (2003). Local, efflux-dependent auxin gradients as a common module for plant organ formation. *Cell, 115*, 591–602.
Betsuyaku, S., Takahashi, F., Kinoshita, A., Miwa, H., Shinozaki, K., Fukuda, H., & Sawa, S. (2011). Mitogen-activated protein kinase regulated by the CLAVATA receptors contributes to shoot apical meristem homeostasis. *Plant & Cell Physiology, 52*, 14–29.
Brand, U., Grünewald, M., Hobe, M., & Simon, R. (2002). Regulation of CLV3 expression by two homeobox genes in Arabidopsis. *Plant Physiology, 129*, 565–575.
Buzas, D. M., Nakamura, M., & Kinoshita, T. (2014). Epigenetic role for the conserved Fe-S cluster biogenesis protein AtDRE2 in Arabidopsis thaliana. *Proceedings of the National Academy of Sciences of the United States of America, 111*, 13565–13570.
Castro, B., Citterico, M., Kimura, S., Stevens, D. M., Wrzaczek, M., & Coaker, G. (2021). Stress-induced reactive oxygen species compartmentalization, perception and signalling. *Nature Plants, 7*, 403–412.
Chen, S. X., & Schopfer, P. (1999). Hydroxyl-radical production in physiological reactions. A novel function of peroxidase. *European Journal of Biochemistry, 260*, 726–735.
Chickarmane, V. S., Gordon, S. P., Tarr, P. T., Heisler, M. G., & Meyerowitz, E. M. (2012). Cytokinin signaling as a positional cue for patterning the apical-basal axis of the growing Arabidopsis shoot meristem. *Proceedings of the National Academy of Sciences of the United States of America, 109*, 4002–4007.
Chitwood, D. H., Nogueira, F. T., Howell, M. D., Montgomery, T. A., Carrington, J. C., & Timmermans, M. C. (2009). Pattern formation via small RNA mobility. *Genes & Development, 23*, 549–554.
Chung, Y., Zhu, Y., Wu, M. F., Simonini, S., Kuhn, A., Armenta-Medina, A., ... Wagner, D. (2019). Auxin Response Factors promote organogenesis by chromatin-mediated repression of the pluripotency gene SHOOTMERISTEMLESS. *Nature Communications, 10*, 886.
Comelli, R. N., & Gonzalez, D. H. (2007). Conserved homeodomain cysteines confer redox sensitivity and influence the DNA binding properties of plant class III HD-Zip proteins. *Archives of Biochemistry and Biophysics, 467*, 41–47.
Cosgrove, D. J. (2018). Nanoscale structure, mechanics and growth of epidermal cell walls. *Current Opinion in Plant Biology, 46*, 77–86.
Coudert, Y., Palubicki, W., Ljung, K., Novak, O., Leyser, O., & Harrison, C. J. (2015). Three ancient hormonal cues co-ordinate shoot branching in a moss. *eLife, 4*, e06808.
Davière, J. M., Wild, M., Regnault, T., Baumberger, N., Eisler, H., Genschik, P., & Achard, P. (2014). Class I TCP-DELLA interactions in inflorescence shoot apex determine plant height. *Current Biology, 24*, 1923–1928.
Després, C., Chubak, C., Rochon, A., Clark, R., Bethune, T., Desveaux, D., & Fobert, P. R. (2003). The Arabidopsis NPR1 disease resistance protein is a novel cofactor that confers redox regulation of DNA binding activity to the basic domain/leucine zipper transcription factor TGA1. *The Plant Cell, 15*, 2181–2191.
Dietz, K. J., Turkan, I., & Krieger-Liszkay, A. (2016). Redox- and reactive oxygen species-dependent signaling into and out of the photosynthesizing chloroplast. *Plant Physiology, 171*(3), 1541–1550.
Donnelly, P. M., Bonetta, D., Tsukaya, H., Dengler, R. E., & Dengler, N. G. (1999). Cell cycling and cell enlargement in developing leaves of Arabidopsis. *Developmental Biology, 215*, 407–419.
Edwards, D., Morris, J. L., Richardson, J. B., & Kenrick, P. (2014). Cryptospores and cryptophytes reveal hidden diversity in early land floras. *The New Phytologist, 202*, 50–78.

Eshed, Y., Baum, S. F., Perea, J. V., & Bowman, J. L. (2001). Establishment of polarity in lateral organs of plants. *Current Biology, 11*, 1251–1260.

Foreman, J., Demidchik, V., Bothwell, J. H., Mylona, P., Miedema, H., Torres, M. A., ... Dolan, L. (2003). Reactive oxygen species produced by NADPH oxidase regulate plant cell growth. *Nature, 422*, 442–446.

Gordon, S. P., Chickarmane, V. S., Ohno, C., & Meyerowitz, E. M. (2009). Multiple feedback loops through cytokinin signaling control stem cell number within the Arabidopsis shoot meristem. *Proceedings of the National Academy of Sciences of the United States of America, 106*, 16529–16534.

Guan, C., Wu, B., Yu, T., Wang, Q., Krogan, N. T., Liu, X., & Jiao, Y. (2017). Spatial auxin signaling controls leaf flattening in arabidopsis. *Current Biology, 27*, 2940–2950.

Gutsche, N., Thurow, C., Zachgo, S., & Gatz, C. (2015). Plant-specific CC-type glutaredoxins: Functions in developmental processes and stress responses. *Biological Chemistry, 396*, 495–509.

Gutsche, N., & Zachgo, S. (2016). The N-terminus of the floral arabidopsis TGA TRANSCRIPTION Factor PERIANTHIA mediates redox-sensitive DNA-binding. *PLoS One, 11*, e0153810.

Hardtke, C. S., Ckurshumova, W., Vidaurre, D. P., Singh, S. A., Stamatiou, G., Tiwari, S. B., ... Berleth, T. (2004). Overlapping and non-redundant functions of the Arabidopsis auxin response factors MONOPTEROS and NONPHOTOTROPIC HYPOCOTYL 4. *Development, 131*, 1089–1100.

Harrison, C. J., & Morris, J. L. (2018). The origin and early evolution of vascular plant shoots and leaves. *Philosophical Transactions of the Royal Society of London, 373*, 20160496.

Hata, Y., & Kyozuka, J. (2021). Fundamental mechanisms of the stem cell regulation in land plants: Lesson from shoot apical cells in bryophytes. *Plant Molecular Biology, 107*, 213–225.

He, H., Van Breusegem, F., & Mhamdi, A. (2018). Redox-dependent control of nuclear transcription in plants. *Journal of Experimental Botany, 69*, 3359–3372.

Hepworth, S. R., Zhang, Y., McKim, S., Li, X., & Haughn, G. W. (2005). BLADE-ON-PETIOLE-dependent signaling controls leaf and floral patterning in Arabidopsis. *The Plant Cell, 17*, 1434–1448.

Hur, Y. S., Um, J. H., Kim, S., Kim, K., Park, H. J., Lim, J. S., ... Cheon, C. I. (2015). Arabidopsis thaliana homeobox 12 (ATHB12), a homeodomain-leucine zipper protein, regulates leaf growth by promoting cell expansion and endoreduplication. *The New Phytologist, 205*, 316–328.

Ichihashi, Y., Kawade, K., Usami, T., Horiguchi, G., Takahashi, T., & Tsukaya, H. (2011). Key proliferative activity in the junction between the leaf blade and leaf petiole of Arabidopsis. *Plant Physiology, 157*, 1151–1162.

Irish, V. F., & Sussex, I. M. (1992). A fate map of the Arabidopsis embryonic shoot apical meristem. *Development, 115*, 745–753.

Ishibashi, Y., Kasa, S., Sakamoto, M., Aoki, N., Kai, K., Yuasa, T., ... Iwaya-Inoue, M. (2015). A role for reactive oxygen species produced by NADPH oxidases in the embryo and aleurone cells in barley seed germination. *PLoS One, 10*, e0143173.

Iyer, N. J., Jia, X., Sunkar, R., Tang, G., & Mahalingam, R. (2012). microRNAs responsive to ozone-induced oxidative stress in Arabidopsis thaliana. *Plant Signaling & Behavior*, 7, 484–491.

Jemmat, A. M., Ranocha, P., Le Ru, A., Neel, M., Jauneau, A., Raggi, S., ... Dunand, C. (2020). Coordination of five class III peroxidase-encoding genes for early germination events of Arabidopsis thaliana. *Plant Science, 298*, 110565.

Juarez, M. T., Kui, J. S., Thomas, J., Heller, B. A., & Timmermans, M. C. (2004). microRNA-mediated repression of rolled leaf1 specifies maize leaf polarity. *Nature, 428*, 84–88.

Jun, J. H., Ha, C. M., & Fletcher, J. C. (2010). BLADE-ON-PETIOLE1 coordinates organ determinacy and axial polarity in arabidopsis by directly activating ASYMMETRIC LEAVES2. *The Plant Cell, 22*, 62–76.

Kalve, S., Sizani, B. L., Markakis, M. N., Helsmoortel, C., Vandeweyer, G., Laukens, K., ... Beemster, G. (2020). Osmotic stress inhibits leaf growth of Arabidopsis thaliana by enhancing ARF-mediated auxin responses. *The New Phytologist, 226*, 1766–1780.

Kawade, K., Horiguchi, G., Usami, T., Hirai, M. Y., & Tsukaya, H. (2013). ANGUSTIFOLIA3 signaling coordinates proliferation between clonally distinct cells in leaves. *Current Biology, 23*, 788–792.

Kelley, D. R., Arreola, A., Gallagher, T. L., & Gasser, C. S. (2012). ETTIN (ARF3) physically interacts with KANADI proteins to form a functional complex essential for integument development and polarity determination in Arabidopsis. *Development, 139*, 1105–1109.

Krogan, N. T., Marcos, D., Weiner, A. I., & Berleth, T. (2016). The auxin response factor MONOPTEROS controls meristem function and organogenesis in both the shoot and root through the direct regulation of PIN genes. *The New Phytologist, 212*, 42–50.

Kuhlemeier, C., & Timmermans, M. C. (2016). The Sussex signal: Insights into leaf dorsiventrality. *Development, 143*, 3230–3237.

Lampugnani, E. R., Khan, G. A., Somssich, M., & Persson, S. (2018). Building a plant cell wall at a glance. *Journal of Cell Science, 131*, jcs207373.

Lantzouni, O., Alkofer, A., Falter-Braun, P., & Schwechheimer, C. (2020). GROWTH-REGULATING FACTORS interact with DELLAs and regulate growth in cold stress. *The Plant Cell, 32*, 1018–1034.

Lee, B. H., Ko, J. H., Lee, S., Lee, Y., Pak, J. H., & Kim, J. H. (2009). The Arabidopsis GRF-INTERACTING FACTOR gene family performs an overlapping function in determining organ size as well as multiple developmental properties. *Plant Physiology, 151*, 655–668.

Lee, B. H., Kwon, S. H., Lee, S. J., Park, S. K., Song, J. T., Lee, S., ... Kim, J. H. (2015). The Arabidopsis thaliana NGATHA transcription factors negatively regulate cell proliferation of lateral organs. *Plant Molecular Biology, 89*, 529–538.

Lee, Y., Rubio, M. C., Alassimone, J., & Geldner, N. (2013). A mechanism for localized lignin deposition in the endodermis. *Cell, 153*, 402–412.

Li, N., Muthreich, M., Huang, L. J., Thurow, C., Sun, T., Zhang, Y., & Gatz, C. (2019). TGACG-BINDING FACTORs (TGAs) and TGA-interacting CC-type glutaredoxins modulate hyponastic growth in Arabidopsis thaliana. *The New Phytologist, 221*, 1906–1918.

Li, Z., Fernie, A. R., & Persson, S. (2016). Transition of primary to secondary cell wall synthesis. *Science Bulletin, 61*, 838–846.

Li, Z., Li, B., Shen, W. H., Huang, H., & Dong, A. (2012). TCP transcription factors interact with AS2 in the repression of class-I KNOX genes in Arabidopsis thaliana. *The Plant Journal, 71*, 99–107.

Liu, J., Shi, X., Chang, Z., Ding, Y., & Ding, C. (2022). Auxin efflux transporters OsPIN1c and OsPIN1d function redundantly in regulating rice (Oryza sativa L.) panicle development. *Plant & Cell Physiology, 63*, 305–316.

Long, J. A., Moan, E. I., Medford, J. I., & Barton, M. K. (1996). A member of the KNOTTED class of homeodomain proteins encoded by the STM gene of Arabidopsis. *Nature, 379*, 66–69.

Lu, D., Wang, T., Persson, S., Mueller-Roeber, B., & Schippers, J. H. (2014). Transcriptional control of ROS homeostasis by KUODA1 regulates cell expansion during leaf development. *Nature Communications, 5*, 3767.

Ludin, A., Gur-Cohen, S., Golan, K., Kaufmann, K. B., Itkin, T., Medaglia, C., ... Lapidot, T. (2014). Reactive oxygen species regulate hematopoietic stem cell self-renewal, migration and development, as well as their bone marrow microenvironment. *Antioxidants & Redox Signaling, 21*, 1605–1619.

Magnani, E., & Barton, M. K. (2011). A per-ARNT-sim-like sensor domain uniquely regulates the activity of the homeodomain leucine zipper transcription factor REVOLUTA in Arabidopsis. *The Plant Cell, 23*, 567–582.

Mangano, S., Denita-Juarez, S. P., Choi, H. S., Marzol, E., Hwang, Y., Ranocha, P., ... Estevez, J. M. (2017). Molecular link between auxin and ROS-mediated polar growth. *Proceedings of the National Academy of Sciences of the United States of America, 114*, 5289–5294.

McConnell, J. R., Emery, J., Eshed, Y., Bao, N., Bowman, J., & Barton, M. K. (2001). Role of PHABULOSA and PHAVOLUTA in determining radial patterning in shoots. *Nature, 411*, 709–713.

Medeiros, D. B., Barros, J., Fernie, A. R., & Araújo, W. L. (2020). Eating away at ROS to regulate stomatal opening. *Trends in Plant Science, 25*, 220–223.

Mei, Q., & Dvornyk, V. (2014). Evolution of PAS domains and PAS-containing genes in eukaryotes. *Chromosoma, 123*, 385–405.

Mittler, R., Kim, Y., Song, L., Coutu, J., Coutu, A., Ciftci-Yilmaz, S., ... Zhu, J. K. (2006). Gain- and loss-of-function mutations in Zat10 enhance the tolerance of plants to abiotic stress. *FEBS Letters, 580*, 6537–6542.

Monshausen, G. B., Messerli, M. A., & Gilroy, S. (2008). Imaging of the Yellow Cameleon 3.6 indicator reveals that elevations in cytosolic Ca2+ follow oscillating increases in growth in root hairs of Arabidopsis. *Plant Physiology, 147*, 1690–1698.

Müller, K., Linkies, A., Vreeburg, R. A., Fry, S. C., Krieger-Liszkay, A., & Leubner-Metzger, G. (2009). In vivo cell wall loosening by hydroxyl radicals during cress seed germination and elongation growth. *Plant Physiology, 150*, 1855–1865.

Neuser, J., Metzen, C. C., Dreyer, B. H., Feulner, C., van Dongen, J. T., Schmidt, R. R., & Schippers, J. (2019). HBI1 Mediates the trade-off between growth and immunity through its impact on apoplastic ROS homeostasis. *Cell Reports, 28*, 1670–1678.

Nosaka-Takahashi, M., Kato, M., Kumamaru, T., & Sato, Y. (2022). Measurements of the number of specified and unspecified cells in the shoot apical meristem during a plastochron in rice (Oryza sativa) reveal the robustness of cellular specification process in plant development. *PLoS One, 17*, e0269374.

Ogawa, E., Yamada, Y., Sezaki, N., Kosaka, S., Kondo, H., Kamata, N., ... Takahashi, T. (2015). ATML1 and PDF2 play a redundant and essential role in arabidopsis embryo development. *Plant & Cell Physiology, 56*, 1183–1192.

O'Malley, R. C., Huang, S. C., Song, L., Lewsey, M. G., Bartlett, A., Nery, J. R., ... Ecker, J. R. (2016). Cistrome and epicistrome features shape the regulatory DNA landscape. *Cell, 165*, 1280–1292.

Orozco-Cárdenas, M. L., Narváez-Vásquez, J., & Ryan, C. A. (2001). Hydrogen peroxide acts as a second messenger for the induction of defense genes in tomato plants in response to wounding, systemin, and methyl jasmonate. *The Plant Cell, 13*, 179–191.

Palatnik, J. F., Allen, E., Wu, X., Schommer, C., Schwab, R., Carrington, J. C., & Weigel, D. (2003). Control of leaf morphogenesis by microRNAs. *Nature, 425*, 257–263.

Pekker, I., Alvarez, J. P., & Eshed, Y. (2005). Auxin response factors mediate Arabidopsis organ asymmetry via modulation of KANADI activity. *The Plant Cell, 17*, 2899–2910.

Piya, S., Liu, J., Burch-Smith, T., Baum, T. J., & Hewezi, T. (2020). A role for Arabidopsis growth-regulating factors 1 and 3 in growth-stress antagonism. *Journal of Experimental Botany, 71*(4), 1402–1417.

Potocký, M., Jones, M. A., Bezvoda, R., Smirnoff, N., & Žárský, V. (2007). Reactive oxygen species produced by NADPH oxidase are involved in pollen tube growth. *The New Phytologist, 174*, 742–751.

Prigge, M. J., Otsuga, D., Alonso, J. M., Ecker, J. R., Drews, G. N., & Clark, S. E. (2005). Class III homeodomain-leucine zipper gene family members have overlapping, antagonistic, and distinct roles in Arabidopsis development. *The Plant Cell, 17*, 61–76.

Py, B., Moreau, P. L., & Barras, F. (2011). Fe-S clusters, fragile sentinels of the cell. *Current Opinion in Microbiology, 14*, 218–223.

Reddy, G. V., & Meyerowitz, E. M. (2005). Stem-cell homeostasis and growth dynamics can be uncoupled in the Arabidopsis shoot apex. *Science, 310*, 663–667.

Reinhardt, D., Frenz, M., Mandel, T., & Kuhlemeier, C. (2005). Microsurgical and laser ablation analysis of leaf positioning and dorsoventral patterning in tomato. *Development, 132*, 15–26.

Rojas-Murcia, N., Hématy, K., Lee, Y., Emonet, A., Ursache, R., Fujita, S., ... Geldner, N. (2020). High-order mutants reveal an essential requirement for peroxidases but not laccases in Casparian strip lignification. *Proceedings of the National Academy of Sciences of the United States of America, 117*, 29166–29177.

Sachs, T. (1969). Regeneration experiments on the determination of the form of leaves. *Israel Journal of Botany, 18*, 21–30.

Sakakibara, K., Nishiyama, T., Deguchi, H., & Hasebe, M. (2008). Class 1 KNOX genes are not involved in shoot development in the moss Physcomitrella patens but do function in sporophyte development. *Evolution & Development, 10*, 555–566.

Sato, H., Takasaki, H., Takahashi, F., Suzuki, T., Iuchi, S., Mitsuda, N., ... Shinozaki, K. (2018). Arabidopsis thaliana NGATHA1 transcription factor induces ABA biosynthesis by activating NCED3 gene during dehydration stress. *Proceedings of the National Academy of Sciences of the United States of America, 115*, E11178–E11187.

Schippers, J. H., Foyer, C. H., & van Dongen, J. T. (2016). Redox regulation in shoot growth, SAM maintenance and flowering. *Current Opinion in Plant Biology, 29*, 121–128.

Schlegel, J., Denay, G., Wink, R., Pinto, K. G., Stahl, Y., Schmid, J., ... Simon, R. G. (2021). Control of Arabidopsis shoot stem cell homeostasis by two antagonistic CLE peptide signalling pathways. *eLife, 10*, e70934.

Schmidt, R., Kunkowska, A. B., & Schippers, J. H. M. (2016). Role of reactive oxygen species during cell expansion in leaves. *Plant Physiology, 172*, 2098–2106.

Schopfer, P. (1996). Hydrogen peroxide-mediated cell-wall stiffening in vitro in maize coleoptiles. *Planta, 199*, 43–49.

Schopfer, P. (2006). Biomechanics of plant growth. *American Journal of Botany, 93*, 1415–1425.

Schweikert, C., Liszkay, A., & Schopfer, P. (2000). Scission of polysaccharides by peroxidase-generated hydroxyl radicals. *Phytochemistry, 53*, 565–570.

Sharples, S. C., Nguyen-Phan, T. C., & Fry, S. C. (2017). Xyloglucan endo-transglucosylase/hydrolases (XTHs) are inactivated by binding to glass and cellulosic surfaces, and released in active form by a heat-stable polymer from cauliflower florets. *Journal of Plant Physiology, 218*, 135–143.

Shumbe, L., Chevalier, A., Legeret, B., Taconnat, L., Monnet, F., & Havaux, M. (2016). Singlet oxygen-induced cell death in arabidopsis under high-light stress is controlled by OXI1 kinase. *Plant Physiology, 170*, 1757–1771.

Siegfried, K. R., Eshed, Y., Baum, S. F., Otsuga, D., Drews, G. N., & Bowman, J. L. (1999). Members of the YABBY gene family specify abaxial cell fate in Arabidopsis. *Development, 126*, 4117–4128.

Steeves, T. A., & Sussex, I. M. (1989). (2nd Edn). *Patterns in Plant Development 338*. Cambridge: Cambridge University Press.

Sussex, I. M. (1951). Experiments on the cause of dorsiventrality in leaves. *Nature, 167*, 651–652.

Tavladoraki, P., Cona, A., & Angelini, R. (2016). Copper-containing amine oxidases and FAD-dependent polyamine oxidases are key players in plant tissue differentiation and organ development. *Frontiers in Plant Science, 7*, 824.

Tsuda, K., Ito, Y., Sato, Y., & Kurata, N. (2011). Positive autoregulation of a KNOX gene is essential for shoot apical meristem maintenance in rice. *The Plant Cell, 23*, 4368–4381.

Tsukagoshi, H., Busch, W., & Benfey, P. N. (2010). Transcriptional regulation of ROS controls transition from proliferation to differentiation in the root. *Cell, 143*, 606–616.

Tsukaya, H. (2006). Mechanism of leaf-shape determination. *Annual Review of Plant Biology, 57*, 477–496.

Tsukaya, H. (2014). Comparative leaf development in angiosperms. *Current Opinion in Plant Biology, 17*, 103–109.

Tsukaya, H. (2021). The leaf meristem enigma: The relationship between the plate meristem and the marginal meristem. *The Plant Cell, 33*, 3194–3206.

Vasco, A., Smalls, T. L., Graham, S. W., Cooper, E. D., Wong, G. K., Stevenson, D. W., ... Ambrose, B. A. (2016). Challenging the paradigms of leaf evolution: Class III HD-Zips in ferns and lycophytes. *The New Phytologist, 212*, 745–758.

Vercruysse, J., Baekelandt, A., Gonzalez, N., & Inzé, D. (2020). Molecular networks regulating cell division during Arabidopsis leaf growth. *Journal of Experimental Botany, 71*, 2365–2378.

Vercruyssen, L., Verkest, A., Gonzalez, N., Heyndrickx, K. S., Eeckhout, D., Han, S. K., ... Inzé, D. (2014). ANGUSTIFOLIA3 binds to SWI/SNF chromatin remodeling complexes to regulate transcription during Arabidopsis leaf development. *The Plant Cell, 26*, 210–229.

Viaene, T., Landberg, K., Thelander, M., Medvecka, E., Pederson, E., Feraru, E., ... Friml, J. (2014). Directional auxin transport mechanisms in early diverging land plants. *Current Biology, 24*, 2786–2791.

Viola, I. L., Güttlein, L. N., & Gonzalez, D. H. (2013). Redox modulation of plant developmental regulators from the class I TCP transcription factor family. *Plant Physiology, 162*, 1434–1447.

Vogt, J. H., & Schippers, J. H. M. (2015). Setting the PAS, the role of circadian PAS domain proteins during environmental adaptation in plants. *Frontiers in Plant Science, 6*, 513.

Wang, X., Li, Q., Yuan, W., Cao, Z., Qi, B., Kumar, S., ... Qian, W. (2016). The cytosolic Fe-S cluster assembly component MET18 is required for the full enzymatic activity of ROS1 in active DNA demethylation. *Scientific Reports, 6*, 26443.

Wang, Y., Salasini, B. C., Khan, M., Devi, B., Bush, M., Subramaniam, R., & Hepworth, S. R. (2019). Clade I TGACG-motif binding basic leucine zipper transcription factors mediate BLADE-ON-PETIOLE-dependent regulation of development. *Plant Physiology, 180*, 937–951.

Wellman, C. H., Osterloff, P. L., & Mohiuddin, U. (2003). Fragments of the earliest land plants. *Nature, 425*, 282–285.

Wiese, A., Christ, M. M., Virnich, O., Schurr, U., & Walter, A. (2007). Spatio-temporal leaf growth patterns of Arabidopsis thaliana and evidence for sugar control of the diel leaf growth cycle. *The New Phytologist, 174*, 752–761.

Wolf, S., Hématy, K., & Höfte, H. (2012). Growth control and cell wall signaling in plants. *Annual Review of Plant Biology, 63*, 381–407.

Wu, F., Chi, Y., Jiang, Z., Xu, Y., Xie, L., Huang, F., ... Pei, Z. M. (2020). Hydrogen peroxide sensor HPCA1 is an LRR receptor kinase in Arabidopsis. *Nature, 578*, 577–581.

Xia, X. J., Zhou, Y. H., Shi, K., Zhou, J., Foyer, C. H., & Yu, J. Q. (2015). Interplay between reactive oxygen species and hormones in the control of plant development and stress tolerance. *Journal of Experimental Botany, 66*, 2839–2856.

Xie, Y., Huhn, K., Brandt, R., Potschin, M., Bieker, S., Straub, D., ... Wenkel, S. (2014). REVOLUTA and WRKY53 connect early and late leaf development in Arabidopsis. *Development, 141*, 4772–4783.

Yadav, R. K., Perales, M., Gruel, J., Girke, T., Jönsson, H., & Reddy, G. V. (2011). WUSCHEL protein movement mediates stem cell homeostasis in the Arabidopsis shoot apex. *Genes & development, 25*, 2025–2030.

Yamauchi, S., Mano, S., Oikawa, K., Hikino, K., Teshima, K. M., Kimori, Y., ... Takemiya, A. (2019). Autophagy controls reactive oxygen species homeostasis in guard cells that is essential for stomatal opening. *Proceedings of the National Academy of Sciences of the United States of America, 116*, 19187–19192.

Yamauchi, T., Yoshioka, M., Fukazawa, A., Mori, H., Nishizawa, N. K., Tsutsumi, N., ... Nakazono, M. (2017). An NADPH oxidase RBOH functions in rice roots during lysigenous aerenchyma formation under oxygen-deficient conditions. *The Plant Cell, 29*, 775–790.

Yanai, O., Shani, E., Dolezal, K., Tarkowski, P., Sablowski, R., Sandberg, G., ... Ori, N. (2005). Arabidopsis KNOXI proteins activate cytokinin biosynthesis. *Current Biology, 15*, 1566–1571.

Yao, X., Wang, H., Li, H., Yuan, Z., Li, F., Yang, L., & Huang, H. (2009). Two types of cis-acting elements control the abaxial epidermis-specific transcription of the MIR165a and MIR166a genes. *FEBS Letters, 583*, 3711–3717.

Zeng, J., Dong, Z., Wu, H., Tian, Z., & Zhao, Z. (2017). Redox regulation of plant stem cell fate. *The EMBO Journal, 36*, 2844–2855.

Zhang, K., Wang, R., Zi, H., Li, Y., Cao, X., Li, D., ... Liu, X. (2018). AUXIN RESPONSE FACTOR3 regulates floral meristem determinacy by repressing cytokinin biosynthesis and signaling. *The Plant Cell, 30*, 324–346.

Zhang, S., Apel, K., & Kim, C. (2014). Singlet oxygen-mediated and EXECUTER-dependent signalling and acclimation of Arabidopsis thaliana exposed to light stress. *Philosophical Transactions of the Royal Society of London. Series B, Biological Sciences, 369*, 20130227.

Zhang, Y., Jiao, Y., Jiao, H., Zhao, H., & Zhu, Y. X. (2017). Two-step functional innovation of the stem-cell factors WUS/WOX5 during plant evolution. *Molecular Biology and Evolution, 34*, 640–653.

Zheng, Y., Zhang, K., Guo, L., Liu, X., & Zhang, Z. (2018). AUXIN RESPONSE FACTOR3 plays distinct role during early flower development. *Plant Signaling & Behavior, 13*, e1467690.

Zhou, G., Meng, S., Li, Y., Ghebre, Y. T., & Cooke, J. P. (2016). Optimal ROS signaling is critical for nuclear reprogramming. *Cell Reports, 15*, 919–925.

Zhou, Q., & Cai, Q. (2018). The superoxide dismutase genes might be required for appropriate development of the ovule after fertilization in Xanthoceras sorbifolium. *Plant Cell Reports, 37*, 727–739.

Zumajo-Cardona, C., Vasco, A., & Ambrose, B. A. (2019). The evolution of the KANADI gene family and leaf development in lycophytes and ferns. *Plants (Basel, Switzerland), 8*, 313.

Further reading

Guo, M., Thomas, J., Collins, G., & Timmermans, M. C. (2008). Direct repression of KNOX loci by the ASYMMETRIC LEAVES1 complex of Arabidopsis. *The Plant Cell, 20*, 48–58.

CHAPTER FIVE

Multiple roles of ROS in flowering plant reproduction

Liang-Zi Zhou*, **and Thomas Dresselhaus***
Cell Biology and Plant Biochemistry, University of Regensburg, Regensburg, Germany
*Corresponding authors. e-mail address: liangzi.zhou@ur.de; thomas.dresselhaus@ur.de

Contents

Abstract

Reactive oxygen species (ROS), which are generated by different cellular compartments through aerobic metabolic pathways, are generally considered to play dual roles in plants. They are toxic compounds at high concentration but are also required to modulate multiple signaling reactions and gene regulatory networks in biotic and abiotic stress responses as well as during plant growth and development. Recent studies revealed that ROS play also critical roles in various aspects of flowering plant reproduction. In male gametophytes, ROS regulate germline specification and differentiation, and they are associated with programmed cell death (PCD) of the tapetum to regulate pollen development. During female gametophyte formation, ROS are involved in cell identity determination, polarity regulation and megaspore PCD. After pollination, ROS are of vital importance for regulation of self-incompatibility responses, for pollen hydration and for successful germination on the stigma. Further roles have been reported during pollen tube growth and guidance in the transmitting

Advances in Botanical Research, Volume 105
ISSN 0065-2296, https://doi.org/10.1016/bs.abr.2022.10.002

tract, pollen tube reception and sperm cell release in the female gametophyte as well as during gamete fusion. In this chapter, we will provide an overview of the multiple functions of ROS signaling during plant reproductive development from germline initiation to double fertilization. We will focus on flowering plants and discuss the generation and elimination of ROS, and the induction and regulation of various downstream signaling pathways.

Nomenclature

ABC	ATP-binding cassette
AGO	ARGONAUTE
AMS	ABORTED MICROSPORES
ANJ-FER	ANJEA-FERONIA
are	*anthocyanin reduced*
BRs	brassinosteroids
BSO	buthionine sulfoximine
cad2–1	*cadmium sensitive2–1*
cGMP	cyclic guanosine monophosphate
ChIP	chromatin immunoprecipitation
CM-H2DCFDA	chloromethyl-2′,7′-dichlorodihydrofluorescein diacetate
CRK	CYSTEINE-RICH RECEPTOR-LIKE KINASES family
CRP	cysteine-rich protein
CrRLK1L	*Catharanthus roseus* RLK1-like
Cu/Zn-SOD	copper-zinc SOD
DAB	diaminobenzidine
DEG	differentially expressed gene
DPI	diphenylene iodonium chloride
DWF	DWARF
EPR	electron paramagnetic resonance
Fe-SOD	iron SOD
FER	FERONIA
GRN	gene regulatory network
GSI	gametophytic self-incompatibility
GRX	glutaredoxin
GSH	glutaredoxin
GSSG	glutathione disulfide
H_2O_2	hydrogen peroxide
IDA	INFLORESCENCE DEFICIENT IN ABSCISSION
IDL	IDA-LIKE
mee33	*maternal effect embryo arrest 33*
MMC	megaspore mother cell
Mn-SOD	manganese SOD
MPK4	MITOGEN-ACTIVATED PROTEIN KINASE 4
MSCA1	MALE STERILE CONVERTED ANTHER 1
NADPH	nicotinamide adenine dinucleotide phosphate
NBT	nitroblue tetrazolium
NO	nitric oxide
OH·	hydroxyl radical

OsHXK1	hexokinase 1
O_2^-	superoxide anion or radical
$^1O_2$	singlet oxygen
PAO	polyamine oxidase
PCD	programmed cell death
PCP-B	POLLEN COAT PROTEIN B-class CRP
PMC	pollen mother cell
PsEND1	*Pisum sativum* ENDOTHECIUM 1
RALF	RAPID ALKALINIZATION FACTOR
RBOH	RESPIRATORY BURST OXIDASE HOMOLOG
RLK	receptor-like kinase
roGFP2	redox-sensitive green fluorescent protein
ROPGEFs	guanine nucleotide exchange factors
ROS	reactive oxygen species
SC	self-compatibility
SCR	S-locus CRPs
SI	self-incompatibility
SODs	superoxide dismutases
SRKs	S-locus receptor kinases
SSI	sporophytic self-incompatibility
SSP	STIGMA-SPECIFIC PEROXIDASE
TF	transcription factor
WT	wild-type

1. Introduction

Flowering plants (angiosperms) represent about 90% of all plant species. With more than 300,000 species, they dominate life on earth and are of vital importance for various ecosystems and the environment (Crepet, 2000). Through sexual and more rarely asexual reproduction, flowering plants produce seeds, which contain and protect the next generation and largely contributed to their evolutionary success. Investigations to elucidate the molecular mechanisms underlying plant reproduction became more and more important in the last century to secure food and feed production for a dramatically growing world population (Pereira & Coimbra, 2019). Moreover, global warming will have a high and negative impact on seed production and biodiversity, thus novel and better adapted plant species and varieties are needed to fulfill the requirements of human and animal nutrition.

Sexual reproduction in flowering plants starts from the initiation of male and female germlines from sporophytic cells in meristems and ends up with the double fertilization process characteristic for angiosperms

(Dresselhaus, Sprunck, & Wessel, 2016), ultimately generating seeds that harbor the next generation. During the life cycle of flowering plants, male and female reproductive organs are generated from flower meristems, initiate differentiation of germlines and finally generate male and female gametes via gametogenesis (Fig. 1). After gamete maturation, pollen grains (male gametophytes) are released from anthers and land on the surface of female stigmatic papilla cells where intensive communication takes place to distinguish between own and alien pollen. After adhesion, compatible pollen grains hydrate and germinate, and produce a pollen tube that serves as a vehicle to transport non-motile sperm cells deep into the female reproductive tissues. After rejection of incompatible pollen, compatible pollen tubes invade into transmitting tracts and grow at a relative high speed towards ovules, which develop into seeds after successful fertilization. Attraction signals from the ovule and embryo sac (female gametophyte) serve as guidance clues and lure pollen tubes towards and inside the female gametophyte. Usually, a single pollen tube penetrates one of the synergid cells, burst at the tip region to release both sperm cells near the egg cell and central cell. After gamete activation, plasmogamy (gamete fusion) takes place to accomplish double fertilization (Dresselhaus & Franklin-Tong, 2013; Higashiyama & Takeuchi, 2015). The whole process requires intensive and precise communication between male and female reproductive tissue and cells. A large number of signaling components and pathways are involved (Mizuta & Higashiyama, 2018; Qu, Li, Lan, & Dresselhaus, 2015; Zhou & Dresselhaus, 2019), often resulting in the regulation of Reactive oxygen species (ROS) levels (Breygina & Klimenko, 2020; Sankaranarayanan, Ju, & Kessler, 2020; Zhang, Zhang, & Gao, 2020).

As an unavoidable by-product of aerobic metabolism or as a consequence of inevitable escape of electrons to oxygen from the electron transport activities, ROS are constantly generated in different plant cellular compartments, including chloroplasts, mitochondria, peroxisomes, endoplasmic reticulum, at the plasma membrane and in the apoplast/cell wall (Mittler, 2017). For biosynthesis of some ROS see Chapters 1 and 2 This Volume in this book. ROS can be subdivided into different groups, including free radicals such as superoxide anion ($O_2^{\cdot -}$) and hydroxyl ($OH^{\cdot}$) as well as nonradical molecules like hydrogen peroxide (H_2O_2) and singlet oxygen ($^1O_2$) (Sharma, Jha, Dubey, & Pessarakli, 2012). These highly reactive molecules are originally thought to be dangerous due to their ability to induce oxidative cell death. However, in recent years, it was also

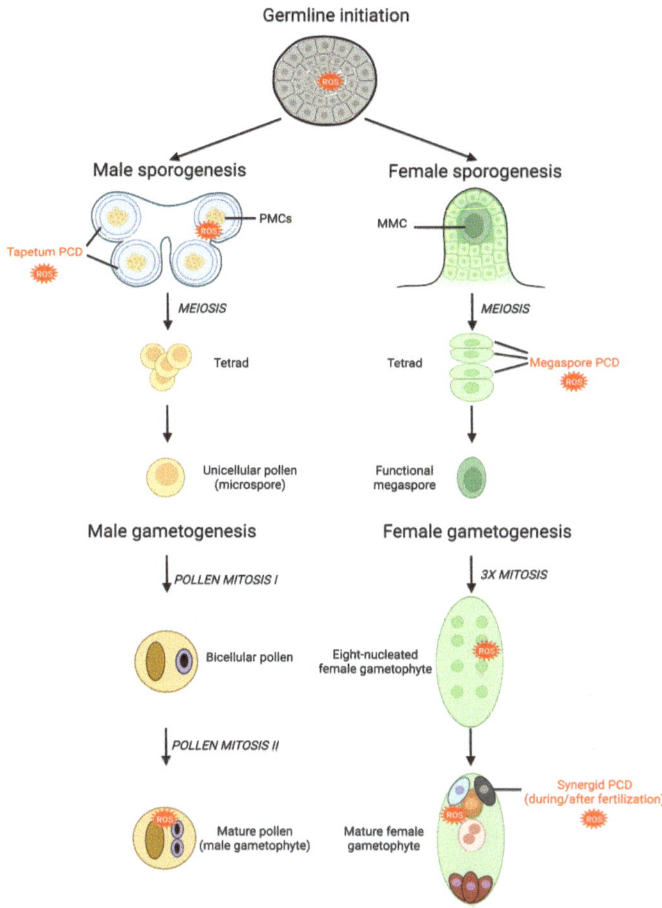

Fig. 1 Schematic model showing the involvement of ROS in plant germline initiation, sporogenesis and gametogenesis. During germline initiation, ROS are of vital importance for archesporial cells formation. In male sporogenesis, ROS regulate pollen mother cell (PMC) differentiation and programmed cell death (PCD) of the tapetum, which is a key step for pollen development and maturation. Accumulation of ROS can be observed in mature pollen grains consisting of the vegetative cell that encloses two sperm cells. After megaspore mother cell (MMC) differentiation and meiosis, ROS are involved in the regulation of PCD of three of the four megaspores during female sporogenesis and are required in female gametogenesis where it is involved in cell division regulation. In the mature embryo sac, accumulation of ROS can be detected at the egg cell and ROS burst occurs in synergid cells, which go through PCD before gamete fusion (receptive synergid) and degenerate after fertilization (persisting synergid), respectively. The large central cell is di-haploid and separates three antipodal cells at the chalazal pole of the embryo sac from the egg apparatus containing the egg and synergid cells at the micropylar pole.

discovered that ROS play critical roles in the control and regulation of various biological processes throughout the plant life cycle, from seed development and germination to root, shoot and flower development, from reproduction to biotic and abiotic stress responses (Mansoor et al., 2022; Mhamdi & Van Breusegem, 2018; Mittler, Vanderauwera, Gollery, & Van Breusegem, 2004; Chapters 4 and 6 This volume). All these studies extended our understanding about the function of ROS in plants and indicated the complex dual roles of ROS. In order to utilize ROS as signaling molecules, plant evolved different mechanisms to balance the production and elimination of ROS using a complex redox network involving several antioxidant enzymes and non-enzymatic reactions (Mittler et al., 2004; Mittler, 2017), which appear to be also highly active during plant reproduction (Martin, Distefano, Zabaleta, & Pagnussat, 2013; Zhang et al., 2020). In this book chapter, we will summarize our current knowledge about the function of ROS at different steps during angiosperm reproduction. We consider and, when possible, compare different plant species and try to build up the current regulatory networks of ROS homeostasis in different reproductive cell types.

2. ROS in germline specification

The concept of germline specification in plants is different from that in animals, as the animal germline is determined already during early embryogenesis and is maintained as isolated germline stem cells through the whole lifetime (Strome & Lehmann, 2007). In contrast, the germline of flowering plants is initiated from undifferentiated meristematic cells of adult plants which produce vegetative tissues and organs. After flowering induction, vegetative meristems are transformed into flowering/inflorescence meristems, which generate male and female reproductive organs that form archesporial cells as precursors of the germlines (Fig. 1). These differentiate into male and female meiocytes (pollen/megaspore mother cells), which, following meiosis, further develop into male and female gametophytes, respectively (Berger & Twell, 2011; Zhou, Juranic, & Dresselhaus, 2017). In flowering plants, both male and female germline specification can be subdivided into two major steps: the first step involves polarized cell divisions depending on positional information during which germ cell lineages are separated from somatic cell lineages. In the second step, germ cell fate is maintained while surrounding somatic cells are prevented from adopting germ cell fate (Berger & Twell, 2011; Strome & Lehmann, 2007). Up to now, the regulation mechanisms of germline specification are still not well understood in plants. Among the limited number of molecular players, ROS was found to be important for germline specification in certain plant species (Kelliher & Walbot, 2012; Whipple, 2012; Zafra, Rodriguez-Garcia, & de Dios Alche, 2010). An overview about genes associated with ROS in reproduction is shown in Table 1.

In male floret primordia of maize (*Zea mays*), for example, archesporial cells are generated as precursors of pollen mother cells (PMCs), representing the origin of the male germline. After meiosis and two additional pollen mitotic divisions, they generate the male gametophyte (pollen/pollen tube) that contains two sperm cells (male gametes; Zhou et al., 2017). *MALE STERILE CONVERTED ANTHER 1* (*MSCA1*) was found to be a key player involved in this process. Mutation of *MSCA1* disturbs male gamete development, generating empty anthers lacking mature pollen. Detailed analysis of the *msca1* phenotype revealed that initiation of stamen primordia was normal in *msca1*, but archesporial cells were not differentiated, indicating that differentiation from a somatic hypodermal to an archesporial cells is blocked (Chaubal et al., 2003). *MSCA1* encodes a glutaredoxin (GRX) protein belonging to CC-type GRXs, which interact with the TGA transcription factor FASCIATED EAR4 in meristems. As a redox regulator, MSCA1 uses glutathione to reduce disulfide bridges of target proteins. Thus, it was speculated that the redox status may play an important role during germ cell fate determination in plants, which represent a distinct mechanism compared with that in animals (Timofejeva et al., 2013; Tsukagoshi, Busch, & Benfey, 2010). In *msca1* flowers, reducing the concentration of oxygen (hypoxia) or H_2O_2 increases the number of germ cells and stimulates superficial germ cell formation in maize. Moreover, the *msca1* mutant phenotype can be rescued. On the other hand, an oxidizing environment causes inhibition of germ cell specification and simultaneously induces ectopic differentiation of cells outside of anther lobes to form archesporials (Kelliher & Walbot, 2012; Whipple, 2012).

Due to technical difficulties, less is known about female germline initiation and the role of ROS is unclear. However, the discovery of MSCA1 and involvement of ROS homeostasis in male germline specification in maize opened a new gate for further investigations and it is reasonable to speculate that similar mechanisms may also exist in female germline initiation in maize and in both germlines in other plant species as well.

Table 1 Overview of molecular players reported in this chapter about the role of ROS in plant reproduction.

Molecular player	Protein family	Functions of molecular players	References
GERMLINE SPECIFICATION			
MSCA1 (1)	CC-type GRX	Archesporial cells differentiation	Chaubal et al. (2003), Kelliher and Walbot (2012), Whipple (2012)
MALE GAMETOPHYTE DEVELOPMENT			
ROXY1/2 (2)	CC-type GRX	Anther lobes or PMC differentiation	Xing and Zachgo (2008)
OsAGO2 (3)	ARGONAUTE	Initiation of tapetum PCD	Zheng et al. (2019)
OsHXK1 (3)	Hexokinase	*oshxk1* can restore *osago2* mutant phenotype	Zheng et al. (2019)
SiIDA (4)	INFLORESCENCE DEFICIENT IN ABSCISSION	Tapetum and septum PCD regulation	Wang et al. (2020)
DWARF (4)	BR biosynthetic gene	Tapetum PCD regulation	Yan et al. (2020)
BZR1 (4)	BR signaling regulator	Tapetum PCD regulation	Yan et al. (2020)
RBOH1 (4)	NADPH oxidase	ROS production/tapetum cell degeneration	Yan et al. (2020)
RBOHE (2)	NADPH oxidase	Tapetum PCD regulation	Chapman et al. (2019)

(continued on next page)

Table 1 *(continued)*

Molecular player	Protein family	Functions of molecular players	References
MYB80 (2)	MYB TF	RBOHE transcriptional level regulation	Chapman et al. (2019)
AMS (2)	ABORTED MICROSPORES TF	RBOHE transcriptional level regulation	Chapman et al. (2019)
PsEND1 (5)	*Pisum sativum* ENDOTHECIUM 1	Tapetum PCD regulation	Hamza et al. (2021)
FEMALE GAMETOPHYTE DEVELOPMENT			
MSD1 (2)	Mn-SOD	Regulates cell division during megagametogenesis	Martin, Distefano, et al. (2013), Martin, Fiol, et al. (2013)
POLLEN ADHESION AND HYDRATION			
KINβγ (2)	Subunit of SNF-related kinase 1	Pollen grain germination on stigma	Halford and Hey (2009)
CATALASE3 (2)	Catalase	ROS reduction	Gao et al. (2016)
ANJEA (2)	*Catharanthus roseus* RLK1-like	Regulates ROS production in papilla cells	Liu et al. (2021), Wang et al. (2017)
FERONIA (2)	*Catharanthus roseus* RLK1-like	Regulates ROS production in papilla cells	Liu et al. (2021), Wang et al. (2017)

(continued on next page)

Table 1 *(continued)*

Molecular player	Protein family	Functions of molecular players	References
RALF23/33 (2)	RAPID ALKALINIZATION FACTORs	Induce ROS production in papilla cells	Liu et al. (2021), Wang et al. (2017)
PCP-Bs (2)	POLLEN COAT PROTEIN B-class CRP peptides	ROS level reduction in papilla cells	Liu et al. (2021), Wang et al. (2017)
SI RESPONSE			
CRK31/41 (2)	CYSTEINE-RICH RECEPTOR-LIKE KINASES	ROS accumulation during SI response	Kodera and Fobis-Loisy (2022)
SC RESPONSE			
SSP (6)	STIGMA-SPECIFIC PEROXIDASE	Recognition between compatible pollen and stigma	McInnis et al. (2006), Zefra et al. (2010), Zinta et al. (2016)
POLLEN GERMINATION			
pad2–1 (2)	GSH deficient mutant	Promotes pollen germination	Parisy et al. (2007), Zechmann et al. (2011)
cad2–1 (2)	GSH deficient mutant	Promotes pollen germination	Garcia-Quiros et al. (2020)
are (4)	flavonoid-3-hydroxylase mutant	Regulates pollen viability, germination and growth	Chapman et al. (2019), Muhlemann et al. (2018)

(continued on next page)

Table 1 *(continued)*

Molecular player	Protein family	Functions of molecular players	References
POLLEN TUBE GROWTH			
AtABCG28 (2)	ATP-binding cassette transporter	Regulates pollen tube growth	Do et al. (2019)
RBOHH/RBOHJ (2)	NADPH oxidase	Regulate pollen tube growth	Kaya et al. (2014), Lassig et al. (2014)
ANXUR1/2 (2)	*Catharanthus roseus* RLK1-like protein	Regulate pollen tube integrity	Boisson-Dernier et al. (2013)
SPERM CELL RELEASE			
MPK4 (2)	MITOGEN-ACTIVATED PROTEIN KINASE	Avoids premature synergid cell PCD	Volz et al. (2022)
FERONIA (2)	*Catharanthus roseus* RLK1-like protein	Pollen tube rupture induction	Huck et al. (2003), Rotman et al. (2003)
ROPGEFs (2)	Guanine nucleotide exchange factors	Interaction partner of FERONIA	Duan et al. (2010)
Plant RHOs (2)	RAC/ROP GTPases	Interaction partner of FERONIA	Duan et al. (2010)

See text for details. Numbers behind molecular players indicate plant species: (1) *Zea mays*; (2) *Arabidopsis thaliana*; (3) *Oryza sativa*; (4) *Solanum lycopersicum*; (5) *Pisum sativum*; (6) *Senecio squalidum*.

3. ROS in gametogenesis

3.1 Male gametophyte development

After male germline initiation, further development of premeiotic archesporial cells results in the formation of PMCs, which will undergo meiosis and each give rise to four haploid microspores (Fig. 1). After two additional rounds of mitosis, mature male gametophytes are generated in anther locules (Chang & Neuffer, 1989). During this process, multiple gene regulatory networks (GRNs) and chemical molecules are involved. The regulation of ROS homeostasis appears as one of the critical aspects of male gametogenesis (Sankaranarayanan et al., 2020).

Redox homeostasis is of vital importance for proper anther and pollen development: not only ROS production systems, but also ROS scavenging mechanisms are required to maintain the redox balance. Spatiotemporal distribution of ROS plays also a critical role during anther development, but the regulatory mechanism of ROS transportation are not well described (Chapman, Muhlemann, Gayomba, & Muday, 2019). Future investigations will show whether mechanisms described below for few species occur also generally in other or all angiosperms.

Glutaredoxins (GRXs) are small oxidoreductases, which are involved in maintaining redox homeostasis (Ray, Huang, & Tsuji, 2012; Shi & Carroll, 2020). They can be subdivided into several subgroups in plants and are of vital importance for oxidative stress responses. Among these subgroups, GRXs containing a CC-motif (CC-type) are plant specific while those containing either CPYC- or CGFS-motifs commonly occur also in prokaryotes and eukaryotes (Meyer, Belin, Delorme-Hinoux, Reichheld, & Riondet, 2012). In *Arabidopsis thaliana*, the CC-type GRXs *ROXY1* and *ROXY2* were found to be involved in anther development. The double mutant *roxy1roxy2* is male sterile and does not produce pollen, while single mutants do not show a difference compared with wild type (WT) plants (Xing & Zachgo, 2008). Interestingly, differentiation of abaxial and adaxial anther lobes were interfered differently: early lobe differentiation was impaired in adaxial lobes, whereas PMC differentiation is defective in abaxial lobes of *roxy1roxy2* plants and PMCs are not capable to enter meiosis. Transcriptomic analysis with WT plants and the double mutant revealed that expression pattern of a large number of anther-specific genes was altered. This indicates that the two GRXs are key regulators during anther development. One conserved glutathione-binding glycine residue is essential for function of ROXY1 suggesting that CC-type GRXs need to interact with glutathione to catalyze biosynthetic reactions. In conclusion, functional analyses of *ROXY1/2* indicates that redox regulation plays critical roles in Arabidopsis male gametogenesis during early PMC differentiation (Xing & Zachgo, 2008).

The rice (*Oryza sativa)* genome encodes 19 ARGONAUTE (AGO) proteins, of which *OsAGO2* is highly expressed in anthers and regulates their development. Mutations of *OsAGO2* leads to overaccumulation of ROS and causes premature initiation of tapetal programmed cell death (PCD) and ultimately in abortion of pollen development. Hexokinase 1 (OsHXK1) was discovered as an interaction partner of OsAGO2, and the expression level of *OsHXK1* is significantly increased in the *ago2–1* mutant. Overexpression of OsHXK1 displays the same phenotype as *ago2–1*, while knockdown of *OsHXK1* in the *ago2–1* background could restore pollen fertility. Direct binding of OsAGO2 to the *OsHXK1* promoter can be detected via chromatin immunoprecipitation (ChIP) assays, suggesting that OsAGO2 regulates the expression of *OsHXK1* via DNA methylation. In summary, OsHXK1 is an important regulator, which controls production of ROS in a spatiotemporal manner leading to proper timing of tapetal PCD, and thus resulting in functional male gametophyte formation in rice (Zheng et al., 2019).

The signaling peptide gene *INFLORESCENCE DEFICIENT IN ABSCISSION* (*IDA*) is widely expressed in *Arabidopsis thaliana*. Together with other members from this gene family, *IDAs* and *IDA-LIKEs* (*IDLs*) are involved in different plant development processes and responses to environmental stimuli, especially during abscission of floral organs after pollination (Aalen, Wildhagen, Sto, & Butenko, 2013). To investigate the function of *IDA* homologs during reproductive development in tomato (*Solanum lycopersicum*), knockout lines of *SiIDA* were generated. *SiIDA* was shown to be expressed in the tapetum, septum and pollen tubes. Mutations of *SiIDA* caused PCD defects in both, tapetum and septum, leading to failure of anther dehiscence and reduction of pollen tube elongation. The phenotypes were linked with insufficient ROS levels, as external application of the mature SiIDA peptide could rescue the pollen tube elongation defect of the mutant and led to enhanced ROS levels in pollen tubes. Moreover, application of ROS scavengers could inhibit the function of *SiIDA*, indicating that *SiIDA* regulates ROS production (Wang et al., 2020). It remained unclear how *SiIDA* regulates ROS production, thus further studies on ROS induction and downstream signaling are necessary for understanding the link between IDA signaling and ROS homeostasis.

It was further shown in tomato that the phytohormone brassinosteroid (BR) is involved in pollen viability and pollen germination through interfering with ROS production and tapetal PCD. Functional analysis of the BR biosynthetic gene *DWARF* (*DWF*) and BR signaling regulator gene *BRASSINAZOLE RESISTANT 1* (*BZR1*) revealed that they interfered with production of ROS and thereby influenced PCD of tapetum cells, resulting in modifications of tapetal cell degeneration. Furthermore, BZR1 can directly bind to the promoter of *RESPIRATORY BURST OXIDASE HOMOLOG 1* (*RBOH1*) in tomato, while RBOH1 is responsible for ROS production at the plasma membrane. All these findings strongly indicate that BZR1-dependent ROS production is of vital importance for tapetum cell degeneration and pollen development (Yan et al., 2020).

RESPIRATORY BURST OXIDASE HOMOLOG E (*RBOHE*), which is expressed in the tapetum of developing anthers, is involved in pollen development in *Arabidopsis thaliana*. During pollen formation, the tapetum needs to be degenerated in order to provide among other lipids, flavonoids and proteins including small cysteine-rich proteins (CRPs) to be deposited on the outer layers of the pollen cell wall (exine). This is required for proper pollen coat development and later functions in pollen-pistil interactions (see 4. below). In the *rbohE* mutant, ROS accumulation is decreased and PCD of tapetum cells is delayed. This results in the formation of an abnormal pollen coat and non-viable pollen. Transcriptional level of *RBOHE* was also found to be regulated by transcription factors *MYB80* and *ABORTED MICROSPORES* (*AMS*), which together regulate tapetum development and pollen formation (Chapman et al., 2019).

ENDOTHECIUM 1 (*PsEND1*) is expressed specifically in pea (*Pisum sativum*) anthers and encodes a protein containing four hemopexin domains, which are responsible for preventing heme-associated oxidative stress in different living creatures. Functional analysis of *PsEND1* revealed that mutation of this gene caused precocious tapetum degradation and impaired pollen development, which are associated with the interference of superoxide anion production and ROS-scavenging enzyme activity alteration. Thus PsEND1 is essential for pollen development by modulating ROS amount during anther development (Hamza et al., 2021).

Olive tree (*Olea europaea*) is an important economic crop for Mediterranean countries and ROS accumulation could also be detected in different reproductive organs. Olive flower development can be divided into five stages, including green bud, white bud, recently opened flower, dehiscent anther and finally, abscised anthers and petals. With histochemical staining analysis, ROS accumulation was detected at the green bud stage as well as at the end of the recently opened flower and dehiscent anther stage. Highest ROS levels were observed in pollen grains, pollen tubes and the stigmatic surface, indicating various functions of ROS during pollen development and during interactions between male and female reproductive tissues (Zafra et al., 2010). The detailed roles of ROS during these processes remained to be elucidated.

In summary, all above findings in different plant species indicated that ROS homeostasis is essential for male gametophyte development. ROS is mainly involved in PMC differentiation and tapetum PCD, which are key steps for successful male gametogenesis. Although investigations were conducted so far only in few plant species, are partly descriptive and identified only few molecular players, we propose that the identified players and mechanisms exist in a similar manner in most (all) other plant species as the above described reproductive processes are highly conserved. Modulation of ROS levels is now necessary to better understand their role (s) for male gametophyte development.

3.2 Female gametophyte development

After initiation of the female germline in ovule primordia, megaspore mother cells (MMCs) undergo meiosis resulting in a linear tetrad of megaspores (Fig. 1). The innermost spore forms a single functional megaspore, while the other three megaspores degenerate via PCD. After three rounds of mitotic nuclear divisions, an embryo sac (female gametophyte) is formed consisting in most angiosperms from a process of cellularization of seven cells (Hater, Nakel, & Gross-Hardt, 2020). In ovules of the model plant *Arabidopsis thaliana*, the haploid female gametophyte is a seven-cell structure including four different cell types, two gametic cells (egg cell and central cell), two synergid cells and three antipodal cells (Webb & Gunning, 1991). The egg cell is located near the micropylar area of the ovule surrounded by two synergid cells forming the egg apparatus, while the di-haploid central cell is located in the middle of the embryo sac and occupies most of the space, while three antipodal cells are located at the chalazal end of the female gametophyte (Drews & Koltunow, 2011). During the process of megagametogenesis (formation of female gametes), ROS is first detected during megaspore PCD. Strong peroxide burst has been reported at the nucellus where dying megaspores are located. Due to the deeply embedded structure of the developing female gametophyte, direct measurement of ROS levels is technically challenging. In plants, superoxide dismutases (SODs), which catalyze the dismutation of superoxide ($O_2^{\cdot -}$), is the first line for cellular defense against ROS. According to their metal cofactors, SOD isozymes can be classified into different groups, iron SODs (Fe-SODs), manganese SODs (Mn-SODs) and copper-zinc SODs (Cu/Zn-SODs; Alscher, Erturk, & Heath, 2002). By using a high throughput forward genetic screen, a female gametophytic mutant, named *maternal effect embryo arrest 33* (*mee33*), was found with defects in female gametophyte transmission (Pagnussat et al., 2005). *mee33* is a transposon insertion line of the Mn-SOD encoding gene *MSD1*, and allelic to the mutant *oiwa* (named from a Japanese ghost story, a woman named Oiwa that was poisoned by her husband and died, which also killed her unborn baby) was recently used for investigating the roles of MSD1 and ROS during female gametogenesis. Notably, *oiwa* was impaired in MSD1 activity and showed abnormally high ROS levels (Martin, Fiol, Sundaresan, Zabaleta, & Pagnussat, 2013). It was shown that this mutation leads to a mitotic arrest during megagametogenesis, as ROS appears to be involved in regulation of cell cycle progression, microtubule organization and cell plate formation, which in the end regulate cell division (Martin, Distefano, et al., 2013; Martin, Fiol, et al., 2013). Thus, ROS homeostasis seems to be essential not only for normal nuclei division, but also for nuclei migration, which in the end guarantees proper patterning and cell fate determination during embryo sac formation.

Female gametogenesis occurs deeply embedded by multiple layers of maternal ovary and ovule tissues, which explains the limited number of investigations on ROS homeostasis during this process. Development of different and novel ROS biosensors (Choi et al., 2021; Erard, Dupre-Crochet, & Nusse, 2018) may strongly improve our understanding of this process in the near future.

4. Multiple functions of ROS during the progamic phase

The progamic phase describes the period between pollination and double fertilization, and is characterized by numerous specific interactions between the male gametophyte with the different maternal tissues of the pistil and ovule, respectively. After anthesis, mature pollen grains land on receptive stigmatic tissue and the interaction between male and female tissue is initiated. This determines whether pollen tubes are formed that are capable of penetrating the stigmatic tissue (Edlund, Swanson, & Preuss, 2004; Hiscock & Allen, 2008). This crosstalk is mediated by intensive signal exchange between pollen grain and stigma. This highly regulated process includes a series of events and occurs in succession including pollen adhesion, hydration, germination, penetration of pollen tubes into female tissues and further growth inside transmitting tracts. The internal signaling mechanism regulating these processes are fairly complex and largely unknown, but a lot of progress has been made in recent years (Cascallares et al., 2020; Dresselhaus & Franklin-Tong, 2013; Zhou & Dresselhaus, 2019). Among all signaling molecules and singling mechanisms involved, the redox network appears to be particularly important.

4.1 Pollen adhesion and hydration

Based on the presence or absence of stigmatic secretion, stigma surface of flowering plants can be classified as wet or dry (Heslop-Harrison & Shivanna, 1977). But so far, most of our knowledge about pollen-stigma interaction was obtained from plants with a dry stigma of the Brassicaceae. After being released from anthers, pollen grains land on stigmatic surfaces and adhere to papilla cells (Fig. 2A). Next, dry pollen take up water (hydration), their metabolism is activated and germination of a pollen tube is initiated (Ma et al., 2012). After the first contact between pollen and female stigma tissue occurred, intensive communication is required to promote compatible, but to prevent incompatible pollen from water uptake. ROS appears to play an important role in this regulatory network.

The first molecular player being identified was KINβγ, a plant-specific subunit of the SNF-related protein kinase 1 complex, which plays critical roles in the regulation of plant development (Halford & Hey, 2009). In Arabidopsis pollen grains, KINβγ is localized to the cytoplasm and nucleus of the vegetative cell, and a mutation of *KINβγ* caused pollen grain germination defects after landing on the stigma. *In vitro* pollen germination of this mutant was not interfered. Further analysis revealed that hydration of mutant pollen grain on stigma was compromised, and this defect can be complemented by external application of water on stigma, indicating that the germination defect was caused by a pollen hydration defect. Morphological analysis showed that mitochondria and peroxisomes were altered in the mutant and the number of these cell organelles was significantly reduced compared with that of WT plants. As mitochondria and peroxisomes are main ROS generation factories in plant cells (see also Chapter 2 in this book), it was speculated that ROS levels were reduced in

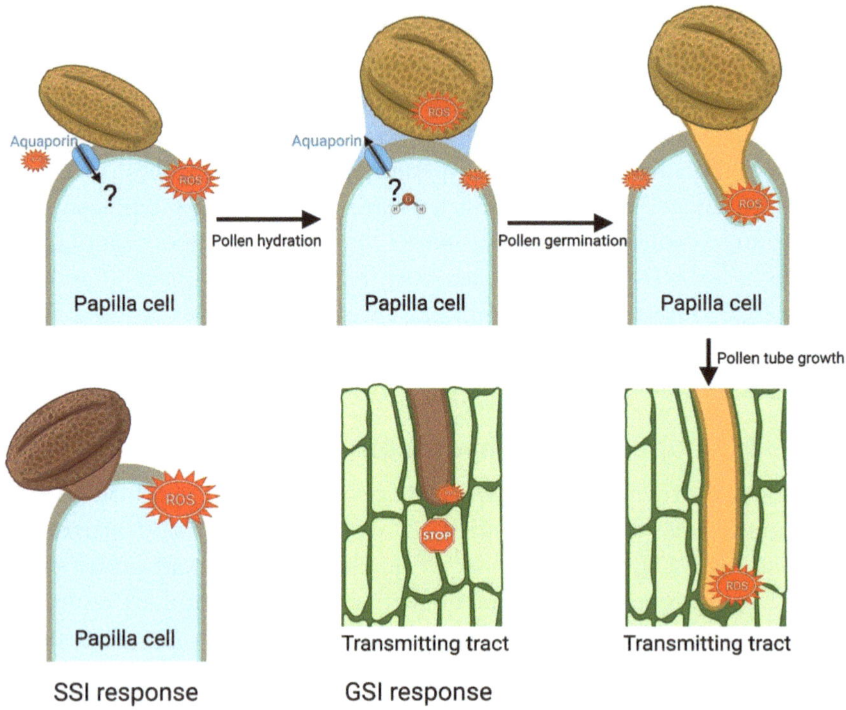

Fig. 2 Schematic model showing ROS during the progamic phase. (A) Stigma papilla cells accumulate ROS during maturation. After pollen adhesion, compatible pollen grains absorb water from the stigma and ROS level are modified (increased/lowered) depending on the plant species. Pollen germinates and penetrates the female tissue between plasma membrane and cell wall of papilla cells and continues growth between transmitting tract cells. ROS accumulation is detectable also in pollen grains and the tip of pollen tubes, respectively. (B) High amount of ROS in stigma papilla cells is involved in the SI response. During the SSI response, ROS level in papilla cells cannot be reduced by incompatible pollination, leading to germination arrest. During the GSI response, ROS at the pollen tube apex play various roles in different plant species ultimately resulting in pollen tube growth arrest inside the transmitting tract.

the *kinβγ* mutant. Indeed, external application of H_2O_2 could partially restore water absorption of mutant pollen grain, while overexpressing of *CATALASE3*, which leads to ROS reduction, resulted in compromised hydration of WT pollen grains on stigmas (Gao et al., 2016). These findings indicate that *KINβγ* is involved in the regulation of ROS levels in pollen grain via mediating the biogenesis of mitochondria and peroxisomes, which are of vital importance for pollen grain hydration on the stigma.

A key role of ROS in pollen hydration was recently elucidated by studying CRPs expressed in pollen grains and papilla cells of the stigma and their interaction with receptor-like kinases of the CrRLK1L (*Catharanthus roseus* RLK1-like) family. In un-pollinated papilla cells of *Arabidopsis thaliana*, the ANJEA-FERONIA (ANJ-FER) receptor kinase complex fulfills the function as a 'stigmatic gatekeeper', which perceives RAPID ALKALINIZATION FACTOR (RALF) peptides RALF23 and RALF33 in an autocrine signaling pathway to induce production of ROS in stigmatic papilla cells. Upon pollination, the POLLEN COAT PROTEIN B-class CRP peptides (PCP-Bs) compete with unrelated CRPs RALF23 and RALF33 on binding with extracellular domains of the ANJ-FER complex, resulting in reduction of stigmatic ROS level and an increase of pollen hydration (Liu et al., 2021; Wang et al., 2017). This investigation provides new insight into how ROS is involved in pollen recognition and hydration in this species. The detailed interaction network of molecular players is illustrated in Zhou, Qu, and Dresselhaus (2021). However, high ROS accumulation in unpollinated stigma was so far only detected in few plant species (Zhang et al., 2020). ROS levels increase after pollination in species like maize (Zhou et al., 2021) and PCP-Bs were reported to exist only in the Brassicaceae. Thus, further clarification is required to investigate whether ROS are widely involved in the regulation/promotion of water uptake after pollen adhesion in flowering plants and whether ROS regulates, for example, the activity of aquaporins (Fig. 2A).

4.2 ROS in self-incompatibility (SI) responses

Most flowering plants produce hermaphroditic flowers containing male and female reproductive organs, which are mature and receptive at the same time. In order to prevent fertilization by alien pollen and to promote outcrossing to individuals of the same species, molecular mechanisms evolved to recognize, accept or reject pollen that lands on stigma. Many angiosperms developed hybridization barriers to prevent fertilization with self-pollen, which is called self-incompatibility (SI). Other plant species evolved mechanism to only accept self-pollen, which makes them self-compatible (SC; Munoz-Sanz, Zuriaga, Cruz-Garcia, McClure, & Romero, 2020).

In the Brassicaceae, most plant species evolved sporophytic self-incompatibility (SSI) at the pollen-stigma interface to avoid inbreeding and to maintain hybrid vigor in the next generation. Recognition between self-pollen and self-stigma in this plant family is often accomplished by interaction between pollen ligand S-locus CRPs (SCRs) and stigma-specific S-locus receptor kinases (SRKs; Leducq et al., 2014; Paetsch, Mayland-Quellhorst, & Neuffer, 2006). The rejection mechanism for self-pollen remained unclear for quite a long time, but recently ROS were found to be involved in this process. Accumulation of ROS in stigma papilla cells can be detected in different Brassicaceae species. After analyzing the expression pattern of fluorescent ROS probes, higher levels of ROS can often be detected at those developmental stages which are considered as 'receptive' to pollen. However, the accumulation pattern of ROS is variable depending on the plant species (Zafra et al., 2010) indicating that this difference could be associated with SI and SC regulation, respectively.

Stigma papilla cells of many Brassicaceae species appear to accumulate large amounts of ROS before pollination (Fig. 2B), which is crucial to accomplish self-incompatibility, to ensure rejection of incompatible self- or related pollen, and to promote compatible pollen acceptance. Transcriptomic analysis of the Brassicaceae SI system revealed that class III peroxidases, which are highly expressed in mature stigmata, might be responsible for ROS homeostasis regulation. In *Arabidopsis thaliana*, the expression levels of two receptor-like kinases from the CYSTEINE-RICH RECEPTOR-LIKE KINASES family (CRK31 and CRK41) were strongly increased upon incompatible pollination (Kodera & Fobis-Loisy, 2022). It was hypothesized that these CRKs belong to the SI response pathway and were associated with ROS accumulation. A similar phenomenon was discovered in Chinese cabbage (*Brassica rapa* ssp. *pekinensis*), as stigmatic ROS levels were reduced after compatible pollination but increased after self-pollination. Application of ROS scavengers or suppressing the expression of RBOHs, which encode plant nicotinamide adenine dinucleotide phosphate (NADPH) oxidases that are responsible for $O_2^{\cdot -}$ production, can both lead to break down of SI (Jimenez-Quesada, Traverso, & Alche Jde, 2016) and, on the contrary, increased ROS levels on stigma resulting in inhibition of germination and penetration of compatible pollen, thus mimicking the incompatible response reaction (Zhang et al., 2021). Thus, in different plant species of the Brassicaceae, ROS accumulates in stigmatic papilla cells and functions in SSI regulation.

While SSI is determined by the sporophytic tissues of the anther and pistil, the response in gametophytic SI (GSI) is determined by the haploid genome of the pollen tube and growth arrest usually occurs in the transmitting tract (Fig. 2B). In both types of SI, PCD of pollen and stigma are of vital importance for pollen rejection (Serrano, Romero-Puertas, Sandalio, & Olmedilla, 2015). In olive (*Olea europaea*), for example, PCD was found to be induced after pollination with self-incompatible pollen. ROS and nitric oxide (NO) are produced, but it remained unclear how they regulate

PCD during self-incompatible pollination. Biochemical studies strongly support the hypothesis that both, $O_2^{\cdot-}$ and NO are essential for inducing PCD in SI responses. Peroxynitrite, a powerful oxidizing and nitrating agent generated by rapid reaction between $O_2^{\cdot-}$ and NO, is produced during pollination and strongly associated with PCD (Serrano, Romero-Puertas, Rodriguez-Serrano, Sandalio, & Olmedilla, 2012). Peroxynitrite mediates PCD during self-incompatible pollination of olive in both, pollen and stigma papilla cells.

During GSI responses in the Solanaceae, the cytotoxic enzyme S-RNase specifically inhibits self-pollen tube elongation. ROS disruption, as a primary target of S-RNase, can be found at the tip of pollen tubes. In self-incompatible pear pollen belonging to the Rosaceae, ROS production was interrupted in mitochondria and the cell wall, which is subsequently followed by termination of Ca^{2+} elevations, depolymerization of the cytoskeleton and later degradation of nuclear DNA (Speranza & Scoccianti, 2012; Wang et al., 2010).

In contrast, in field poppy (*Papaver rhoeas*), which is also self-incompatible, pollination with incompatible pollen grain induces rapid and transient increases of ROS together with NO in pollen tubes and this increase is triggered by upstream Ca^{2+} elevation (Franklin-Tong & Bosch, 2021; Wilkins et al., 2011). Taken together, different ROS levels are involved in similar reproductive responses indicating the diversity of underlying regulatory mechanisms and likely molecular players in different species.

In the Nymphaeaceae, commonly called water lily, hybrid breeding is challenging due to the low compatibility between pollen and stigma of related species/varieties. Transcriptomic analysis of samples at different time points after pollination showed that differentially expressed genes (DEGs) are mainly involved in ROS metabolism, biogenesis of flavonoids and cell wall organization or biogenesis (Sun et al., 2019). Although the precise role of ROS metabolism remained unclear, this study further improved our knowledge of the role of ROS in incompatibility reactions in highly diverse angiosperm families. It is generally thought that the knowledge can be used in the future for overcoming hybridization barriers and to improve plant breeding.

4.3 ROS during self-compatibility (SC) responses

In SI pollination systems, interaction mechanisms between pollen and stigma have been well studied at both, the cellular and molecular level.

However, the interaction of pollen and stigma in self-compatible (SC) pollination is less understood. *Arabidopsis thaliana*, a SC species, provides an excellent system for detailed functional analysis (Gao et al., 2016). ROS and NO together appear to regulate pollen-pistil interaction in this and related plant species (Duan et al., 2020; McInnis, Desikan, Hancock, & Hiscock, 2006; Prado, Colaço, Moreno, Silva, & Feijó, 2008). Before pollen grains lands on stigma, higher level of ROS in papilla cells enhance the activity of STIGMA-SPECIFIC PEROXIDASE (SSP) in *Senecio squalidus* (Asteraceae), while pollen retain high NO levels. After pollination, interaction between pollen and papilla cells lead to a dramatic decline of both, ROS levels and SSP activity, thus promoting recognition between compatible pollen and pistils (McInnis et al., 2006; Zafra et al., 2010; Zinta, Khan, AbdElgawad, Verma, & Srivastava, 2016).

High level of peroxidase activity can also be detected on angiosperm stigmas when they are receptive to pollen. In *Senecio squalidus*, high amount of ROS, especially H_2O_2, can be observed at stigmatic papilla cells and its amount is reduced after getting into contact with compatible pollen. The regulation of this process requires cooperation with NO derived from pollen, as treatment of stigma with NO resulted in reduced amount of stigmatic ROS/H_2O_2 (McInnis et al., 2006). In summary, pollen and stigma are primed already before pollination to quickly respond to compatible pollen and ROS/NO levels appear to play an important role in the associated cellular response.

4.4 Pollen germination

ROS is involved in tissue loosening at the onset of pollen germination and pollen tube emergence. The microfibrillar network of the pollen intine requires loosening of its cell wall material to form a bulge outward, which allows pollen tube protrusion. This process is similar with cell wall loosening during growth extension of other plant tissues, but more spatially controlled (Cascallares et al., 2020; Smirnova, Matveyeva, & Yermakov, 2014).

During pollen germination, the *phytoalexin-deficient* (*pad2–1*) mutant was found in *Arabidopsis thaliana* to be involved in this process as the mutant shows a poor pollen germination ratio. Previous studies showed that GSH is essential for detoxification of ROS/H_2O_2, which is required for survival of living creatures. As a glutathione (GSH) deficient mutant, the phenotype of *pad2–1* indicates that GSH is essential for proper pollen germination. On the other hand, external application of buthionine sulfoximine (BSO), an inhibitor of GSH synthesis, also impaired the pollen germination ratio *in vitro*, while application of both, reduced GSH with BSO, can restore the pollen germination ratio and GSH level (Parisy et al., 2007; Zechmann, Koffler, & Russel, 2011). This demonstrated that GSH synthesis inhibition is responsible for the defect of pollen germination *in vitro*. A similar defect on pollen germination and pollen tube elongation can be observed in another glutathione-deficient *cad2–1* (cadmium sensitive2–1) mutant. Expressing roGFP2 (redox-sensitive green fluorescent protein) reporter in WT and *cad2–1* mutant plants clearly showed reduced glutathione (GSH) and increased glutathione disulfide (GSSG) in mutant flowers compared with those of WT plants. Reproductive organs of the *cad2–1* mutant, including stigma, style, anther, pollen grains and pollen tubes, were more oxidized than WT flower organs. Reduced pollen germination ratio and earlier pollen tube arrest can be detected in *cad2–1* indicating that the accumulation of glutathione and high GSH/GSSG ratio is of vital importance for proper pollen germination and pollen tube growth (Garcia-Quiros, Alche, Karpinska, & Foyer, 2020).

In *Nicotiana tabacum*, stigma exudates were collected at different developmental stages and compared. By using electron paramagnetic resonance (EPR) and colorimetric analyses methods, it was found that $O_2^{\cdot-}$ and H_2O_2 levels were remarkably decreased during pistil maturation, while the activity of SOD was increased. The SOD enzyme exists in three different isozymes, Fe-SOD, Cu/Zn-SOD and Mn-SOD. ROS balance was shifted because of selective suppression of Cu/Zn-SOD activity, and this shift reduced pollen germination ratio and fertilization efficiency. This indicates that H_2O_2 in stigma exudate is necessary for accelerating pollen germination and ensures successful fertilization. Moreover, NADPH oxidase inhibition can decrease $O_2^{\cdot-}$ production and pollen germination was slowed down, but seed set was not interfered (Breygina, Schekaleva, Klimenko, & Luneva, 2022). Thus, the role of $O_2^{\cdot-}$ is not of vital importance in this case, indicating that there exist various functions of the different ROS during pollen germination.

Flavonols and their glycosylated derivatives accumulate in plant reproductive organs, particularly in pollen, anther and stigma (Muhlemann, Younts, & Muday, 2018; Stracke et al., 2010). Recent studies in different plant species indicated that flavonols may act as regulators of ROS-dependent processes. In the tomato mutant *anthocyanin reduced* (*are*), flavonol accumulation was reduced, and mutants produced less pollen with decreased viability, bad germination, slow growth and more fragile pollen tubes. ROS measurement with fluorescence probes showed that ROS levels were inversely proportional to flavonol levels, indicating that flavonol could act as ROS scavenger in pollen (Chapman et al., 2019; Muhlemann et al., 2018).

In kiwifruit, very early ROS production after pollen grain re-activation can be detected, and ROS were shown to accumulate at the germination apertures, suggesting that ROS are closely associated with pollen tube emergence (Speranza, Crinelli, Scoccianti, & Geitmann, 2012). Similar regulatory mechanism can also be found in tobacco, indicating conserved function of ROS during pollen tube emergence (Smirnova, Matveeva, Polesskaia, & Ermakov, 2009).

ROS interact with different chemical molecules during regulation of pollen germination. Selenium is a trace element, which is also essential for human nutrition, but it is not clear whether selenium also serves as essential micronutrient in plants. However, it has been reported to play a role during pollen germination (Tedeschini et al., 2015). Selenium can enhance stress tolerance via regulating the production and elimination of ROS, and thereby also influences ROS levels required for the crosstalk between pollen and stigma, and promotion of pollen tube growth. It is not clear how selenium promotes pollen germination while serving as a ROS inhibitor. To investigate the underlying mechanism(s), two olive cultivars (*Arbequina* and *Maurino*) were used for external selenium application. It was found that selenium enrichment in non-stressed plants triggered generation of more viable pollen, which also germinated better, while in non-stressed plants this effect was less significant (Del Pino et al., 2019; Tedeschini et al., 2015). This knowledge is thought to be usable to improve the reproduction rate and to achieve a higher yield of olive in the future.

As described above, together with ROS, NO appears as a key regulator of pollen tube growth in angiosperms, but the roles of both molecules during reproduction in gymnosperms remained unclear. Investigations of pollen tube elongation using Arizona cypress (*Cupressus arizonica Greene*) detected both, ROS and NO in nuclei of generative cells and accumulate at the tip region of pollen tubes. Treatment with a NO scavenger and NO synthesis inhibitor, respectively, reduced the pollen germination ratio and pollen tube growth. Additionally, the morphology of the pollen tube was severely interfered, generating shorter pollen tubes with abnormal branching and/or swollen pollen tube tips. The similar phenomenon can be observed after treatment using an NADPH oxidase inhibitor: the actin cytoskeleton pattern was disrupted in pollen tubes (Pasqualini et al., 2015), indicating that ROS and NO play important roles as signaling molecules in pollen germination and pollen tube growth in gymnosperms. Similar to cypress, ROS were also found to be involved in pollen germination and pollen tube growth regulation in blue spruce. ROS are secreted from pollen grain of blue spruce during early stage of activation, and the different ROS showed varying localization patterns in pollen tubes and appear to play also different roles during this process. Intracellular ROS, especially $O_2^{\cdot-}$, is necessary for pollen germination, while extracellular ROS, for instance $OH^{\cdot}$, is not required for this process. External application of H_2O_2 does not have an effect on germination efficiency but accelerates pollen tube growth in a concentration-dependent manner. Treatment with different inhibitors further demonstrated that ROS reduction leads to depolarization of the pollen tube plasma membrane and a block of pollen germination (Maksimov, Evmenyeva, Breygina, & Yermakov, 2018).

4.5 Pollen tube growth and integrity maintenance

Besides pollen tube emergence, ROS are also intensively involved in pollen tube elongation (Fig. 3). Pollen tubes belong to the cell types with a polar growth behavior: they possess a typical tip structure and polar cell expansion is limited to the tip region. To keep cell (wall) integrity, cell expansion needs to be precisely regulated in order to generate a correct shape, size and to withstand the immense internal turgor pressure.

Previous studies showed that apoplastic ROS homeostasis is modulated by NADPH oxidases as well as by secreted type III peroxidases and has a significant influence on cell wall property and cell expansion. Besides, ROS fulfills its function by regulating oscillation of Ca^{2+} and pH (Mangano, Juarez, & Estevez, 2016), and thus together regulates pollen tube polarity growth and integrity (Fig. 3).

The involvement of ROS in tip-growth processes is accomplished either by accumulation at the tip region, where it functions together with calcium channel activation, or by interaction with other key molecules such as NO (Potocky, Jones, Bezvoda, Smirnoff, & Zarsky, 2007). In *Lilium longiflorum*, NO is produced at the sub-apical region of pollen tubes and functions as a negative regulator of pollen tube growth (Prado, Porterfield, & Feijo, 2004). NO is generally known to be involved in stress responses in different plants. In *Camellia sinensis*, for example, NO regulates pollen germination and pollen tube growth via cGMP (cyclic guanosine monophosphate) under low-temprature stress (Wang et al., 2012). Changes in the transcript and miRNA composition of pollen tubes were discovered under cold and NO treatment,

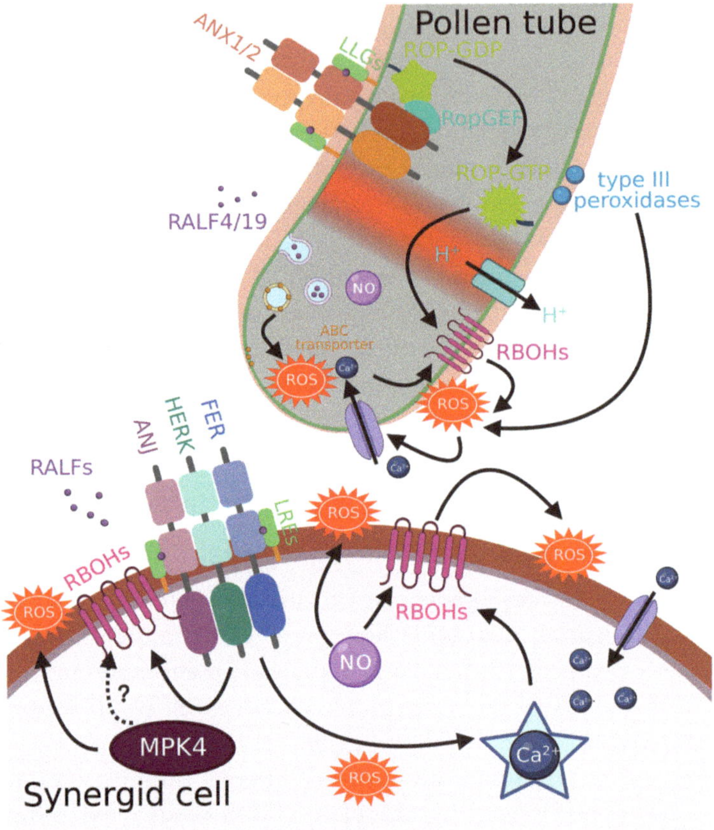

Fig. 3 Model of the ROS regulatory network during pollen tube growth and reception at synergid cells. (A) ROS homeostasis is required for pollen tube integrity and polarity regulation during cell elongation (polar tip growth). Apoplastic ROS levels are generated and modified by RBOHs and type III peroxidases. Calcium signaling along a tip-gradient is required for fine-tuning RBOH activity. Cooperation of ROS and NO near the sub-apical region is important for pollen tube growth. ABC transporters located at tip localized vesicles are involved in regulating ROS production in the tip and an alkaline band appears in the subapical pollen tube region generated by proton pumps. CrRLK1L receptor-like kinases ANX1/2 form complexes with LLG GPI-anchor proteins together with pollen tube secreted ligands from the RALF family, thereby activating a ROPGEFs-ROP/RAC signaling pathway that likely regulates RBOH activity. (B) After pollen tube arrival, synergid cells will go through PCD to prevent multiple pollen tube penetration into one embryo sac. In this process, MPK4 and NO are involved in ROS level regulation, while RBOHs generating ROS are likely regulated by CrRLK1L receptor-like kinase complexes consisting of FER/ANJ/HERK and the co-receptor LRE (a GPI-anchor protein of the LLG family). RALFs from arriving pollen tubes modulate activity of the complex and thus ROS levels directly, while downstream calcium spiking indirectly regulates ROS synthesis.

which further demonstrated that NO plays critical roles during pollen tube growth regulation in *Camellia sinensis* (Pan et al., 2016; Xu et al., 2021). NO production was also found in pollen grains of olive (*Olea europaea L.*), which correlates with pollen viability. Exogenous application of NO inhibits pollen germination and pollen tube growth, while treatment of NO donors or NO scavengers leads to morphology change of pollen tubes (Jimenez-Quesada et al., 2017). In addition, ROS and NO also cooperate with each other at the interface between male and female cells during the progamic phase, where pollen-derived NO counteracts abundant ROS from the stigma resulting in pollen tube penetration into female tissues (Hiscock, Bright, McInnis, Desikan, & Hancock, 2007). Recent studies also revealed that ROS and NO were regulated by other signaling pathways. In tobacco, pollen maturation, germination and pollen tube elongation were also found to be altered by exogenous application of polyamines via regulating the metabolism of both, NO and ROS (Benko, Jee, Kaszler, Feher, & Gemes, 2020). This indicated that a highly complex signaling mechanism is required for maintaining the cooperation between the different signaling molecules.

In the plant kingdom, tip-accumulation of ROS can be found in different tip growing cells including pollen tubes and root hairs (Mangano et al., 2016). This tip-accumulation pattern, particularly that of H_2O_2, activates downstream signaling to regulate directional growth·H_2O_2 is more stable than other ROS and therefore can also be used as a signaling molecule. The formation of such tip-localized pattern of H_2O_2 in polar growing plant cells has been discussed for a long time, as chemical properties of H_2O_2 allow rapid diffusion (Bienert, Schjoerring, & Jahn, 2006). In pollen tubes, tip-localization of ROS cannot be established via subcellular localization of its producing enzymes like NADPH oxidase, which catalyzes the production of $O_2^{\cdot-}$ (Ushio-Fukai, 2009), but instead pollen tubes involve polyamine oxidase (PAO) for ROS generation, which uses polyamines as substrate to produce H_2O_2 (Wu et al., 2010). Transportation of polyamines in pollen tubes occurs via ABC (ATP-binding cassette) transporters, which are generally involved in protecting cells by removing toxic molecules. Members of the ABCG subfamily of ABC transporters are expressed in pollen tubes and may act there as polyamine transporters thus supporting H_2O_2 production (Do et al., 2019). The ABC transporter *AtABCG28*, which is expressed in pollen grains and pollen tubes, was found to be involved in ROS homeostasis via regulating ROS status in *Arabidopsis thaliana*. AtABCG28 localizes to secretory vesicles in the tip region of growing pollen tubes containing polyamines. Mutation of

AtABCG28 caused pollen tube growth defects and male sterility, as tip accumulation of polyamine and ROS was impaired. Ectopic expression of *AtABCG28* in root hairs was shown to promote tip-localization of ROS under high-pH conditions (Do et al., 2019). This indicated that the gene is directly responsible for ROS homeostasis regulation in pollen tubes via polyamines and that the spatial organization mechanism of ROS is regulated by an ABC transporters.

Enzymatic ROS production is involved in pollen tube tip growth, but a ROS gradient in pollen tubes is required for pollen tube elongation. In *Arabidopsis thaliana*, pollen tube tip growth of the NADPH oxidase double mutants *rbohH rbohJ* was severely impaired due to defects of ROS accumulation in pollen tubes. During this regulation process, calcium signaling is also involved, as both RBOHH and RBOHJ contain EF-hand motifs, which are capable of binding calcium. Point mutation of this motif causes defects of Ca^{2+}-induced ROS production and failure of mutant phenotype complementation. Thus, ROS production by RBOHH and RBOHJ is essential for pollen tube tip growth, and Ca^{2+} signaling is involved in modulation or fine-tuning of this regulatory process (Kaya et al., 2014). Recent studies showed that post-translational modifications (PTMs) of proteins are also involved in regulating ROS production together with NO during plant development and immune responses (Begara-Morales et al., 2014; Palma et al., 2020; Yun et al., 2011). Although evidence that PTM regulates ROS production via NO during plant reproduction is missing, it is likely that similar regulatory mechanism also exist during plant reproduction.

Tip growth of pollen tubes occurs in an oscillatory manner, which requires precise coordination between cell expansion and exocytosis-mediated new cell wall material deposition (Chebli, Kaneda, Zerzour, & Geitmann, 2012). Defects in this coordination leads to reduction of pollen tube growth and cell wall integrity, ultimately resulting in premature pollen tube burst. In the above described *rbohH rbohJ* double mutant, oscillation pattern of pollen tube growth was impaired, generating faster growing pollen tubes that finally burst (Lassig, Gutermuth, Bey, Konrad, & Romeis, 2014). Such growth behavior of the double mutant indicated that ROS can reduce the maximal growth speed of pollen tubes, allowing deposition of sufficient new cell wall material via exocytosis. Both, RBOHH and RBOHJ are regulated by CrRLK1L receptor-like kinases ANXUR1 and ANXUR2, which are co-localized at the plasma membrane of the pollen tube tip region. Interaction between them generates a machinery that coordinates pollen tube growth and pollen tube integrity (Boisson-Dernier et al., 2013).

In tobacco, tip-localized ROS were detected in pollen tubes via the general ROS dye chloromethyl-2′,7′-dichlorodihydrofluorescein diacetate (CM-H2DCFDA), while $O_2^{\cdot -}$ production can specifically be detected by nitroblue tetrazolium (NBT) and H_2O_2 levels by diaminobenzidine (DAB; Potocky et al., 2007). To further analyze the origin of ROS in pollen tubes, knockdown of a pollen-specific NADPH oxidase was generated, resulting in pollen tube growth inhibition in tobacco. Application of ROS scavengers and the NADPH oxidase inhibitor diphenylene iodonium chloride (DPI) were further shown to inhibit pollen tube growth and ROS production. External application of H_2O_2 could rescue growth inhibition, while external $CaCl_2$ application caused NBT reduction at the tip of pollen tubes, suggesting that Ca^{2+} can increase the activity of pollen NADPH oxidases (Potocky et al., 2007). In conclusion, tip-localized ROS, which are produced by NADPH oxidases, is in concert with other small molecules like NO, H^+ and Ca^{2+} of key importance to secure normal pollen tube growth.

5. Function of ROS during double fertilization

5.1 Pollen tube guidance

Guidance of pollen tubes towards and inside ovules has been well studied during the past two decades, but the correlation with ROS homeostasis remained unclear. Little is known about the involvement of ROS during guidance through the transmitting tract and towards ovules by sporophytic attraction signals. In the haploid female gametophyte, synergid cells of the embryo sac serve as glandular cells and are responsible for pollen tube attraction at the micropylar region of the ovule (Zhou & Dresselhaus, 2019). After sperm cell release and fertilization accomplishment, the guidance cues are removed and the persistent synergid cell will go through PCD and/or fuse with the large central cell to avoid polytubey (growth of multiple pollen tubes inside the female gametophyte) to indirectly prevent polyspermy (Maruyama et al., 2015; Yu et al., 2021). Thus, to achieve successful fertilization, tight coordination of synergid mediated pollen tube attraction, receptive synergid degeneration, and elimination of the persistent synergid is required. In *Arabidopsis thaliana*, *MITOGEN-ACTIVATED PROTEIN KINASE 4* (*MPK4*) is involved in cell death regulation. During this process, synergid preservation and ROS homeostasis were shown to be closely linked and maintained by *MPK4*. Mutation of *MPK4* caused increased ROS levels in synergid cells, and as a consequence, synergid cells prematurely undergo PCD before pollen tube reception. Treatment of ROS scavengers can restore ROS homeostasis in the *mpk4* mutant, leading to synergid maintenance and normal pollen tube perception (Volz, Harris, Hirt, & Lee, 2022).

5.2 Pollen tube reception and sperm cell release

Upon arrival, pollen tube needs to receive signals from female gametophyte to induce tip burst and to release sperm cells (pollen tube reception) for fertilization (for detailed molecular interactions see Fig. 3). In *Arabidopsis thaliana*, the CrRLK1L family member FERONIA (FER) is involved in pollen tube rupture induction. The *fer* mutant showed pollen tube overgrowth inside the female gametophyte due to failure(s) to induce pollen tube rupture (Huck, Moore, Federer, & Grossniklaus, 2003; Rotman et al., 2003). In roots, it has been demonstrated that FER interacts with guanine nucleotide exchange factors (ROPGEFs) and activates RAC/ROP GTPases (plant RHOs). It's also known that FER functions as a cell surface regulator in FER-ROPGEF-RAC/ROP complexes to induce NADPH oxidase mediated ROS production, which in the end regulates polar growth of root hairs (Duan, Kita, Li, Cheung, & Wu, 2010). Apparently, the FER-ROPGEF-RAC/ROP-NADPH pathway exists also in the embryo sac to regulate pollen tube perception. FER signaling establishes high level of ROS at the entrance of the embryo sac, which are required for pollen tube rupture and sperm cell release in a Ca^{2+} dependent manner involving Ca^{2+} channel activation (Duan et al., 2014). Upon contact of pollen tubes with synergid cells, Ca^{2+} spiking is induced in synergid cells and its amplitude correlates with PCD (Denninger et al., 2014; Hamamura et al., 2014). RBOH-dependent ROS production in synergid cells can be inhibited by Diphenyleneiodonium preventing NADHP-dependent ROS biosynthesis (Chapman et al., 2019). As synergid cells are responsible for pollen tube attraction at the micropylar region and are also involved in pollen tube burst and sperm cell release, it is very likely that ROS produced in synergid cells by RBOHs play a key role for pollen tube perception. Whether all and/or specific ROS are involved in these processes, and how ROS mediates sperm cell discharge is a task for further experimentation.

Little is known about the role of ROS after sperm cell release and its involvement in gamete fusion and fertilization-induced activation of embryo, as well as endosperm development (Dresselhaus & Jurgens, 2021). In rice, it was shown that egg cells contain large amounts of ROS and that these levels progressively decrease in zygotes after fertilization. Inhibition of

ROS production led to an arrest of zygote and early embryo development, respectively (Rattanawong et al., 2021), indicating that ROS likely play also an important role after fertilization.

6. Conclusions and perspectives

In conclusion, we have summarized that ROS play multiple roles during reproduction in plants starting from male and female germline initiation, followed by male and female gametogenesis, pollen-stigma interaction, progamic phase, until fertilization. Spatiotemporal dynamic changes in ROS levels, regulation of antioxidant molecules and the corresponding biosynthesis/decay enzyme activities control many aspects of plant reproduction (Figs. 1–3). We have learned that ROS production and scavenging via different enzymes needs to be tightly coordinated to generate a stable, but also rapidly altered ROS status, which can modulate all of the above-described processes. ROS are clearly associated with reproductive PCD including degeneration of tapetal cell layer in anthers, degeneration of non-functional megaspores, degeneration of the receptive synergid cell during pollen tube burst and disintegration of the persistent synergid cell after successful fertilization. ROS are also key players in cell differentiation during reproduction and cell wall integrity regulation during pollen tube invasion and further growth through the transmitting tract.

How ROS-regulated redox signaling is regulated in reproductive cells is largely unknown and only few downstream players like CC-GPXs have been studied in few reproductive processes. Similarly, few SODs have been studied so far and reports about the many other known ROS enzymes like apoplastic type III peroxidases and 5-lipoxygenases are scarce. Quantification of ROS levels during reproduction appears already as a limiting factor for further studies, especially as female gametogenesis, the progamic phase and fertilization occur deeply embedded inside maternal plant tissues. Traditional ROS detectors (including above mentioned DAB, NBT, CM-H2DCFA and other fluorescent probes) are not precise enough to study the dynamics of ROS changes. During the last decade, several ROS biosensors were developed for more detailed analysis of ROS levels in living plant and animal cells and tissues (Bogdanova, Schultz, & Belousov, 2017; Choi et al., 2021; Erard et al., 2018; Ortega-Villasante, Buren, Baron-Sola, Martinez, & Hernandez, 2016; Ortega-Villasante, Buren, Blazquez-Castro, Baron-Sola, & Hernandez, 2018). These cytoplasmic and novel apoplastic biosensors will significantly contribute to a more precise understanding of ROS during plant reproduction and provide possibilities to quantify and study the dynamics of different ROS at every step during reproduction. A major future challenge will be to precisely elucidate ROS target molecules and clarify their functional mechanisms, which will finally give rise to a comprehensive ROS regulated network for plant reproduction. During plant development, TFs seem to represent major targets in the cytoplasm/nucleoplasm (see Chapter 4 This Volume), but as outlined above during reproduction apoplastic ROS appears especially important. The potential to use ROS in agriculture has also been demonstrated recently when modulated ROS levels in pollen tubes led to the induction of haploid plants (Jiang et al., 2022). This example indicates that a better understanding of the ROS network in reproduction has a great potential to promote plant breeding in the future.

Acknowledgments

Work in the Dresselhaus lab on CRP signaling during reproduction is supported by the German Research Foundation DFG via Collaborative Research Center SFB924 and on ROS via DFG Research Unit ICIPS.

References

Aalen, R. B., Wildhagen, M., Sto, I. M., & Butenko, M. A. (2013). *IDA: A peptide ligand regulating cell separation processes in Arabidopsis. Journal of Experimental Botany, 64*, 5253–5261.

Alscher, R. G., Erturk, N., & Heath, L. S. (2002). Role of superoxide dismutases (SODs) in controlling oxidative stress in plants. *Journal of Experimental Botany, 53*, 1331–1341.

Begara-Morales, J. C., Sanchez-Calvo, B., Chaki, M., Valderrama, R., Mata-Perez, C., Lopez-Jaramillo, J., ... Barroso, J. B. (2014). Dual regulation of cytosolic ascorbate peroxidase (APX) by tyrosine nitration and S-nitrosylation. *Journal of Experimental Botany, 65*(2), 527–538. https://doi.org/10.1093/jxb/ert396

Benko, P., Jee, S., Kaszler, N., Feher, A., & Gemes, K. (2020). Polyamines treatment during pollen germination and pollen tube elongation in tobacco modulate reactive oxygen species and nitric oxide homeostasis. *Journal of Plant Physiology, 244*, 153085. https://doi.org/10.1016/j.jplph.2019.153085

Berger, F., & Twell, D. (2011). Germline specification and function in plants. *Annual Review of Plant Biology, 62*, 461–484.

Bienert, G. P., Schjoerring, J. K., & Jahn, T. P. (2006). Membrane transport of hydrogen peroxide. *Biochimica et Biophysica Acta, 1758*, 994–1003.

Bogdanova, Y. A., Schultz, C., & Belousov, V. V. (2017). Local generation and imaging of hydrogen peroxide in living cells. *Current Protocols in Chemical Biology, 9*, 117–127.

Boisson-Dernier, A., Lituiev, D. S., Nestorova, A., Franck, C. M., Thirugnanarajah, S., & Grossniklaus, U. (2013). ANXUR receptor-like kinases coordinate cell wall integrity with growth at the pollen tube tip via NADPH oxidases. *PLoS Biology, 11*, e1001719.

Breygina, M., & Klimenko, E. (2020). ROS and ions in cell signaling during sexual plant reproduction. *International Journal of Molecular Sciences, 21*.

Breygina, M., Schekaleva, O., Klimenko, E., & Luneva, O. (2022). The balance between different ROS on tobacco stigma during flowering and its role in pollen germination. *Plants (Basel), 11*.

Cascallares, M., Setzes, N., Marchetti, F., Lopez, G. A., Distefano, A. M., Cainzos, M., ... Pagnussat, G. C. (2020). A complex journey: Cell wall remodeling, interactions, and integrity during pollen tube growth. *Frontiers in Plant Science, 11*, 599247.

Chang, M. T., & Neuffer, M. G. (1989). A simple method for staining nuclei of mature and germinated maize pollen. *Stain Technology, 64*, 181–184.

Chapman, J. M., Muhlemann, J. K., Gayomba, S. R., & Muday, G. K. (2019). RBOH-dependent ROS synthesis and ROS scavenging by plant specialized metabolites to modulate plant development and stress responses. *Chemical Research in Toxicology, 32*, 370–396.

Chaubal, R., Anderson, J. R., Trimnell, M. R., Fox, T. W., Albertsen, M. C., & Bedinger, P. (2003). The transformation of anthers in the msca1 mutant of maize. *Planta, 216*, 778–788.

Chebli, Y., Kaneda, M., Zerzour, R., & Geitmann, A. (2012). The cell wall of the Arabidopsis pollen tube—Spatial distribution, recycling, and network formation of polysaccharides. *Plant Physiology, 160*, 1940–1955.

Choi, D. W., Roh, Y. J., Kim, S., Lee, H. M., Kim, M., Shin, D., ... Lee, B. C. (2021). Development of a novel fluorescent biosensor for dynamic monitoring of metabolic methionine redox status in cells and tissues. *Biosensors and Bioelectronics, 178*, 113031.

Crepet, W. L. (2000). Progress in understanding angiosperm history, success, and relationships: Darwin's abominably "perplexing phenomenon". *Proceedings of the National Academy of Sciences of the United States of America, 97*, 12939–12941.

Del Pino, A. M., Regni, L., D'Amato, R., Tedeschini, E., Businelli, D., Proietti, P., & Palmerini, C. A. (2019). Selenium-enriched pollen grains of Olea europaea L.: Ca(2+) signaling and germination under oxidative stress. *Frontiers in Plant Science, 10*, 1611.

Denninger, P., Bleckmann, A., Lausser, A., Vogler, F., Ott, T., Ehrhardt, D. W., ... Grossmann, G. (2014). Male–female communication triggers calcium signatures during fertilization in *Arabidopsis*. *Nature Communications, 5*, 4645.

Do, T. H. T., Choi, H., Palmgren, M., Martinoia, E., Hwang, J. U., & Lee, Y. (2019). Arabidopsis ABCG28 is required for the apical accumulation of reactive oxygen species in growing pollen tubes. *Proceedings of the National Academy of Sciences of the United States of America, 116*, 12540–12549.

Dresselhaus, T., & Franklin-Tong, N. (2013). Male-female crosstalk during pollen germination, tube growth and guidance, and double fertilization. *Molecular Plant, 6*, 1018–1036.

Dresselhaus, T., & Jurgens, G. (2021). Comparative embryogenesis in angiosperms: Activation and patterning of embryonic cell lineages. *Annual Review of Plant Biology, 72*, 641–676.

Dresselhaus, T., Sprunck, S., & Wessel, G. M. (2016). Fertilization mechanisms in flowering plants. *Current Biology, 26*, R125–R139.

Drews, G. N., & Koltunow, A. M. (2011). The female gametophyte. *Arabidopsis Book, 9*, e0155.

Duan, Q., Kita, D., Johnson, E. A., Aggarwal, M., Gates, L., Wu, H. M., & Cheung, A. Y. (2014). Reactive oxygen species mediate pollen tube rupture to release sperm for fertilization in Arabidopsis. *Nature Communications, 5*, 3129.

Duan, Q., Kita, D., Li, C., Cheung, A. Y., & Wu, H. M. (2010). FERONIA receptor-like kinase regulates RHO GTPase signaling of root hair development. *Proceedings of the National Academy of Sciences of the United States of America, 107*, 17821–17826.

Duan, Q., Liu, M.-C. J., Kita, D., Jordan, S. S., Yeh, F.-L. J., Yvon, R., ... Cheung, A. Y. (2020). FERONIA controls pectin- and nitric oxide-mediated male–female interaction. *Nature, 579*(7800), 561–566.

Edlund, A. F., Swanson, R., & Preuss, D. (2004). Pollen and stigma structure and function: The role of diversity in pollination. *Plant Cell, 16*(*Suppl*), S84–S97.

Erard, M., Dupre-Crochet, S., & Nusse, O. (2018). Biosensors for spatiotemporal detection of reactive oxygen species in cells and tissues. *The American Journal of Physiology-Regulatory, Integrative and Comparative Physiology, 314*, R667–R683.

Franklin-Tong, N., & Bosch, M. (2021). Plant biology: Stigmatic ROS decide whether pollen is accepted or rejected. *Current Biology, 31*, R904–R906.

Gao, X. Q., Liu, C. Z., Li, D. D., Zhao, T. T., Li, F., Jia, X. N., ... Zhang, X. S. (2016). The Arabidopsis KINbetagamma subunit of the SnRK1 complex regulates pollen hydration on the stigma by mediating the level of reactive oxygen species in pollen. *PLoS Genetics, 12*, e1006228.

Garcia-Quiros, E., Alche, J. D., Karpinska, B., & Foyer, C. H. (2020). Glutathione redox state plays a key role in flower development and pollen vigour. *Journal of Experimental Botany, 71*(2), 730–741. https://doi.org/10.1093/jxb/erz376

Halford, N. G., & Hey, S. J. (2009). Snf1-related protein kinases (SnRKs) act within an intricate network that links metabolic and stress signalling in plants. *Biochemical Journal, 419*, 247–259.

Hamamura, Y., Nishimaki, M., Takeuchi, H., Geitmann, A., Kurihara, D., & Higashiyama, T. (2014). Live imaging of calcium spikes during double fertilization in *Arabidopsis*. *Nature Communications, 5*, 4722.

Hamza, R., Roque, E., Gomez-Mena, C., Madueno, F., Beltran, J. P., & Canas, L. A. (2021). PsEND1 is a key player in pea pollen development through the modulation of redox homeostasis. *Frontiers in Plant Science, 12*, 765277.

Hater, F., Nakel, T., & Gross-Hardt, R. (2020). Reproductive multitasking: The female gametophyte. *Annual Review of Plant Biology, 71*, 517–546.

Heslop-Harrison, Y., & Shivanna, K. (1977). The receptive surface of the angiosperm stigma. *Annals of Botany (Lond), 41*, 1233–1258.

Higashiyama, T., & Takeuchi, H. (2015). The mechanism and key molecules involved in pollen tube guidance. *Annual Review of Plant Biology, 66*, 393–413.

Hiscock, S. J., & Allen, A. M. (2008). Diverse cell signalling pathways regulate pollen-stigma interactions: The search for consensus. *New Phytologist, 179*, 286–317.

Hiscock, S. J., Bright, J., McInnis, S. M., Desikan, R., & Hancock, J. T. (2007). Signaling on the stigma: Potential new roles for ROS and NO in plant cell signaling. *Plant Signaling & Behavior, 2*, 23–24.

Huck, N., Moore, J. M., Federer, M., & Grossniklaus, U. (2003). The Arabidopsis mutant feronia disrupts the female gametophytic control of pollen tube reception. *Development, 130*, 2149–2159.

Jimenez-Quesada, M. J., Carmona, R., Lima-Cabello, E., Traverso, J. A., Castro, A. J., Claros, M. G., & Alche, J. D. (2017). Generation of nitric oxide by olive (Olea europaea L.) pollen during in vitro germination and assessment of the S-nitroso- and nitro-proteomes by computational predictive methods. *Nitric Oxide, 68*, 23–37. https://doi.org/10.1016/j.niox.2017.06.005

Jimenez-Quesada, M. J., Traverso, J. A., & Alche Jde, D. (2016). NADPH oxidase-dependent superoxide production in plant reproductive tissues. *Frontiers in Plant Science, 7*, 359.

Kaya, H., Nakajima, R., Iwano, M., Kanaoka, M. M., Kimura, S., Takeda, S., ... Kuchitsu, K. (2014). Ca2+-activated reactive oxygen species production by Arabidopsis RbohH and RbohJ is essential for proper pollen tube tip growth. *Plant Cell, 26*, 1069–1080.

Kelliher, T., & Walbot, V. (2012). Hypoxia triggers meiotic fate acquisition in maize. *Science, 337*, 345–348.

Kodera, C., & Fobis-Loisy, I. (2022). Actors of ROS homeostasis in stigmatic cells essential for plant reproduction. *Archives of Molecular Biology and Genetics, 1*, 9.

Lassig, R., Gutermuth, T., Bey, T. D., Konrad, K. R., & Romeis, T. (2014). Pollen tube NAD(P)H oxidases act as a speed control to dampen growth rate oscillations during polarized cell growth. *The Plant Journal, 78*, 94–106.

Leducq, J. B., Gosset, C. C., Gries, R., Calin, K., Schmitt, E., Castric, V., & Vekemans, X. (2014). Self-incompatibility in Brassicaceae: Identification and characterization of SRK-like sequences linked to the S-locus in the tribe Biscutelleae. *G3 (Bethesda), 4*, 983–992.

Liu, C., Shen, L., Xiao, Y., Vyshedsky, D., Peng, C., Sun, X., ... Li, C. (2021). Pollen PCP-B peptides unlock a stigma peptide-receptor kinase gating mechanism for pollination. *Science, 372*, 171–175.

Ma, J. F., Liu, Z. H., Chu, C. P., Hu, Z. Y., Wang, X. L., & Zhang, X. S. (2012). Different regulatory processes control pollen hydration and germination in Arabidopsis. *Sexual Reproduction in Plants, 25*, 77–82.

Maksimov, N., Evmenyeva, A., Breygina, M., & Yermakov, I. (2018). The role of reactive oxygen species in pollen germination in Picea pungens (blue spruce). *Reproduction in Plants, 31*, 357–365.

Mangano, S., Juarez, S. P., & Estevez, J. M. (2016). ROS regulation of polar growth in plant cells. *Plant Physiology, 171*, 1593–1605.

Mansoor, S., Ali Wani, O., Lone, J. K., Manhas, S., Kour, N., Alam, P., ... Ahmad, P. (2022). Reactive oxygen species in plants: from source to sink. *Antioxidants (Basel), 11*.

Martin, M. V., Distefano, A. M., Zabaleta, E. J., & Pagnussat, G. C. (2013). New insights into the functional roles of reactive oxygen species during embryo sac development and fertilization in Arabidopsis thaliana. *Plant Signaling & Behavior, 8*. https://doi.org/10.4161/psb 25714

Martin, M. V., Fiol, D. F., Sundaresan, V., Zabaleta, E. J., & Pagnussat, G. C. (2013). oiwa, a female gametophytic mutant impaired in a mitochondrial manganese-superoxide dismutase, reveals crucial roles for reactive oxygen species during embryo sac development and fertilization in Arabidopsis. *Plant Cell, 25*, 1573–1591.

Maruyama, D., Volz, R., Takeuchi, H., Mori, T., Igawa, T., Kurihara, D., ... Higashiyama, T. (2015). Rapid elimination of the persistent synergid through a cell fusion mechanism. *Cell, 161*, 907–918.

McInnis, S. M., Desikan, R., Hancock, J. T., & Hiscock, S. J. (2006). Production of reactive oxygen species and reactive nitrogen species by angiosperm stigmas and pollen: Potential signalling crosstalk? *New Phytologist, 172*, 221–228.

Meyer, Y., Belin, C., Delorme-Hinoux, V., Reichheld, J. P., & Riondet, C. (2012). Thioredoxin and glutaredoxin systems in plants: Molecular mechanisms, crosstalks, and functional significance. *Antioxidants & Redox Signaling, 17*, 1124–1160.

Mhamdi, A., & Van Breusegem, F. (2018). Reactive oxygen species in plant development. *Development, 145*.

Mittler, R. (2017). ROS are good. *Trends in Plant Science, 22*, 11–19.

Mittler, R., Vanderauwera, S., Gollery, M., & Van Breusegem, F. (2004). Reactive oxygen gene network of plants. *Trends in Plant Science, 9*, 490–498.

Mizuta, Y., & Higashiyama, T. (2018). Chemical signaling for pollen tube guidance at a glance. *Journal of Cell Science, 131*.

Muhlemann, J. K., Younts, T. L. B., & Muday, G. K. (2018). Flavonols control pollen tube growth and integrity by regulating ROS homeostasis during high-temperature stress. *Proceedings of the National Academy of Sciences of the United States of America, 115*, E11188–E11197.

Munoz-Sanz, J. V., Zuriaga, E., Cruz-Garcia, F., McClure, B., & Romero, C. (2020). Self-(In)compatibility systems: Target traits for crop-production, plant breeding, and biotechnology. *Frontiers in Plant Science, 11*, 195.

Ortega-Villasante, C., Buren, S., Baron-Sola, A., Martinez, F., & Hernandez, L. E. (2016). In vivo ROS and redox potential fluorescent detection in plants: Present approaches and future perspectives. *Methods, 109*, 92–104.

Ortega-Villasante, C., Buren, S., Blazquez-Castro, A., Baron-Sola, A., & Hernandez, L. E. (2018). Fluorescent in vivo imaging of reactive oxygen species and redox potential in plants. *Free Radical Biology and Medicine, 122*, 202–220.

Paetsch, M., Mayland-Quellhorst, S., & Neuffer, B. (2006). Evolution of the self-incompatibility system in the Brassicaceae: Identification of S-locus receptor kinase (SRK) in self-incompatible Capsella grandiflora. *Heredity (Edinb), 97*, 283–290.

Pagnussat, G. C., Yu, H. J., Ngo, Q. A., Rajani, S., Mayalagu, S., Johnson, C. S., ... Sundaresan, V. (2005). Genetic and molecular identification of genes required for female gametophyte development and function in Arabidopsis. *Development, 132*, 603–614.

Palma, J. M., Mateos, R. M., Lopez-Jaramillo, J., Rodriguez-Ruiz, M., Gonzalez-Gordo, S., Lechuga-Sancho, A. M., & Corpas, F. J. (2020). Plant catalases as NO and H2S targets. *Redox Biology, 34*, 101525. https://doi.org/10.1016/j.redox.2020.101525

Pan, J., Wang, W., Li, D., Shu, Z., Ye, X., Chang, P., & Wang, Y. (2016). Gene expression profile indicates involvement of NO in Camellia sinensis pollen tube growth at low temperature. *BMC Genomics, 17*(1), 809. https://doi.org/10.1186/s12864-016-3158-4

Parisy, V., Poinssot, B., Owsianowski, L., Buchala, A., Glazebrook, J., & Mauch, F. (2007). Identification of PAD2 as a gamma-glutamylcysteine synthetase highlights the importance of glutathione in disease resistance of Arabidopsis. *The Plant Journal, 49*, 159–172.

Pasqualini, S., Cresti, M., Casino, C., Faleri, C., Frenguelli, G., Tedeschini, E., & Ederli, L. (2015). Roles for NO and ROS signalling in pollen germination and pollen-tube elongation in Cupressus arizonica. *Biologia Plantarum, 59*, 735–744.

Pereira, A. M., & Coimbra, S. (2019). Advances in plant reproduction: From gametes to seeds. *Journal of Experimental Botany, 70*, 2933–2936.

Potocky, M., Jones, M. A., Bezvoda, R., Smirnoff, N., & Zarsky, V. (2007). Reactive oxygen species produced by NADPH oxidase are involved in pollen tube growth. *New Phytologist, 174*, 742–751.

Prado, A. M., Colaço, R., Moreno, N., Silva, A. C., & Feijó, J. A. (2008). Targeting of pollen tubes to ovules is dependent on nitric oxide (NO) signaling. *Molecular Plant, 1*(4), 703–714. https://doi.org/10.1093/mp/ssn034

Prado, A. M., Porterfield, D. M., & Feijo, J. A. (2004). Nitric oxide is involved in growth regulation and re-orientation of pollen tubes. *Development, 131*, 2707–2714.

Qu, L. J., Li, L., Lan, Z., & Dresselhaus, T. (2015). Peptide signalling during the pollen tube journey and double fertilization. *Journal of Experimental Botany, 66*, 5139–5150.

Rattanawong, K., Koiso, N., Toda, E., Kinoshita, A., Tanaka, M., Tsuji, H., & Okamoto, T. (2021). Regulatory functions of ROS dynamics via glutathione metabolism and glutathione peroxidase activity in developing rice zygote. *The Plant Journal, 108*, 1097–1115.

Ray, P. D., Huang, B. W., & Tsuji, Y. (2012). Reactive oxygen species (ROS) homeostasis and redox regulation in cellular signaling. *Cellular Signalling, 24*, 981–990.

Rotman, N., Rozier, F., Boavida, L., Dumas, C., Berger, F., & Faure, J. E. (2003). Female control of male gamete delivery during fertilization in Arabidopsis thaliana. *Current Biology, 13*, 432–436.

Sankaranarayanan, S., Ju, Y., & Kessler, S. A. (2020). Reactive oxygen species as mediators of gametophyte development and double fertilization in flowering plants. *Frontiers in Plant Science, 11*, 1199.

Serrano, I., Romero-Puertas, M. C., Rodriguez-Serrano, M., Sandalio, L. M., & Olmedilla, A. (2012). Peroxynitrite mediates programmed cell death both in papillar cells and in self-incompatible pollen in the olive (Olea europaea L.). *Journal of Experimental Botany, 63*, 1479–1493.

Serrano, I., Romero-Puertas, M. C., Sandalio, L. M., & Olmedilla, A. (2015). The role of reactive oxygen species and nitric oxide in programmed cell death associated with self-incompatibility. *Journal of Experimental Botany, 66*, 2869–2876.

Sharma, P., Jha, A. B., Dubey, R. S., & Pessarakli, M. (2012). Reactive oxygen species, oxidative damage, and antioxidative defense mechanism in plants under stressful conditions. *Journal of Botany, 2012*, 1–26.

Shi, Y., & Carroll, K. S. (2020). Activity-based sensing for site-specific proteomic analysis of cysteine oxidation. *Accounts of Chemical Research, 53*, 20–31.
Smirnova, A. V., Matveeva, N. P., Polesskaia, O. G., & Ermakov, I. P. (2009). Formation of reactive oxygen species during pollen grain germination. *Ontogenez, 40*, 425–434.
Smirnova, A. V., Matveyeva, N. P., & Yermakov, I. P. (2014). Reactive oxygen species are involved in regulation of pollen wall cytomechanics. *Plant Biology (Stuttg), 16*, 252–257.
Speranza, A., & Scoccianti, V. (2012). New insights into an old story Pollen ROS also play a role in hay fever. *Plant Signaling & Behavior, 7*, 5.
Speranza, A., Crinelli, R., Scoccianti, V., & Geitmann, A. (2012). Reactive oxygen species are involved in pollen tube initiation in kiwifruit. *Plant Biology (Stuttg), 14*, 64–76.
Stracke, R., Jahns, O., Keck, M., Tohge, T., Niehaus, K., Fernie, A. R., & Weisshaar, B. (2010). Analysis of PRODUCTION OF FLAVONOL GLYCOSIDES-dependent flavonol glycoside accumulation in Arabidopsis thaliana plants reveals MYB11-, MYB12- and MYB111-independent flavonol glycoside accumulation. *New Phytologist, 188*, 985–1000.
Strome, S., & Lehmann, R. (2007). Germ versus soma decisions: Lessons from flies and worms. *Science, 316*, 392–393.
Sun, C. Q., Chen, F. D., Teng, N. J., Yao, Y. M., Shan, X., & Dai, Z. L. (2019). Transcriptomic and proteomic analysis reveals mechanisms of low pollen-pistil compatibility during water lily cross breeding. *BMC Plant Biology, 19*, 542.
Tedeschini, E., Proietti, P., Timorato, V., D'Amato, R., Nasini, L., Dei Buono, D., ... Frenguelli, G. (2015). Selenium as stressor and antioxidant affects pollen performance in Olea europaea. *Flora – Morphology, Distribution, Functional Ecology of Plants, 215*, 16–22.
Timofejeva, L., Skibbe, D. S., Lee, S., Golubovskaya, I., Wang, R., Harper, L., ... Cande, W. Z. (2013). Cytological characterization and allelism testing of anther developmental mutants identified in a screen of maize male sterile lines. *G3 (Bethesda), 3*, 231–249.
Tsukagoshi, H., Busch, W., & Benfey, P. N. (2010). Transcriptional regulation of ROS controls transition from proliferation to differentiation in the root. *Cell, 143*, 606–616.
Ushio-Fukai, M. (2009). Compartmentalization of redox signaling through NADPH oxidase-derived ROS. *Antioxidants & Redox Signaling, 11*, 1289–1299.
Volz, R., Harris, W., Hirt, H., & Lee, Y. H. (2022). ROS homeostasis mediated by MPK4 and SUMM2 determines synergid cell death. *Nature Communications, 13*, 1746.
Wang, C. L., Wu, J., Xu, G. H., Gao, Y. B., Chen, G., Wu, J. Y., ... Zhang, S. L. (2010). S-RNase disrupts tip-localized reactive oxygen species and induces nuclear DNA degradation in incompatible pollen tubes of Pyrus pyrifolia. *Journal of Cell Science, 123*, 4301–4309.
Wang, L., Clarke, L. A., Eason, R. J., Parker, C. C., Qi, B., Scott, R. J., & Doughty, J. (2017). PCP-B class pollen coat proteins are key regulators of the hydration checkpoint in Arabidopsis thaliana pollen-stigma interactions. *New Phytologist, 213*, 764–777.
Wang, R., Shi, C., Wang, X., Li, R., Meng, Y., Cheng, L., ... Li, T. (2020). Tomato SlIDA has a critical role in tomato fertilization by modifying reactive oxygen species homeostasis. *The Plant Journal, 103*, 2100–2118.
Wang, Y. H., Li, X. C., Zhu-Ge, Q., Jiang, X., Wang, W. D., Fang, W. P., ... Li, X. H. (2012). Nitric oxide participates in cold-inhibited Camellia sinensis pollen germination and tube growth partly via cGMP in vitro. *PLoS One, 7*(12), e52436. https://doi.org/10.1371/journal.pone.0052436
Webb, M. C., & Gunning, B. E. (1991). The microtubular cytoskeleton during development of the zygote, proembryo and free-nuclear endosperm in Arabidopsis thaliana (L.) Heynh. *Planta, 184*, 187–195.
Whipple, C. (2012). Plant science. Defining the plant germ line—Nature or nurture? *Science, 337*, 301–302.
Wilkins, K. A., Bancroft, J., Bosch, M., Ings, J., Smirnoff, N., & Franklin-Tong, V. E. (2011). Reactive oxygen species and nitric oxide mediate actin reorganization and

programmed cell death in the self-incompatibility response of papaver. *Plant Physiology, 156*, 404–416.

Wu, J., Shang, Z., Wu, J., Jiang, X., Moschou, P. N., Sun, W., ... Zhang, S. (2010). Spermidine oxidase-derived H(2)O(2) regulates pollen plasma membrane hyperpolarization-activated Ca(2+) -permeable channels and pollen tube growth. *The Plant Journal, 63*, 1042–1053.

Xing, S., & Zachgo, S. (2008). ROXY1 and ROXY2, two Arabidopsis glutaredoxin genes, are required for anther development. *The Plant Journal, 53*, 790–801.

Xu, X., Wang, W., Sun, Y., Xing, A., Wu, Z., Tian, Z., ... Wang, Y. (2021). MicroRNA omics analysis of camellia sinesis pollen tubes in response to low-temperature and nitric oxide. *Biomolecules, 11*(7), https://doi.org/10.3390/biom11070930

Yan, M. Y., Xie, D. L., Cao, J. J., Xia, X. J., Shi, K., Zhou, Y. H., ... Yu, J. Q. (2020). Brassinosteroid-mediated reactive oxygen species are essential for tapetum degradation and pollen fertility in tomato. *The Plant Journal, 102*, 931–947.

Yu, X., Zhang, X., Zhao, P., Peng, X., Chen, H., Bleckmann, A., ... Sun, M. X. (2021). Fertilized egg cells secrete endopeptidases to avoid polytubey. *Nature, 592*, 433–437.

Yun, B. W., Feechan, A., Yin, M., Saidi, N. B., Le Bihan, T., Yu, M., ... Loake, G. J. (2011). S-nitrosylation of NADPH oxidase regulates cell death in plant immunity. *Nature, 478*(7368), 264–268. https://doi.org/10.1038/nature10427

Zafra, A., Rodriguez-Garcia, M. I., & de Dios Alche, J. (2010). Cellular localization of ROS and NO in olive reproductive tissues during flower development. *BMC Plant Biology, 10*, 14.

Zechmann, B., Koffler, B. E., & Russel, S. D. (2011). Glutathione synthesis is essential for pollen germination in vitro. *BMC Plant Biology, 11*, 11.

Zhang, L., Huang, J., Su, S., Wei, X., Yang, L., Zhao, H., ... Duan, Q. (2021). FERONIA receptor kinase-regulated reactive oxygen species mediate self-incompatibility in Brassica rapa. *Current Biology, 31*, 3004–3016 e3004.

Zhang, M. J., Zhang, X. S., & Gao, X. Q. (2020). ROS in the male-female interactions during pollination: Function and regulation. *Frontiers in Plant Science, 11*, 177.

Zheng, S., Li, J., Ma, L., Wang, H., Zhou, H., Ni, E., ... Zhuang, C. (2019). OsAGO2 controls ROS production and the initiation of tapetal PCD by epigenetically regulating OsHXK1 expression in rice anthers. *Proceedings of the National Academy of Sciences of the United States of America, 116*, 7549–7558.

Zhou, L. Z., & Dresselhaus, T. (2019). Friend or foe: Signaling mechanisms during double fertilization in flowering seed plants. *Current Topics in Developmental Biology, 131*, 453–496.

Zhou, L. Z., Juranic, M., & Dresselhaus, T. (2017). Germline development and fertilization mechanisms in maize. *Molecular Plant, 10*, 389–401.

Zhou, L. Z., Qu, L. J., & Dresselhaus, T. (2021). Stigmatic ROS: Regulator of compatible pollen tube perception? *Trends in Plant Science, 26*, 993–995.

Zinta, G., Khan, A., AbdElgawad, H., Verma, V., & Srivastava, A. K. (2016). Unveiling the redox control of plant reproductive development during abiotic stress. *Frontiers in Plant Science, 7*, 700.

CHAPTER SIX

ROS in seed germination

Christophe Bailly*
Sorbonne Université, IBPS, CNRS, UMR 7622 Biologie du Développement, Paris, France
*Corresponding author. e-mail address: christophe.bailly@sorbonne-universite.fr

Contents

Abstract

Seed germination is a critical event in higher plant life because its success ensures the establishment of a new generation and survival of the species. It is regulated by a myriad of regulatory components and modulated by the conditions of seed development on the mother plant and subsequent conditions of seed storage. Reactive Oxygen Species (ROS) are key regulators of the germination process by itself, as well as play a role in all stages of seed life, including the seed maturation program and the physiological changes that can occur in dry mature seeds, such as dormancy alleviation and loss of viability. Their homeostasis during seed germination also regulates the decision to germinate. ROS play a critical role in sensing the environmental conditions that prevail at the time of sowing and they regulate seed dormancy. The source of ROS production in seeds varies with the stage of seed life and during their progression towards germination, resulting in a complex oxidative network controlling seed germination. Oxidative signaling during seed germination triggers a large variety of downstream signaling events ranging from crosstalk with plant hormones to changes in gene expression and modification of cell wall properties. This chapter summarizes the current knowledge on the role of ROS in seed germination and provides novel insights, such as the concepts of oxidative memory and oxidative

Advances in Botanical Research, Volume 105
ISSN 0065-2296, https://doi.org/10.1016/bs.abr.2022.10.003

network, that bring a better understanding on their function in this key developmental process.

1. Introduction

Seeds are time and space vector of genetic inheritance between generations of higher plants. Seeds of many plant species are desiccation tolerant and exhibit a low moisture content when mature, which allows them to remain viable for long periods of time. They ensure the transmission of genetic information from the mother plant to the progeny and are central for plant reproduction, dispersal, and survival of plant species. In addition, they are also critical for human and animal nutrition, and they account for ca. 70% of energy intake of the human population. Germination, which occurs after rehydration of mature dry quiescent seed, results in the elongation of the embryonic axis, and emergence of the radicle is a crucial developmental transition in the plant life cycle because it permits the establishment of a new generation. Seed germination can appear as a simple developmental event, when compared to other developmental processes such as flowering, for example, because it only relies on cell elongation. However, this is a very complex process whose success is controlled by a wide range of interconnected and additive molecular and signaling events that are modulated by conditions of seed development on the mother plant, seed storage and seed imbibition. In addition, germination must occur in environmental conditions that allow successful growth of the produced seedling and plant survival. Thus the "decision to germinate" is tightly regulated by external factors but also by internal factors, such as seed dormancy. Dormancy is a blockage of germination that controls the timing of germination to ensure development of the emerging seedling in appropriate conditions (Bewley, 1997).

These special features of seed life, from formation to seedling emergence, require a network of cell signaling components that can sense environmental conditions and control seed dormancy. The plant hormones abscisic acid (ABA), gibberellins (GAs) and ethylene, play a key role in the regulation of seed germination. Dormancy, in particular, depends mainly on an intrinsic balance of ABA and GAs. GAs have been described as an internal regulator involved in the induction of germination, whereas ABA is involved in the establishment and maintenance of seed dormancy (Bewley, 1997; Finch-Savage & Leubner-Metzger, 2006). In addition, the role of reactive oxygen species (ROS) in the regulation of seed germination and dormancy progressively emerged in the 2 last decades, and ROS are now widely acknowledged as key actors of this developmental processes. However, their mechanisms of production and modes of action during the germination processes are far from being fully resolved. The objective of this chapter is to provide an up-to-date perspective on the state of knowledge on the role of ROS in seed germination.

2. Seed life and the concept of an oxidative memory

2.1 From fertilization to germination

Germination requires a synchronized control of molecular and cellular signaling event whose integration ultimately provides the decision to proceed, or not. In this wide array of regulatory pathways ROS are key regulatory players, along with plant hormones, and increasing evidence demonstrate that their role is probably even more important than originally thought. Their controlled production is possible in every single cell where oxygen and an active electron transfer chain are present, but they can also be produced in anhydrobiosis through non enzymatic autoxidative reactions, which makes them very good candidates for being involved in every step of seed life, from development on the mother plant to germination. Indeed, one must emphasize that, in contrast to other higher plants organs, seeds display a unique pattern of development. Seeds emerge from fertilization and they first develop on the mother plant (Zhou & Dresselhaus, 2022, This Volume). After a phase of cell division and morphogenesis, embryos accumulate reserves during the seed maturation phase, before entering an abrupt and dramatic desiccation phase leading to mature and dehydrated orthodox seed. Primary dormancy is induced during seed maturation on the mother plant (Finch-Savage & Leubner-Metzger, 2006). The progressive loss of water during seed maturation induces the formation of a so-called cellular "glassy state", that restricts molecular mobility and metabolism and allows mature orthodox seeds to survive in a quiescent state for long periods of time (Leprince, Pellizzaro, Berriri, & Buitink, 2017). Dry seeds then rehydrate in the presence of water, ensuring active metabolism resumption and, if environmental conditions are appropriate and if seeds are non-dormant, the completion of all cellular events that lead to the elongation of the embryonic axis, the first visible symptom of seed germination.

The 3 steps of seed life (formation, quiescence, germination) cannot be considered as independent phases since each individual phase has a dramatic effect on the subsequent one, even though they may appear as being physiologically disconnected. For example, in *Arabidopsis thaliana*, cool temperatures during seed maturation induce strong primary dormancy (Chiang et al., 2011; Donohue, Heschel, Chiang, Butler, & Barua, 2007) while high temperatures reduce dormancy (Huang, Footitt, & Finch-Savage, 2014). In sunflower, it has been shown that faster seed desiccation during the late maturation phase induced lower seed dormancy (Lachabrouilli, Rigal, Corbineau, & Bailly, 2021), and that drought stress during seed maturation induced higher seed vigor and lower dormancy (Vancostenoble, Blanchet, Langlade, & Bailly, 2022). The conditions of seed maturation and seed desiccation on the plant have also a marked effect on dry seed longevity (Leprince et al., 2017). In the context of global warming, the understanding of the mechanisms involved in maternal effect on the offspring is an important issue. Subsequently, duration and conditions of storage also have a pronounced effect on seed germination. Seed dormancy release occurs during dry storage in a process called after-ripening that allows the transition of a dormant to a non-dormant state. Longer storage period, or inappropriate conditions of storage (e.g., high temperature, high relative humidity), progressively alter seed vigor, i.e., the ability to germinate in a wide range of environmental conditions, and ultimately lead to loss of seed viability.

The observed phenotype of germination results from additive effects of seed development and storage conditions, as well as from the environmental conditions at the time of seed imbibition. This indicates that a message is transmitted from developing seeds to dry seeds and then from dry to germinating seeds. Here I suggest that this message mostly consists of oxidative processes, as it will be further explained here, and I propose a novel concept of "oxidative memory" to explain the transmission of information throughout the different steps of seed life.

2.2 Oxidative memory: sending the message from the developing to the germinating seed

ROS formation and homeostasis in developing seeds is poorly documented (Bailly, 2004). In the whole process of seed development, the seed desiccation phase, which consists in dramatic removal of water from cells, is a stressful condition generating ROS, as it has been reported for water stress in other plant organs (Choudhury, Rivero, Blumwald, & Mittler, 2017;

Noctor, Mhamdi, & Foyer, 2014). Among other mechanisms (reviewed by Leprince et al. (2017) and Smolikova, Leonova, Vashurina, Frolov, and Medvedev (2020)) acquisition of dessiccation tolerance requires an activation of antioxidant systems to avoid oxidative damage (Kranner & Birtić, 2005). For example, Chen et al. (2014) showed that ascorbate peroxidase activity was critical for the maturation-drying phase in Arabidopsis. It has been shown that protein carbonylation, an oxidative modification of proteins, increased during the natural developmental desiccation stage in seeds of *M. truncatula* (Satour et al., 2018). Interestingly, Kaur et al. (2015) showed that OsHSP18.2, a molecular chaperone that restricts ROS accumulation during maturation-associated drying, improved seed vigor and longevity. Similarly, an uncharacterized member of the Fumarylacetoacetate hydrolase (FAH) superfamily, named AtFAHD1a, was shown to participate in seed redox regulation late seed maturation and desiccation, and play a role in seed dormancy and longevity (Gerna et al., 2021). There are therefore several lines of evidence indicating that controlling seed desiccation rate, by modulating the oxidative status of the seed tissue, would in turn alter seed dormancy and longevity. Thus, the first oxidative memory timepoint corresponds to the oxidation status of the dry mature seeds (Fig. 1). The higher the oxidation level and ROS content in the dry seeds, the faster the physiological changes occurring during dry storage.

ROS generation indeed continuously occurs in seeds during dry storage. This accumulation can be first beneficial, e.g., allowing seed dormancy alleviation through after ripening, and then becomes progressively deleterious and associated with loss of viability. As previously stated, dry orthodox seeds are in a quiescent state where high viscosity prevents metabolic activity, whereas seed desiccation on the mother plant has created an oxidative environment within tissues of the mature seed. In such conditions, however, and in the presence of oxygen, non-enzymatic oxidation of lipids, in particular, may serve as a source of free radicals. Interestingly, Gerna et al. (2022) has recently brought experimental evidence that gaseous oxygen contributed to lipid peroxidation and seed ageing when the cytoplasm was glassy. The effect of oxygen on oxidative reactions in dry seeds has also been emphasized by storing seeds under "Elevated Partial Pressure of Oxygen", a method that accelerates dry after-ripening storage and ageing (Buijs, Kodde, Groot, & Bentsink, 2018; Groot, Surki, de Vos, & Kodde, 2012; Nagel et al., 2016). Evidence that dormancy release during dry after-ripening resulted from ROS generation was brought by Oracz et al. (2007) in sunflower seeds and

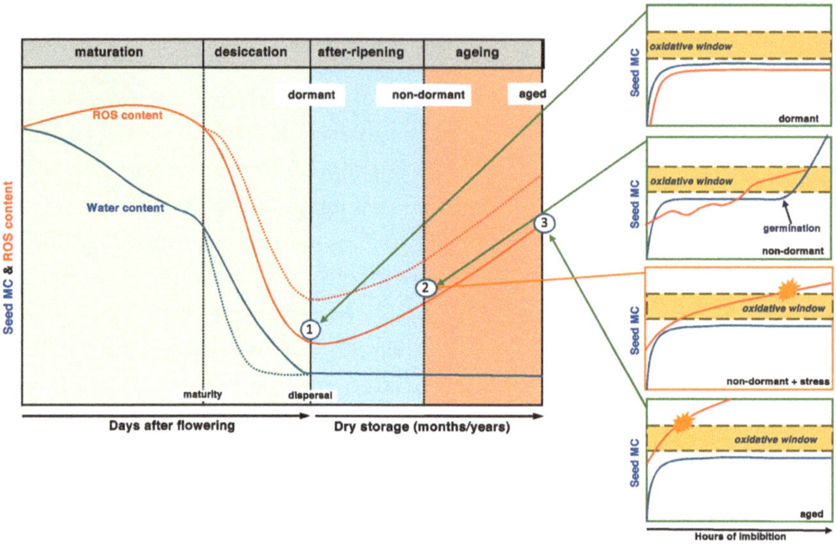

Fig. 1 Schematic representation of the concepts of oxidative memory and oxidative window in the regulation of seed germination. Changes in seed moisture content (MC, blue line) and ROS content (red line) during seed development, seed storage (left-hand side) and seed germination (right-hand side) are represented. Dotted lines show the effects of an excessive desiccation rate on seed MC and ROS content during seed development. In normal developmental conditions (plain lines), seeds are dormant at harvest (1). Their germination under appropriate conditions (green arrows) does not allow germination because ROS content remains below the oxidative window. Dormancy is alleviated during dry storage (2). If non-dormant seeds are imbibed in favorable conditions (green arrow) their ROS content is within the oxidative window and they germinate. If they are imbibed under stressful conditions (water stress, thermal stress, hypoxia.; orange arrow) their ROS content is above the ROS window, and they do not germinate. Prolonged storage of seeds (3) is associated with seed ageing and ROS accumulation. Even when aged seed are imbibed under favorable conditions (green arrow), they do not germinate because their ROS content is above the oxidative window. 1, 2 and 3 are key oxidative memory timepoints (see text). *Adapted from Graeber, K., Nakabayashi, K., & Leubner-Metzger, G. (2017). Development of dormancy. In B. Thomas, B.G. Murray, & D.J. Murphy (Éds.), Encyclopedia of applied plant sciences (2nd ed., pp. 483–489). Academic Press. https://doi.org/10.1016/B978-0-12-394807-6.00209-4 (Graeber, Nakabayashi, & Leubner-Metzger, 2017).*

then confirmed by Bazin et al. (2011) who demonstrated that a subset of mRNAs became specifically oxidized during sunflower seed storage, and that this mechanism allowed dormancy release. After-ripening of sunflower embryonic axes is also accompanied by a shift in the thiol-based cellular redox environment towards more oxidizing conditions (Morscher, Kranner, Arc, Bailly, & Roach, 2015). The oxidative hypothesis of seed dormancy release during dry storage is now widely accepted and is the only one that can explain the transition from a dormant to non-dormant status without an active metabolism (El-Maarouf-Bouteau, Meimoun, Job, Job, & Bailly, 2013; Katsuya-Gaviria, Caro, Carrillo-Barral, & Iglesias-Fernández, 2020; Matilla, 2020). The mechanism of ROS generation in anhydrobiosis is nevertheless not timely restricted and is likely to occur continuously during seed storage, progressively leading to unregulated oxidative damage that can trigger seed ageing and loss of viability. The role of ROS in seed ageing is very well known and has been extensively reviewed elsewhere (Bailly, El-Maarouf-Bouteau, & Corbineau, 2008; Bailly, 2004; Kurek, Plitta-Michalak, & Ratajczak, 2019; Zhang, Zhang, Sun, Meng, & Tao, 2021) and is not presented here. In addition, whether ROS act by themselves or through their oxidative derivates resulting from the oxidation of biomolecules, such as peroxyl, alkoxyl or hydroperoxyl radicals, is not clearly established. The reactive aldehyde products derived from lipid peroxidation, such as malondialdehyde (MDA) for example, are stable and can oxidize proteins (Carbone, Doorn, Kiebler, & Petersen, 2005; Sun, Wu, & Mao, 2022). It must be emphasized that the antioxidant enzymatic machinery is inefficient in a glassy cytoplasm and that the only antioxidant activity in dry seeds relies on non-enzymatic ROS scavengers such as tocopherols and tocochromanols, glutathione or ascorbate (Kranner, Minibayeva, Beckett, & Seal, 2010), that progressively become oxidized and inefficient. The amount of ROS and the related oxidative modifications of cellular components and biomolecules after a storage period represent the second oxidative memory point within the seed life, and it will directly control the subsequent seed ability to germinate after rehydration (Fig. 1).

3. ROS as key regulators of seed germination

3.1 The process of seed germination

As previously evoked, the seed ability to germinate is strongly influenced by a combinational interaction of biological events that precede seed imbibition and that strongly rely on ROS metabolism. ROS, however, are not the only signals that can be transmitted from the previous phases of seed life to the germinating seed. In particular conditions of seed development on the mother plant have an effect on hormone metabolism that regulates seed maturation and dormancy (Gutierrez, Van Wuytswinkel, Castelain, & Bellini, 2007). Among these hormones ABA is playing a leading role because it controls the induction of dormancy (Chen et al., 2020; Nonogaki, 2019). For example, maternal temperature variations have a strong effect on ABA content and therefore on subsequent seed dormancy (Sano & Marion-Poll, 2021). In addition, genetic and epigenetic components are also involved in inheritance of seed traits (Alvarez, Bleich, & Donohue, 2021; Donohue, 2009). In contrast, in dry resting seeds, oxidative processes are primarily active and potentially able to further transmit information to the germinating seed.

Seed germination starts with water uptake by the dry seed, thus allowing a gradual resumption of metabolism (Nonogaki, Bassel, & Bewley, 2010), and is followed by a period of limited water uptake (Phase II of the germination) that allows the completion of a range of cellular events that will in turn lead to radicle protrusion. After germination, a novel increase in water uptake occurs and is associated with Phase III of the process leading to seedling growth. Seed imbibition is associated with a rapid loss of cytoplasmic glassy state and a decrease in viscosity allowing metabolism resumption. In particular, energy metabolism is reactivated rapidly, and ATP accumulation and oxygen uptake have been shown to occur within the first minutes of imbibition (Nietzel et al., 2020). This is concomitant with *de novo* transcriptional activity (Nakabayashi, Okamoto, Koshiba, Kamiya, & Nambara, 2005). Then, a large set of synchronized and timely regulated molecular events occur during phase II of the germination process. Their successful realization allows cell elongation and radicle protrusion and this phase, which is tightly regulated by endogenous and exogenous regulatory factors, is considered as being the critical phase of the whole process. A large range of molecular actors involved in seed germination have been identified so far (Carrera-Castaño, Calleja-Cabrera, Pernas, Gómez, & Oñate-Sánchez, 2020) such as *DELAY OF GERMINATION 1* (*DOG1*; Alonso-Blanco, Bentsink, Hanhart, Blankestijn-de Vries, & Koornneef, 2003; Bentsink, Jowett, Hanhart, & Koornneef, 2006) and *REDUCED DORMANCY 5* (*RDO5*; Xiang et al., 2014). Germination also requires a general restarting of previously paused protein translation using long lived-stored mRNAs (Nonogaki et al., 2010) and the degradation of stored mRNAs accumulated during seed maturation and associated with dormancy (Basbouss-Serhal, Soubigou-Taconnat, Bailly, & Leymarie, 2015; Basbouss-Serhal, Pateyron, Cochet, Leymarie, & Bailly, 2017).

The transition to phase III is driven by the antagonist relationship between ABA and GA but other hormones, such as ethylene or brassinosteroids, are also playing a role in seed germination (Graeber, Nakabayashi, Miatton, Leubner-Metzger, & Soppe, 2012). In addition, reprogramming of the epigenome (including DNA methylation and higher order chromatin organization) is also involved in seed germination (Narsai et al., 2017) and a large array of chromatin modifiers and remodelers have been associated with dormancy or germination (Carrera-Castaño et al., 2020). Ultimately, cell elongation occurring at the end of phase II requires specific modifications facilitating cell wall loosening to promote embryo growth (Steinbrecher & Leubner-Metzger, 2018).

3.2 ROS homeostasis and seed germination

Metabolism resumption during seed imbibition is associated with active ROS production and completion of the whole germination process requires a regulated endogenous production of ROS. ROS production is a prerequisite for germination completion, providing its content remained enclosed within the so-called oxidative window, as defined by Bailly et al. (2008). Beside the concept of oxidative memory, previously stated, the concept of oxidative window suggests that the ability of a seed to germinate is directly dependent on ROS homeostasis (Fig. 1). Seed germination occurs only when seed ROS content reaches a sufficient amount for allowing ROS signaling but is prevented when this amount of ROS is too low (e.g., dormant seeds) or too elevated (aged seeds, seeds imbibed under inappropriate conditions for germination; Fig. 1). Bailly (2019) has recently referenced the studies that reported changes in seed ROS homeostasis in permissive and non-permissive conditions of seed germination. These studies underline the key regulatory role of oxidative window for the completion of germination. The occurrence of this oxidative window is particularly clear in the transition from dormant to non-dormant states as it has been demonstrated in seeds of various species such as Arabidopsis (Leymarie et al., 2012), barley (Ishibashi et al., 2017), sunflower (Oracz et al., 2009), apple (Ciacka, Tyminski, Gniazdowska, & Krasuska, 2022) or rice (Bhattacharjee, Chakrabarty, Kora, & Roy, 2022).

3.3 ROS sources and ROS trafficking: towards an oxidative network of seed germination

The transition from a dry quiescent seed to an imbibed seed is associated with a transition of ROS production mode from non-enzymatic and unregulated mechanisms to enzymatic and regulated production. In imbibed seeds, superoxide anion ($O_2^{\cdot -}$), hydrogen peroxide and hydroxyls radical have been frequently detected (Apel & Hirt, 2004; Bailly, 2004; Bailly, 2019; Tiwari et al., 2017). The major roles of ROS in seed germination have been attributed to hydrogen peroxide (H_2O_2), the most stable ROS which is frequently mentioned as being involved in cell signaling in plants (Mignolet-Spruyt et al., 2016; Mittler, Zandalinas, Fichman, & Van Breusegem, 2022; Petrov & Van Breusegem, 2012). The rapid reactivation of mitochondrial metabolism at the onset of seed imbibition (Nietzel et al., 2020) suggests that mitochondria are likely to be the main site of active and regulated ROS production. Electron leakage from complex I and III of the mitochondrial electron transport chain harbors enough energy to reduce oxygen and form superoxide radicals. Two pathways of oxygen reduction have been identified to date. The first pathway relies on O_2 consumption via the cytochrome oxidase to produce water and the second depends on the direct reduction of O_2 to $O_2^{\cdot -}$ in the NADPH dehydrogenase segment of the respiratory chain (Jezek & Hlavatá, 2005; Murphy, 2009). The generated superoxide radicals are subject to dismutation via the superoxide dismutase (SOD) to give rise to H_2O_2 (Murphy, 2009). The key role of ROS production by the mitochondrial electron transport chain (mtETC) during early seed imbibition has been recently document by (Jurdak, Launay-Avon, Paysant-Le Roux, & Bailly, 2021). Using cell imaging approaches these authors demonstrated that, until 6 h imbibition, mitochondria were the major intracellular source of ROS of both dormant and non-dormant Arabidopsis seeds, but that dormancy release treatments (e.g., ethylene, gibberellins, stratification and light) magnified this production, which shed a new light on the possible role of this organelle in seed germination. Bailly and Merendino (2021) suggested that mitochondria and the respiration process were not only serving as energy suppliers but that they could be playing a role in the sensing of environmental conditions permissive for germination, as also evidenced by Ma et al. (2019). NADPH oxidases (NOXs), also known as respiratory burst oxidase homologues (RBOHs), are one the major ROS producers in plants and they have often been proposed to be involved in ROS production during seed germination. NOXs localize at the plasma membrane and induce the production of $O_2^{\cdot -}$ in the cell wall by transferring electrons from the cytosolic NADH or NADPH to apoplastic oxygen. In barley (Ishibashi, Tawaratsumida, Zheng, Yuasa, & Iwaya-Inoue, 2010) and rice (Li et al., 2017) it was proposed that NOXs would regulate radicle and root elongation during seed germination. In lettuce, using molecular tools, Yang et al. (2020) reported that the NADPH-oxidase LsRbohC1 was involved in the radicle elongation growth and endosperm weakening during lettuce seed germination. In Arabidopsis, seeds of the NADPH oxidase mutants (*rbohD*) were shown as being dormant (Leymarie et al., 2012) but (Jurdak et al., 2021, 2022) showed that the efficiency of a subset of ROS-related dormancy release treatments did required NOX activity. We propose that NOXs are involved in the late events of seed germination, i.e., that they play a role in cell wall weakening prior to cell elongation. Together with apoplastic copper-containing amine oxidases and polyamine oxidases, they could generate ROS within the cell wall and participate to the cleavage of cell wall polysaccharides (Fry, 1998; Liszkay, Kenk, & Schopfer, 2003; Müller, Linkies, et al., 2009). The hydroxyl radical produced in the apoplastic space of plant tissues is likely to play a role in the cell wall loosening processes underlying cell expansion (Podgórska, Burian, & Szal, 2017). In plant cells, peroxisomes also produce ROS as a result of their normal metabolism and they contain various enzymatic systems involved in H_2O_2 production (Corpas, González-Gordo, & Palma, 2020). In seeds they have been mostly reported as being involved in fatty acid β-oxidation to sustain the early steps of seedling growth (Kao, Gonzalez, & Bartel, 2018). However it has been recently proposed that they were also involved in ROS production prior to radicle protrusion in Arabidopsis seeds (Jurdak et al., 2022), but their potential role as germination regulators is not documented yet. Lastly, in germinating Arabidopsis seeds, ROS have also been shown to localize in the nucleus at the onset of radicle protrusion. This appears to be a prerequisite to cell elongation since this was demonstrated in various physiological conditions, thus suggesting that nuclear ROS would play a key role in the late events of germination (Jurdak et al., 2021, 2022; Leymarie et al., 2012). There is no clear evidence showing ROS production systems in the nucleus, even though it has already been proposed (Alkhalfioui et al., 2007), rather it is suspected that ROS could be transferred from other cellular compartments to the nucleus (Martins et al., 2018).

ROS homeostasis in seed cells does not rely only on ROS production sources but also depends on a balance between ROS generation and scavenging though enzymatic and non-enzymatic antioxidants processes (already reviewed by Bailly, 2004). However, the signaling role of ROS during seed germination probably results more from regulated ROS production than from the function of detoxication mechanisms. ROS scavenging systems would be mainly involved in the prevention of oxidative damage, allowing ROS content to be enclosed within the oxidative window when appropriate conditions for germination are present. This is probably the reason why high seed vigor has been associated with an efficient ROS scavenging system. For example sunflower seed vigor has been associated with high catalase activity and efficient capacity of H_2O_2 detoxication (Bailly, Benamar, Corbineau, & Côme, 1998) and Arabidopsis seeds of vitamin E mutants have lower longevity (Sattler, Gilliland, Magallanes-Lundback, Pollard, & DellaPenna, 2004). Accordingly, freshly harvested seeds of catalase (*cat2–1*) and vitamin E (*vte1–1*) mutants were less dormant because their ROS content was high in mature seeds, even without any storage period (after-ripening), in accordance with the concept of oxidative memory previously stated. Lastly, peroxisomes, that were evoked previously as a site of ROS generation, also contain enzymes such as catalases, ascorbate peroxidases and several types of peroxiredoxins, involved in H_2O_2 scavenging (Su, Li, Wang, & Ma, 2019), so that they could fine tune ROS levels in plant cells during germination, but this role has not be clearly shown yet.

In addition to the identification of ROS production sites during germination, the role of ROS trafficking during seed germination has recently emerged. Indeed, one cannot consider the different sites of ROS production as being fully independent but one should rather consider each of them as a component of the oxidative network controlling seed germination (Fig. 2). Seed germination is a dynamic process that relies on a range of timely synchronized molecular and cellular events ranging from reactivation of mitochondrial activity and metabolism to transcriptional regulation and cell wall weakening. It is therefore unrealistic to consider independently ROS production by the mitochondria, for example, from ROS accumulation within the nucleus or NOX activation. Recently (Jurdak et al., 2021, 2022) have brought the first evidence of such an oxidative network by demonstrating that the positive effect of ethylene on Arabidopsis seed germination involved a mitochondrial retrograde response (MRR) from mitochondria to nucleus. In MRR, mitochondria generate an oxidative signal that is transmitted to the nucleus and modify gene expression (Ng et al., 2014). Transcription factors from the NAC (No apical meristem, ATAF1/2 and CUC2 cup-shaped cotyledon) family, notably ANAC013, which localizes to the ER and translocates to the nucleus (De Clercq et al., 2013), have been demonstrated to participate to MRR in germinating Arabidopsis seeds (Jurdak et al., 2021; Fig. 2). In another context, i.e., dormancy alleviation by light, gibberellins and stratification, Jurdak et al. (2022) demonstrated peroxisomes could serve as a relay between mitochondria and nucleus. Cell imaging approaches and gene expression studies showed that peroxisomes could be a major source

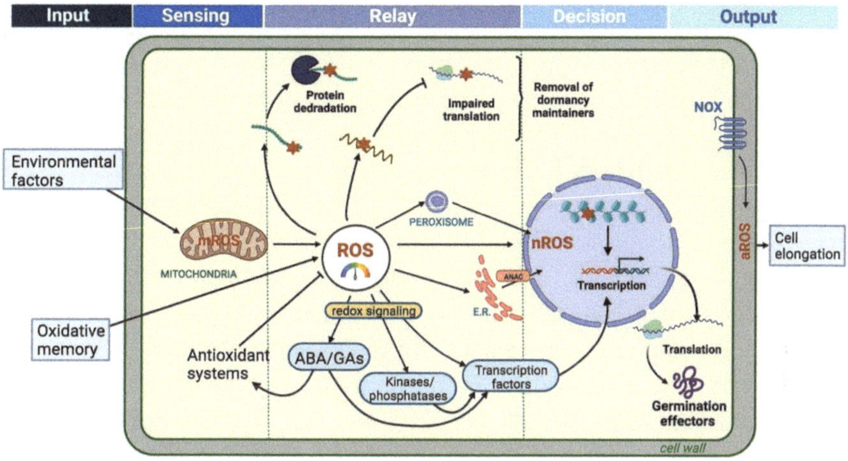

Fig. 2 The oxidative network of seed germination. Germination regulators (the inputs) control mitochondrial ROS (mROS) production (sensing) and subsequently modify cytosolic ROS homeostasis. This leads to the regulation of a subset of ROS related mechanisms that relay the information to the nucleus, including redox regulation of hormone biosynthesis and signaling pathways involving kinases, phosphatases, and of transcription factors. ROS can also directly oxidize dormancy-associated proteins and mRNAs. Organelles, such as endoplasmic reticulum (ER) and peroxisomes can also serve as transmitters of the oxidative signal. Nuclear ROS (nROS) are generated within the nucleus allowing chromatin decompaction and activation of gene expression. Changes in gene expression, which can also be modified by other factors than ROS, make the decision to germinate. Lastly, apoplastic ROS (aROS), that can be generated by NOX activity, allows cell wall weakening which favor cell elongation, the first symptom of seed germination. ANAC, transcription factor from the NAC (No apical meristem, ATAF1/2 and CUC2 cup-shaped cotyledon) family. Created with BioRender.com.

of ROS at the mid-time point to germination, relaying the initial ROS production by mitochondria but preceding the nuclear ROS generation (Jurdak et al., 2022; Fig. 2). The role of other subcellular compartments in ROS generation during seed germination remains to be elucidated. For example, the possible role of plastid-derived ROS during seed germination has not been investigated yet, although it could be important especially when germination occurs under light. Moreover, the most important phytohormones involved in seed germination, i.e. ABA and GAs, are also synthesized in plastids. Recently a specific class of plastid thioredoxins (Trxs), the y-type Trxs, which can fulfill antioxidant functions, have been demonstrated to be involved in Arabidopsis seed longevity and vigor (Née et al., 2021), and it has been proposed that their function in seed germination was linked to ABA and GA metabolisms. In summary, the available data suggest that the oxidative nuclear status can be considered as the central hub that controls the decision to germinate. These findings allow us to propose a comprehensive model of ROS trafficking in germinating seeds in which ROS systems are integrated in an oxidative network regulating seed germination (Fig. 2). Lastly it has to be emphasized that ROS production is not homogenous in the various embryonic tissues during seed germination and that it follows a dynamic spatiotemporal localization. At the tissue level it was shown that ROS production was first more important in the meristem area of the radicle but that it progressively re-localized in the elongating zone of the hypocotyl during further steps of seed germination. This reveals diverse functions of ROS during the germination process, from early sensing of environmental conditions (early oxidative burst at the radicle tip) to late cell wall weakening (late ROS production in the hypocotyl area; Bailly, 2019).

4. Mechanisms of action of ROS

The downstream effects of ROS on cell signaling has been widely studied and reviewed in a large array of physiological contexts (e.g., Foyer & Noctor, 2013; Mittler, Vanderauwera, Gollery, & Van Breusegem, 2004; Mittler, 2017; Noctor, Reichheld, & Foyer, 2018; Petrov & Van Breusegem, 2012; Waszczak, Carmody, & Kangasjärvi, 2018). However information about the mode of action of ROS during seed germination remain sparse.

4.1 Redox signaling and seed germination

Proteins are one of the most sensitive molecules to in vivo oxidation (Shacter, 2000). Redox regulation of thiol groups of cysteine has been extensively studied during seed germination (Arc et al., 2011; Buchanan & Balmer, 2005). ROS production during seed germination is associated with a redox change of protein thiols, that is likely to play a role in cell signaling, probably by interacting with hormone signaling pathways, and cell metabolism (Bykova, Hoehn, Rampitsch, Banks, et al., 2011; Bykova, Hoehn, Rampitsch, Hu, et al., 2011). However, it has also been demonstrated that during the first steps of seed imbibition, the resumption of active metabolism would induce a reduction in disulfide bonds of metabolic proteins (Nietzel et al., 2020) which can appear as contradictory with ROS production.

However, Meyer et al. (2021) proposed that this reduction step is a prerequisite for ROS signaling pathways to be activated during seed germination. Indeed, dry mature seeds are in an oxidative environment, as detailed below, but for ROS to be efficient in regulating proteins containing Cys targets, these proteins have first to be reduced to reset the redox system. Cellular redox agents, such as thioredoxins or glutathione/glutaredoxin participate inh redox regulation of target proteins. Thioredoxins (TRx), for example, have been shown to reduce oxidized proteins during seed imbibition and thus allow the progression of metabolism towards germination (Alkhalfioui et al., 2007; Arc et al., 2011; Wong et al., 2004), thereby initiating the molecular switch leading to germination. Redox regulation during seed germination is not playing only a metabolic role, as many enzymes are regulated by this process, but might also be important for controlling gene expression through kinases, phosphatases or redox sensitive transcription factors (He, Van Breusegem, & Mhamdi, 2018).

Protein carbonylation is another oxidative attack that has been frequently reported in germinating seeds. Carbonylation is irreversible and targets Lys, Arg, Pro, and Thr and induces formation of carbonyl groups (aldehydes and ketones; Nyström, 2005). Carbonyl groups are induced by ROS into the side chain of amino acids via the Fenton reaction or by non-enzymatic reactions with lipid-derived aldehydes resulting from lipid peroxidation (Tola, Jaballi, & Missihoun, 2021). Protein carbonylation has been found during germination of Arabidopsis (Job, Rajjou, Lovigny, Belghazi, & Job, 2005; Müller, Carstens, Linkies, Torres, & Leubner-Metzger, 2009), sunflower (Oracz et al., 2007), rice (Zhang, He, Li, & Yang, 2017), apple (Debska, Krasuska, Budnicka, Bogatek, & Gniazdowska, 2013), and pea (Barba-espín et al., 2011). The carbonylation of reserve proteins has often been observed and it was proposed that this mechanism would favor their mobilization (Job et al., 2005). However, it also been proposed that carbonylation could target negative regulators of seed germination and thus play a role in seed dormancy alleviation (Oracz et al., 2007).

Cellular response to ROS can also involve reversible protein phosphorylation (Baxter, Mittler, & Suzuki, 2014). The shift towards an oxidative status of cytosolic redox potential activates *MITOGEN ACTIVATED PROTEIN KINASE* (MAPK) signaling (Jalmi & Sinha, 2015; Liu & He, 2017) and the ROS-responsive MAPKKK, MEKK1, MPK3, MPK4, and MPK6 have been identified as playing specific roles in ROS signaling (Colcombet & Hirt, 2008). In addition, H_2O_2 has also been reported to activate MAPK by inactivating MAPK repressors (Liu & He, 2017). For example, in Arabidopsis seedlings it was shown that mitochondrial ROS production, previously evoked here as being important for seed germination, activated MAPK signaling (Chang, Jang, Branco-Price, Nghiem, & Bailey-Serres, 2012). Although MAPK are very good candidates for being involved in ROS signaling during seed germination, their role in this specific context is poorly documented. The role of *MPK1* and *MPK6* in the regulation of sunflower seed dormancy has been reported by Bazin et al. (2011) and Oracz et al. (2009), respectively. The phosphorylation of the transcription factor MYB44 by MPK3 and MPK6 was shown to regulate seed germination (Nguyen et al., 2012) and MPK8 promoted seed germination through its interaction with Teosinte Branched1/Cycloidea/Proliferating cell factor 14 (TCP14; Zhang et al., 2019), but in both cases the relationship with ROS homeostasis was not established. Several transcription factors act downstream of MAPKs in ROS signaling (Liu & He, 2017), and therefore further works will be required to bring additional insights about this mechanism during seed germination.

ROS are known to ultimately lead to changes in gene expression, which can rely on the activation of ROS responsive transcription factors. Furthermore, a complex transcription factor network that regulates gene expression during germination has recently been identified (Narsai et al., 2017). ROS responsive transcription factors can be activated by direct oxidation (Gill & Tuteja, 2010; Neill, Desikan, Clarke, Hurst, & Hancock, 2002), through the change in redox cellular status acting on redox-sensitive Cys residue (Dietz, 2014) or by MAPK cascades (He et al., 2018), as stated previously. The transcription factors activated by ROS can belong to different families including MYB, DREB, ZAT, bHLH, WRKY, bZIP and NAC for example (He et al., 2018; Khan, Li, Wang, & Yin, 2018) but they have been poorly studied during seed germination. Albertos et al. (2021) have recently demonstrated that the redox-sensitive transcription factor ANAC089 was repressing ABA synthesis and signaling during seed germination. In contrast, Jurdak et al. (2021) demonstrated that ANAC013 was positively involved in the response to ethylene and in MRR during dormancy alleviation of Arabidopsis seeds (Fig. 2).

4.2 Crosstalk between ROS and hormones

Given the paramount importance of hormone metabolism and signaling in the control of seed germination and dormancy (Finch-Savage & Leubner-Metzger, 2006), attention has been paid to the possible relationship between ROS and hormones. The relationship between ABA and

ROS is probably the best documented. It has been widely studied during ABA induction of stomatal closure which induce a signaling cascade that relies on ROS production thus demonstrating a synergetic effect between ABA and ROS (Postiglione & Muday, 2020). In contrast, in seed tissues, the relationship between ABA and ROS appears antagonistic. Many studies have demonstrated that ROS production during seed germination was associated with ABA degradation (Anand, Kumari, Thakur, & Koul, 2019; El-Maarouf-Bouteau et al., 2015; Ishibashi et al., 2017; Liu, Ye, Liu, Chen, & Zhang, 2010), and, in contrast, ABA inhibits ROS production in seeds (Bi et al., 2017; El-Maarouf-Bouteau et al., 2015; Lariguet et al., 2013). Since ABA regulates seed germination by its antagonistic interaction with GAs, it makes sense to also consider the relationship between ROS and GAs. ROS have been shown to act in concert with GAs synthesis and signaling to promote germination. They stimulate GAs biosynthesis (Anand et al., 2019; Bahin et al., 2011; Kai et al., 2016) and GAs stimulate ROS production (Cembrowska-Lech, Koprowski, & Kepczynski, 2015; El-Maarouf-Bouteau et al., 2015; Jurdak et al., 2022). Ethylene is another plant hormone that has been reported to stimulate germination and dormancy alleviation (Corbineau, Xia, Bailly, & El-Maarouf-Bouteau, 2014). The synergistic relationship between ethylene and ROS in seed dormancy release has been demonstrated in various species such as *Sorbus pohuashanensis* (Wang, Tang, Wang, Shen, & Yang, 2021), wild cardoon (Huarte, Puglia, Prjibelski, & Raccuia, 2020) *Brassica olereace* (Sami, Riaz, Zhou, Zhu, & Zhou, 2019) or sunflower (El-Maarouf-Bouteau et al., 2015). It was recently shown that release of Arabidopsis seed dormancy by ethylene triggered ROS production by the mitochondrial electron transfer chain, thus inducing retrograde signaling and nuclear ROS production (Jurdak et al., 2021).

4.3 From transcriptome to epigenome

Transcriptional reprogramming during seed germination is undoubtedly linked to a reorganization of the chromatin landscape. While dry seeds are characterized by small nuclei with highly condensed chromatin, seed imbibition induces drastic nuclear and chromatin reorganization leading to progressive increase in nuclear size and chromatin decondensation (Narsai et al., 2017; van Zanten et al., 2011; van Zanten, Tessadori, Peeters, & Fransz, 2012). Nuclear architecture and chromatin states can also undergo marked changes in response to environmental signals which makes them good candidates for regulating seed germination (reviewed in Bourbousse,

Benhamed, & Barneche, 2018; Probst & Mittelsten Scheid, 2015; Samo, Ebert, Kopka, & Mozgová, 2021). Nuclear reorganization could also play a role in the control of seed dormancy as suggested for example by the effect of COMPASS-like complex, a major actor in the deposition of H3K4me3 chromatin transcription hallmark, which regulates the DOG1 seed dormancy master regulatory gene in cis (Fiorucci et al., 2019). Recent evidences suggest that ROS could have either direct and indirect effects on chromatin and it has been proposed that the cell redox status could induce epigenetic modifications in plants under abiotic stress conditions and during developmental processes (see Ramakrishnan et al., 2022 for review). Cell redox status can intervene at multiple levels in controlling epigenetics modifications, such as post-translational modification of regulatory proteins, DNA methylation, histone acetylation and methylation, regulation of miRNA or chromatin remodeling (Bhat, Hora, Pal, Jha, & Taneja, 2017; Kietzmann, Petry, Shvetsova, Gerhold, & Görlach, 2017; Kreuz & Fischle, 2016; Ramakrishnan et al., 2022; Shen, Issakidis-Bourguet, & Zhou, 2016). Yet, while a redox-dependent modulation of chromatin organization has been proposed (Doyle & Fitzpatrick, 2010; Saravana Kumar et al., 2020), evidence of chromatin oxidation in the context of germination remains poorly documented. Jurdak et al. (2022) recently demonstrated that nuclear ROS accumulation, which precedes radicle protrusion, resulted in heterochromatin decompaction and reorganization. Therefore ROS-related changes in transcriptional activity during seed germination could involve chromatin decompaction and increase accessibility to DNA of the transcriptional machinery.

5. Concluding remarks & future perspectives

Significant progress has been made in understanding the role of ROS in seed germination. The spatio-temporal regulation of ROS production plays a key role in the decision and the ability of seeds to germinate. ROS homeostasis translates internal and external messages into a germinative response, and the oxidative memory transfers message from the developing seed to the resting seed and to the germinating seed. ROS downstream mode of action, which is better understood now, has to be considered with regards to the other signaling pathways regulating seed germination, among which hormones are the most important. Outstanding questions remain however unsolved. The specificity of cellular response towards the various

ROS is not known. The potential role of mitochondria as the master sensor of environmental and endogenous cues during seed germination should be properly addressed, as well as the relative parts of the other ROS producing systems. The importance of ROS trafficking during seed germination has just been discovered and is poorly known. It would be particularly interesting to identify the systems leading to ROS accumulation within the nucleus. Lastly ROS could be the missing link between chromatin organization, redox signaling and transcriptional reprogramming but this direct causal relationship has not been established yet.

References

Albertos, P., Tatematsu, K., Mateos, I., Sánchez-Vicente, I., Fernández-Arbaizar, A., Nakabayashi, K., ... Lorenzo, O. (2021). Redox feedback regulation of ANAC089 signaling alters seed germination and stress response. *Cell Reports, 35*(11), 109263. https://doi.org/10.1016/j.celrep.2021.109263

Alkhalfioui, F., Renard, M., Vensel, W. H., Wong, J., Tanaka, C. K., Hurkman, W. J., ... Montrichard, F. (2007). Thioredoxin-linked proteins are reduced during germination of Medicago truncatula seeds. *Plant Physiology, 144*(3), 1559–1579. https://doi.org/10.1104/pp.107.098103

Alonso-Blanco, C., Bentsink, L., Hanhart, C. J., Blankestijn-de Vries, H., & Koornneef, M. (2003). Analysis of natural allelic variation at seed dormancy loci of Arabidopsis thaliana. *Genetics, 164*(2), 711–729.

Alvarez, M., Bleich, A., & Donohue, K. (2021). Genetic differences in the temporal and environmental stability of transgenerational environmental effects. *Evolution, 75*(11), 2773–2790. https://doi.org/10.1111/evo.14367

Anand, A., Kumari, A., Thakur, M., & Koul, A. (2019). Hydrogen peroxide signaling integrates with phytohormones during the germination of magnetoprimed tomato seeds. *Scientific Reports, 9*(1), 8814. https://doi.org/10.1038/s41598-019-45102-5

Arc, E., Galland, M., Cueff, G., Godin, B., Lounifi, I., Job, D., & Rajjou, L. (2011). Reboot the system thanks to protein post translational modifications and proteome diversity. How quiescent seeds restart their metabolism to prepare seedling establishment. *Proteomics, 11*(9), 1606–1618. https://doi.org/10.1002/pmic.201000641

Bahin, E., Bailly, C., Sotta, B., Kranner, I., Corbineau, F., & Leymarie, J. (2011). Crosstalk between reactive oxygen species and hormonal signalling pathways regulates grain dormancy in barley. *Plant, Cell & Environment, 34*(6), 980–993. https://doi.org/10.1111/j.1365-3040.2011.02298.x

Bailly, C. (2004). Active oxygen species and antioxidants in seed biology. *Seed Science Research, 14*(2), 93–107. https://doi.org/10.1079/SSR2004159

Bailly, C. (2019). The signalling role of ROS in the regulation of seed germination and dormancy. *Biochemical Journal, 476*(20), 3019–3032. https://doi.org/10.1042/BCJ20190159

Bailly, C., & Merendino, L. (2021). Oxidative signalling in seed germination and early seedling growth: An emerging role for ROS trafficking and inter-organelle communication. *Biochemical Journal, 478*(10), 1977–1984. https://doi.org/10.1042/BCJ20200934

Bailly, C., Benamar, A., Corbineau, F., & Côme, D. (1998). Free radical scavenging as affected by accelerated ageing and subsequent priming in sunflower seeds. *Physiologia Plantarum, 104*(4), 646–652. https://doi.org/10.1034/j.1399-3054.1998.1040418.x

Bailly, C., El-Maarouf-Bouteau, H., & Corbineau, F. (2008). From intracellular signaling networks to cell death: The dual role of reactive oxygen species in seed physiology. *Comptes Rendus Biologies, 331*(10), 806–814. https://doi.org/10.1016/j.crvi.2008.07.022

Barba-espín, G., Diaz-vivancos, P., Job, D., Belghazi, M., Job, C., & Hernández, J. A. (2011). Understanding the role of H2O2 during pea seed germination: A combined proteomic and hormone profiling approach. *Plant, Cell & Environment, 34*(11), 1907–1919. https://doi.org/10.1111/j.1365-3040.2011.02386.x

Basbouss-Serhal, I., Pateyron, S., Cochet, F., Leymarie, J., & Bailly, C. (2017). 5′ to 3′ mRNA decay contributes to the regulation of arabidopsis seed germination by dormancy. *Plant Physiology, 173*(3), 1709–1723. https://doi.org/10.1104/pp.16.01933

Basbouss-Serhal, I., Soubigou-Taconnat, L., Bailly, C., & Leymarie, J. (2015). Germination potential of dormant and nondormant arabidopsis seeds is driven by distinct recruitment of messenger RNAs to polysomes. *Plant Physiology, 168*(3), 1049–1065. https://doi.org/10.1104/pp.15.00510

Baxter, A., Mittler, R., & Suzuki, N. (2014). ROS as key players in plant stress signalling. *Journal of Experimental Botany, 65*(5), 1229–1240. https://doi.org/10.1093/jxb/ert375

Bazin, J., Langlade, N., Vincourt, P., Arribat, S., Balzergue, S., El-Maarouf-Bouteau, H., & Bailly, C. (2011). Targeted mRNA oxidation regulates sunflower seed dormancy alleviation during dry after-ripening. *The Plant Cell, 23*(6), 2196–2208. https://doi.org/10.1105/tpc.111.086694

Bentsink, L., Jowett, J., Hanhart, C. J., & Koornneef, M. (2006). Cloning of DOG1, a quantitative trait locus controlling seed dormancy in Arabidopsis. *Proceedings of the National Academy of Sciences of the United States of America, 103*(45), 17042–17047. https://doi.org/10.1073/pnas.0607877103

Bewley, J. D. (1997). Seed germination and dormancy. *The Plant Cell, 9*(7), 1055–1066. https://doi.org/10.1105/tpc.9.7.1055

Bhat, A. V., Hora, S., Pal, A., Jha, S., & Taneja, R. (2017). Stressing the (epi)genome: Dealing with reactive oxygen species in cancer. *Antioxidants & Redox Signaling*. https://doi.org/10.1089/ars.2017.7158

Bhattacharjee, S., Chakrabarty, A., Kora, D., & Roy, U. K. (2022). Hydrogen peroxide induced antioxidant-coupled redox regulation of germination in rice: Redox metabolic, transcriptomic and proteomic evidences. *Journal of Plant Growth Regulation*. https://doi.org/10.1007/s00344-022-10615-3

Bi, C., Ma, Y., Wu, Z., Yu, Y.-T., Liang, S., Lu, K., & Wang, X.-F. (2017). Arabidopsis ABI5 plays a role in regulating ROS homeostasis by activating CATALASE 1 transcription in seed germination. *Plant Molecular Biology, 94*(1–2), 197–213. https://doi.org/10.1007/s11103-017-0603-y

Bourbousse, C., Benhamed, M., & Barneche, F. (2018). *Profiling developmentally and environmentally controlled chromatin reprogramming. Plant Chromatin Dynamics*. New York, NY: Humana Press, 3–30. https://doi.org/10.1007/978-1-4939-7318-7_1

Buchanan, B. B., & Balmer, Y. (2005). Redox regulation: A broadening horizon. *Annual Review of Plant Biology, 56*, 187–220. https://doi.org/10.1146/annurev.arplant.56.032604.144246

Buijs, G., Kodde, J., Groot, S. P. C., & Bentsink, L. (2018). Seed dormancy release accelerated by elevated partial pressure of oxygen is associated with DOG loci. *Journal of Experimental Botany, 69*(15), 3601–3608. https://doi.org/10.1093/jxb/ery156

Bykova, N. V., Hoehn, B., Rampitsch, C., Banks, T., Stebbing, J.-A., Fan, T., & Knox, R. (2011). Redox-sensitive proteome and antioxidant strategies in wheat seed dormancy control. *Proteomics, 11*(5), 865–882. https://doi.org/10.1002/pmic.200900810

Bykova, N. V., Hoehn, B., Rampitsch, C., Hu, J., Stebbing, J.-A., & Knox, R. (2011). Thiol redox-sensitive seed proteome in dormant and non-dormant hybrid genotypes of wheat. *Phytochemistry, 72*(10), 1162–1172. https://doi.org/10.1016/j.phytochem.2010.12.021

Carbone, D. L., Doorn, J. A., Kiebler, Z., & Petersen, D. R. (2005). Cysteine modification by lipid peroxidation products inhibits protein disulfide isomerase. *Chemical Research in Toxicology, 18*(8), 1324–1331. https://doi.org/10.1021/tx050078z

Carrera-Castaño, G., Calleja-Cabrera, J., Pernas, M., Gómez, L., & Oñate-Sánchez, L. (2020). An updated overview on the regulation of seed germination. *Plants, 9*(6), 703. https://doi.org/10.3390/plants9060703

Cembrowska-Lech, D., Koprowski, M., & Kepczynski, J. (2015). Germination induction of dormant Avena fatua caryopses by KAR(1) and GA(3) involving the control of reactive oxygen species (H2O2 and O-2(center dot-)) and enzymatic antioxidants (superoxide dismutase and catalase) both in the embryo and the aleurone layers. *Journal of Plant Physiology, 176*, 169–179. https://doi.org/10.1016/j.jplph.2014.11.010

Chang, R., Jang, C. J. H., Branco-Price, C., Nghiem, P., & Bailey-Serres, J. (2012). Transient MPK6 activation in response to oxygen deprivation and reoxygenation is mediated by mitochondria and aids seedling survival in Arabidopsis. *Plant Molecular Biology, 78*(1), 109–122. https://doi.org/10.1007/s11103-011-9850-5

Chen, C., Letnik, I., Hacham, Y., Dobrev, P., Ben-Daniel, B.-H., Vanková, R., ... Miller, G. (2014). ASCORBATE PEROXIDASE6 protects arabidopsis desiccating and germinating seeds from stress and mediates cross talk between reactive oxygen species, abscisic acid, and auxin. *Plant Physiology, 166*(1), 370–383. https://doi.org/10.1104/pp.114.245324

Chen, K., Li, G.-J., Bressan, R. A., Song, C.-P., Zhu, J.-K., & Zhao, Y. (2020). Abscisic acid dynamics, signaling, and functions in plants. *Journal of Integrative Plant Biology, 62*(1), 25–54. https://doi.org/10.1111/jipb.12899

Chiang, G. C. K., Bartsch, M., Barua, D., Nakabayashi, K., Debieu, M., Kronholm, I., ... De Meaux, J. (2011). DOG1 expression is predicted by the seed-maturation environment and contributes to geographical variation in germination in Arabidopsis thaliana. *Molecular Ecology, 20*(16), 3336–3349. https://doi.org/10.1111/j.1365-294X.2011.05181.x

Choudhury, F. K., Rivero, R. M., Blumwald, E., & Mittler, R. (2017). Reactive oxygen species, abiotic stress and stress combination. *The Plant Journal, 90*(5), 856–867. https://doi.org/10.1111/tpj.13299

Ciacka, K., Tyminski, M., Gniazdowska, A., & Krasuska, U. (2022). Cold stratification-induced dormancy removal in apple (Malus domestica Borkh.) seeds is accompanied by an increased glutathione pool in embryonic axes. *Journal of Plant Physiology, 274*, 153736. https://doi.org/10.1016/j.jplph.2022.153736

Colcombet, J., & Hirt, H. (2008). Arabidopsis MAPKs: A complex signalling network involved in multiple biological processes. *Biochemical Journal, 413*(2), 217. https://doi.org/10.1042/BJ20080625

Corbineau, F., Xia, Q., Bailly, C., & El-Maarouf-Bouteau, H. (2014). Ethylene, a key factor in the regulation of seed dormancy. *Frontiers in Plant Science, 5*, 539. https://doi.org/10.3389/fpls.2014.00539

Corpas, F. J., González-Gordo, S., & Palma, J. M. (2020). Plant peroxisomes: A factory of reactive species. *Frontiers in Plant Science, 11*. https://www.frontiersin.org/articles/10.3389/fpls.2020.00853

De Clercq, I., Vermeirssen, V., Van Aken, O., Vandepoele, K., Murcha, M. W., Law, S. R., ... Van Breusegem, F. (2013). The membrane-bound NAC transcription factor ANAC013 functions in mitochondrial retrograde regulation of the oxidative stress response in Arabidopsis. *The Plant Cell, 25*(9), 3472–3490. https://doi.org/10.1105/tpc.113.117168

Debska, K., Krasuska, U., Budnicka, K., Bogatek, R., & Gniazdowska, A. (2013). Dormancy removal of apple seeds by cold stratification is associated with fluctuation in H_2O_2, NO production and protein carbonylation level. *Journal of Plant Physiology, 170*(5), 480–488. https://doi.org/10.1016/j.jplph.2012.11.018

Dietz, K.-J. (2014). Redox regulation of transcription factors in plant stress acclimation and development. *Antioxidants & Redox Signaling, 21*(9), 1356–1372. https://doi.org/10.1089/ars.2013.5672

Donohue, K. (2009). Completing the cycle: Maternal effects as the missing link in plant life histories. *Philosophical Transactions of the Royal Society B: Biological Sciences, 364*(1520), 1059–1074. https://doi.org/10.1098/rstb.2008.0291

Donohue, K., Heschel, M. S., Chiang, G. C. K., Butler, C. M., & Barua, D. (2007). Phytochrome mediates germination responses to multiple seasonal cues. *Plant, Cell & Environment, 30*(2), 202–212. https://doi.org/10.1111/j.1365-3040.2006.01619.x

Doyle, K., & Fitzpatrick, F. A. (2010). Redox signaling, alkylation (carbonylation) of conserved cysteines inactivates class I histone deacetylases 1, 2, and 3 and antagonizes their transcriptional repressor function*. *Journal of Biological Chemistry, 285*(23), 17417–17424. https://doi.org/10.1074/jbc.M109.089250

El-Maarouf-Bouteau, H., Meimoun, P., Job, C., Job, D., & Bailly, C. (2013). Role of protein and mRNA oxidation in seed dormancy and germination. *Frontiers in Plant Science, 4*. https://doi.org/10.3389/fpls.2013.00077

El-Maarouf-Bouteau, H., Sajjad, Y., Bazin, J., Langlade, N., Cristescu, S. M., Balzergue, S., ... Bailly, C. (2015). Reactive oxygen species, abscisic acid and ethylene interact to regulate sunflower seed germination. *Plant Cell and Environment, 38*(2), 364–374. https://doi.org/10.1111/pce.12371

Finch-Savage, W. E., & Leubner-Metzger, G. (2006). Seed dormancy and the control of germination. *New Phytologist, 171*(3), 501–523. https://doi.org/10.1111/j.1469-8137.2006.01787.x

Fiorucci, A.-S., Bourbousse, C., Concia, L., Rougée, M., Deton-Cabanillas, A.-F., Zabulon, G., ... Barneche, F. (2019). Arabidopsis S2Lb links AtCOMPASS-like and SDG2 activity in H3K4me3 independently from histone H2B monoubiquitination. *Genome Biology, 20*(1), 100. https://doi.org/10.1186/s13059-019-1705-4

Foyer, C. H., & Noctor, G. (2013). Redox signaling in plants. *Antioxidants & Redox Signaling, 18*(16), 2087–2090. https://doi.org/10.1089/ars.2013.5278

Fry, S. C. (1998). Oxidative scission of plant cell wall polysaccharides by ascorbate-induced hydroxyl radicals. *Biochemical Journal, 332*(2), 507–515. https://doi.org/10.1042/bj3320507

Gerna, D., Arc, E., Holzknecht, M., Roach, T., Jansen-Dürr, P., Weiss, A. K. H., & Kranner, I. (2021). AtFAHD1a: A new player influencing seed longevity and dormancy in arabidopsis? Art. 6 *International Journal of Molecular Sciences, 22*(6), https://doi.org/10.3390/ijms22062997

Gerna, D., Ballesteros, D., Arc, E., Stöggl, W., Seal, C. E., Marami-Zonouz, N., ... Roach, T. (2022). Does oxygen affect ageing mechanisms of Pinus densiflora seeds? A matter of cytoplasmic physical state. *Journal of Experimental Botany, 73*(8), 2631–2649. https://doi.org/10.1093/jxb/erac024

Gill, S. S., & Tuteja, N. (2010). Reactive oxygen species and antioxidant machinery in abiotic stress tolerance in crop plants. *Plant Physiology and Biochemistry, 48*(12), 909–930. https://doi.org/10.1016/j.plaphy.2010.08.016

Graeber, K., Nakabayashi, K., & Leubner-Metzger, G. (2017). Development of dormancy. In B. Thomas, B. G. Murray, & D. J. Murphy (Eds.). *Encyclopedia of applied plant sciences* (pp. 483–489). (2nd ed.). Academic Press. https://doi.org/10.1016/B978-0-12-394807-6.00209-4

Graeber, K., Nakabayashi, K., Miatton, E., Leubner-Metzger, G., & Soppe, W. J. J. (2012). Molecular mechanisms of seed dormancy. *Plant, Cell & Environment, 35*(10), 1769–1786. https://doi.org/10.1111/j.1365-3040.2012.02542.x

Groot, S. P. C., Surki, A. A., de Vos, R. C. H., & Kodde, J. (2012). Seed storage at elevated partial pressure of oxygen, a fast method for analysing seed ageing under dry conditions. *Annals of Botany, 110*(6), 1149–1159. https://doi.org/10.1093/aob/mcs198

Gutierrez, L., Van Wuytswinkel, O., Castelain, M., & Bellini, C. (2007). Combined networks regulating seed maturation. *Trends in Plant Science, 12*(7), 294–300. https://doi.org/10.1016/j.tplants.2007.06.003

He, H., Van Breusegem, F., & Mhamdi, A. (2018). Redox-dependent control of nuclear transcription in plants. *Journal of Experimental Botany, 69*(14), 3359–3372. https://doi.org/10.1093/jxb/ery130

Huang, Z., Footitt, S., & Finch-Savage, W. E. (2014). The effect of temperature on reproduction in the summer and winter annual Arabidopsis thaliana ecotypes Bur and Cvi. *Annals of Botany, 113*(6), 921–929. https://doi.org/10.1093/aob/mcu014

Huarte, H. R., Puglia, G. D., Prjibelski, A. D., & Raccuia, S. A. (2020). Seed transcriptome annotation reveals enhanced expression of genes related to ROS homeostasis and ethylene metabolism at alternating temperatures in wild cardoon. *Plants-Basel, 9*(9), 1225. https://doi.org/10.3390/plants9091225

Ishibashi, Y., Aoki, N., Kasa, S., Sakamoto, M., Kai, K., Tomokiyo, R., ... Iwaya-Inoue, M. (2017). The interrelationship between abscisic acid and reactive oxygen species plays a key role in barley seed dormancy and germination. *Frontiers in Plant Science, 8*, 275. https://doi.org/10.3389/fpls.2017.00275

Ishibashi, Y., Tawaratsumida, T., Zheng, S.-H., Yuasa, T., & Iwaya-Inoue, M. (2010). NADPH oxidases act as key enzyme on germination and seedling growth in barley (Hordeum vulgare L.). *Plant Production Science, 13*(1), 45–52. https://doi.org/10.1626/pps.13.45

Jalmi, S., & Sinha, A. (2015). ROS mediated MAPK signaling in abiotic and biotic stress-striking similarities and differences. *Frontiers in Plant Science, 6*. https://www.frontiersin.org/article/10.3389/fpls.2015.00769

Jezek, P., & Hlavatá, L. (2005). Mitochondria in homeostasis of reactive oxygen species in cell, tissues, and organism. *The International Journal of Biochemistry & Cell Biology, 37*(12), 2478–2503. https://doi.org/10.1016/j.biocel.2005.05.013

Job, C., Rajjou, L., Lovigny, Y., Belghazi, M., & Job, D. (2005). Patterns of protein oxidation in Arabidopsis seeds and during germination. *Plant Physiology, 138*(2), 790–802. https://doi.org/10.1104/pp.105.062778

Jurdak, R., Launay-Avon, A., Paysant-Le Roux, C., & Bailly, C. (2021). Retrograde signalling from the mitochondria to the nucleus translates the positive effect of ethylene on dormancy breaking of Arabidopsis thaliana seeds. *New Phytologist, 229*(4), 2192–2205. https://doi.org/10.1111/nph.16985

Jurdak, R., Rodrigues, G., de, A. G., Chaumont, N., Schivre, G., Bourbousse, C., ... Bailly, C. (2022). Intracellular reactive oxygen species trafficking participates in seed dormancy alleviation in Arabidopsis seeds. *New Phytologist, 234*(3), 850–866. https://doi.org/10.1111/nph.18038

Kai, K., Kasa, S., Sakamoto, M., Aoki, N., Watabe, G., Yuasa, T., ... Ishibashi, Y. (2016). Role of reactive oxygen species produced by NADPH oxidase in gibberellin biosynthesis during barley seed germination. *Plant Signaling & Behavior, 11*(5), e1180492. https://doi.org/10.1080/15592324.2016.1180492

Kao, Y.-T., Gonzalez, K. L., & Bartel, B. (2018). Peroxisome function, biogenesis, and dynamics in plants1[OPEN]. *Plant Physiology, 176*(1), 162–177. https://doi.org/10.1104/pp.17.01050

Katsuya-Gaviria, K., Caro, E., Carrillo-Barral, N., & Iglesias-Fernández, R. (2020). Reactive oxygen species (ROS) and nucleic acid modifications during seed dormancy. Art. 6 *Plants, 9*(6), https://doi.org/10.3390/plants9060679

Kaur, H., Petla, B. P., Kamble, N. U., Singh, A., Rao, V., Salvi, P., ... Majee, M. (2015). Differentially expressed seed aging responsive heat shock protein OsHSP18.2 implicates in seed vigor, longevity and improves germination and seedling establishment under abiotic stress. *Frontiers in Plant Science, 6*, 713. https://doi.org/10.3389/fpls.2015.00713

Khan, S.-A., Li, M.-Z., Wang, S.-M., & Yin, H.-J. (2018). Revisiting the role of plant transcription factors in the battle against abiotic stress. *International Journal of Molecular Sciences, 19*(6), https://doi.org/10.3390/ijms19061634

Kietzmann, T., Petry, A., Shvetsova, A., Gerhold, J. M., & Görlach, A. (2017). The epigenetic landscape related to reactive oxygen species formation in the cardiovascular system. *British Journal of Pharmacology, 174*(12), 1533–1554. https://doi.org/10.1111/bph.13792

Kranner, I., & Birtić, S. (2005). A modulating role for antioxidants in desiccation tolerance. *Integrative and Comparative Biology, 45*(5), 734–740. https://doi.org/10.1093/icb/45.5.734

Kranner, I., Minibayeva, F. V., Beckett, R. P., & Seal, C. E. (2010). What is stress? Concepts, definitions and applications in seed science. *New Phytologist, 188*(3), 655–673. https://doi.org/10.1111/j.1469-8137.2010.03461.x

Kreuz, S., & Fischle, W. (2016). Oxidative stress signaling to chromatin in health and disease. *Epigenomics, 8*(6), 843–862. https://doi.org/10.2217/epi-2016-0002

Kurek, K., Plitta-Michalak, B., & Ratajczak, E. (2019). Reactive oxygen species as potential drivers of the seed aging process. *Plants, 8*(6), 174. https://doi.org/10.3390/plants8060174

Lachabrouilli, A.-S., Rigal, K., Corbineau, F., & Bailly, C. (2021). Effects of agroclimatic conditions on sunflower seed dormancy at harvest. *European Journal of Agronomy, 124*, 126209. https://doi.org/10.1016/j.eja.2020.126209

Lariguet, P., Ranocha, P., De Meyer, M., Barbier, O., Penel, C., & Dunand, C. (2013). Identification of a hydrogen peroxide signalling pathway in the control of light-dependent germination in Arabidopsis. *Planta, 238*(2), 381–395. https://doi.org/10.1007/s00425-013-1901-5

Leprince, O., Pellizzaro, A., Berriri, S., & Buitink, J. (2017). Late seed maturation: Drying without dying. *Journal of Experimental Botany, 68*(4), 827–841. https://doi.org/10.1093/jxb/erw363

Leymarie, J., Vitkauskaite, G., Hoang, H. H., Gendreau, E., Chazoule, V., Meimoun, P., ... Bailly, C. (2012). Role of reactive oxygen species in the regulation of arabidopsis seed dormancy. *Plant and Cell Physiology, 53*(1), 96–106. https://doi.org/10.1093/pcp/pcr129

Li, W.-Y., Chen, B.-X., Chen, Z.-J., Gao, Y.-T., Chen, Z., & Liu, J. (2017). Reactive oxygen species generated by NADPH oxidases promote radicle protrusion and root elongation during rice seed germination. *International Journal of Molecular Sciences, 18*(1), 110. https://doi.org/10.3390/ijms18010110

Liszkay, A., Kenk, B., & Schopfer, P. (2003). Evidence for the involvement of cell wall peroxidase in the generation of hydroxyl radicals mediating extension growth. *Planta, 217*(4), 658–667. https://doi.org/10.1007/s00425-003-1028-1

Liu, Y., & He, C. (2017). A review of redox signaling and the control of MAP kinase pathway in plants. *Redox Biology, 11*, 192–204. https://doi.org/10.1016/j.redox.2016.12.009

Liu, Y., Ye, N., Liu, R., Chen, M., & Zhang, J. (2010). H2O2 mediates the regulation of ABA catabolism and GA biosynthesis in Arabidopsis seed dormancy and germination. *Journal of Experimental Botany, 61*(11), 2979–2990. https://doi.org/10.1093/jxb/erq125

Ma, W., Guan, X., Li, J., Pan, R., Wang, L., Liu, F., ... Zhang, T. (2019). Mitochondrial small heat shock protein mediates seed germination via thermal sensing. *Proceedings of the National Academy of Sciences of the United States of America, 116*(10), 4716–4721. https://doi.org/10.1073/pnas.1815790116

Matilla, A. J. (2020). Seed dormancy: Molecular control of its induction and alleviation. Art. 10 *Plants, 9*(10), https://doi.org/10.3390/plants9101402

Meyer, A. J., Dreyer, A., Ugalde, J. M., Feitosa-Araujo, E., Dietz, K.-J., & Schwarzländer, M. (2021). Shifting paradigms and novel players in Cys-based redox regulation and ROS signaling in plants—And where to go next. *Biological Chemistry, 402*(3), 399–423. https://doi.org/10.1515/hsz-2020-0291

Mignolet-Spruyt, L., Xu, E., Idänheimo, N., Hoeberichts, F. A., Mühlenbock, P., Brosché, M., ... Kangasjärvi, J. (2016). Spreading the news: Subcellular and organellar reactive oxygen species production and signalling. *Journal of Experimental Botany, 67*(13), 3831–3844. https://doi.org/10.1093/jxb/erw080

Mittler, R. (2017). ROS are good. *Trends in Plant Science, 22*(1), 11–19. https://doi.org/10.1016/j.tplants.2016.08.002

Mittler, R., Vanderauwera, S., Gollery, M., & Van Breusegem, F. (2004). Reactive oxygen gene network of plants. *Trends in Plant Science, 9*(10), 490–498. https://doi.org/10.1016/j.tplants.2004.08.009

Mittler, R., Zandalinas, S. I., Fichman, Y., & Van Breusegem, F. (2022). Reactive oxygen species signalling in plant stress responses. *Nature Reviews Molecular Cell Biology, 23*(10), 663–679. https://doi.org/10.1038/s41580-022-00499-2

Morscher, F., Kranner, I., Arc, E., Bailly, C., & Roach, T. (2015). Glutathione redox state, tocochromanols, fatty acids, antioxidant enzymes and protein carbonylation in sunflower seed embryos associated with after-ripening and ageing. *Annals of Botany, 116*(4), 669–678. https://doi.org/10.1093/aob/mcv108

Müller, K., Carstens, A. C., Linkies, A., Torres, M. A., & Leubner-Metzger, G. (2009). The NADPH-oxidase AtrbohB plays a role in Arabidopsis seed after-ripening. *New Phytologist, 184*(4), 885–897. https://doi.org/10.1111/j.1469-8137.2009.03005.x

Müller, K., Linkies, A., Vreeburg, R. A. M., Fry, S. C., Krieger-Liszkay, A., & Leubner-Metzger, G. (2009). In vivo cell wall loosening by hydroxyl radicals during cress seed germination and elongation growth. *Plant Physiology, 150*(4), 1855–1865. https://doi.org/10.1104/pp.109.139204

Murphy, M. P. (2009). How mitochondria produce reactive oxygen species. *Biochemical Journal, 417*(Pt 1), 1–13. https://doi.org/10.1042/BJ20081386

Nagel, M., Kodde, J., Pistrick, S., Mascher, M., Borner, A., & Groot, S. P. C. (2016). Barley seed aging: Genetics behind the dry elevated pressure of oxygen aging and moist controlled deterioration. *Frontiers in Plant Science, 7*, 388. https://doi.org/10.3389/fpls.2016.00388

Nakabayashi, K., Okamoto, M., Koshiba, T., Kamiya, Y., & Nambara, E. (2005). Genome-wide profiling of stored mRNA in Arabidopsis thaliana seed germination: Epigenetic and genetic regulation of transcription in seed. *Plant Journal, 41*(5), 697–709. https://doi.org/10.1111/j.1365-313X.2005.02337.x

Narsai, R., Gouil, Q., Secco, D., Srivastava, A., Karpievitch, Y. V., Liew, L. C., ... Whelan, J. (2017). Extensive transcriptomic and epigenomic remodelling occurs during Arabidopsis thaliana germination. *Genome Biology, 18*, 172. https://doi.org/10.1186/s13059-017-1302-3

Née, G., Châtel-Innocenti, G., Meimoun, P., Leymarie, J., Montrichard, F., Satour, P., ... Issakidis-Bourguet, E. (2021). A new role for plastid thioredoxins in seed physiology in relation to hormone regulation. Art. 19 *International Journal of Molecular Sciences, 22*(19), https://doi.org/10.3390/ijms221910395

Neill, S. J., Desikan, R., Clarke, A., Hurst, R. D., & Hancock, J. T. (2002). Hydrogen peroxide and nitric oxide as signalling molecules in plants. *Journal of Experimental Botany, 53*(372), 1237–1247.

Ng, S., De Clercq, I., Van Aken, O., Law, S. R., Ivanova, A., Willems, P., ... Whelan, J. (2014). Anterograde and retrograde regulation of nuclear genes encoding mitochondrial proteins during growth, development, and stress. *Molecular Plant, 7*(7), 1075–1093. https://doi.org/10.1093/mp/ssu037

Nguyen, X. C., Hoang, M. H. T., Kim, H. S., Lee, K., Liu, X.-M., Kim, S. H., ... Chung, W. S. (2012). Phosphorylation of the transcriptional regulator MYB44 by mitogen activated protein kinase regulates Arabidopsis seed germination. *Biochemical and Biophysical Research Communications, 423*(4), 703–708. https://doi.org/10.1016/j.bbrc.2012.06.019

Nietzel, T., Mostertz, J., Ruberti, C., Née, G., Fuchs, P., Wagner, S., ... Schwarzländer, M. (2020). Redox-mediated kick-start of mitochondrial energy metabolism drives resource-efficient seed germination. *Proceedings of the National Academy of Sciences of the United States of America, 117*(1), 741–751. https://doi.org/10.1073/pnas.1910501117

Noctor, G., Mhamdi, A., & Foyer, C. H. (2014). The roles of reactive oxygen metabolism in drought: Not so cut and dried. *Plant Physiology, 164*(4), 1636–1648. https://doi.org/10.1104/pp.113.233478

Noctor, G., Reichheld, J.-P., & Foyer, C. H. (2018). ROS-related redox regulation and signaling in plants. *Seminars in Cell & Developmental Biology, 80*, 3–12.

Nonogaki, H. (2019). Chapter Five—ABA responses during seed development and germination. In M. Seo, & A. Marion-Poll (Vol. Eds.), *Advances in botanical research: Vol. 92*, (pp. 171–217). Academic Press. https://doi.org/10.1016/bs.abr.2019.04.005

Nonogaki, H., Bassel, G. W., & Bewley, J. D. (2010). Germination-still a mystery. *Plant Science, 179*(6), 574–581. https://doi.org/10.1016/j.plantsci.2010.02.010

Nyström, T. (2005). Role of oxidative carbonylation in protein quality control and senescence. *The EMBO Journal, 24*(7), 1311–1317. https://doi.org/10.1038/sj.emboj.7600599

Oracz, K., Bouteau, H. E.-M., Farrant, J. M., Cooper, K., Belghazi, M., Job, C., ... Bailly, C. (2007). ROS production and protein oxidation as a novel mechanism for seed dormancy alleviation. *Plant Journal, 50*(3), 452–465. https://doi.org/10.1111/j.1365-313X.2007.03063.x

Oracz, K., El-Maarouf-Bouteau, H., Kranner, I., Bogatek, R., Corbineau, F., & Bailly, C. (2009). The mechanisms involved in seed dormancy alleviation by hydrogen cyanide unravel the role of reactive oxygen species as key factors of cellular signaling during germination. *Plant Physiology, 150*(1), 494–505. https://doi.org/10.1104/pp.109.138107

Petrov, V. D., & Van Breusegem, F. (2012). Hydrogen peroxide-a central hub for information flow in plant cells. *AoB PLANTS, 2012*, pls014. https://doi.org/10.1093/aobpla/pls014

Podgórska, A., Burian, M., & Szal, B. (2017). Extra-cellular but extra-ordinarily important for cells: Apoplastic reactive oxygen species metabolism. *Frontiers in Plant Science, 8*. https://www.frontiersin.org/article/10.3389/fpls.2017.01353

Postiglione, A. E., & Muday, G. K. (2020). The role of ROS homeostasis in ABA-induced guard cell signaling. *Frontiers in Plant Science, 11*. https://www.frontiersin.org/articles/10.3389/fpls.2020.00968

Probst, A. V., & Mittelsten Scheid, O. (2015). Stress-induced structural changes in plant chromatin. *Current Opinion in Plant Biology, 27*(Suppl. C), 8–16. https://doi.org/10.1016/j.pbi.2015.05.011

Ramakrishnan, M., Papolu, P. K., Satish, L., Vinod, K. K., Wei, Q., Sharma, A., ... Zhou, M. (2022). Redox status of the plant cell determines epigenetic modifications under abiotic stress conditions and during developmental processes. *Journal of Advanced Research*. https://doi.org/10.1016/j.jare.2022.04.007

Sami, A., Riaz, M. W., Zhou, X., Zhu, Z., & Zhou, K. (2019). Alleviating dormancy in Brassica oleracea seeds using NO and KAR1 with ethylene biosynthetic pathway, ROS and antioxidant enzymes modifications. *BMC Plant Biology, 19*(1), 577. https://doi.org/10.1186/s12870-019-2118-y

Samo, N., Ebert, A., Kopka, J., & Mozgová, I. (2021). Plant chromatin, metabolism and development – An intricate crosstalk. *Current Opinion in Plant Biology, 61*, 102002. https://doi.org/10.1016/j.pbi.2021.102002

Sano, N., & Marion-Poll, A. (2021). ABA metabolism and homeostasis in seed dormancy and germination. Art. 10 *International Journal of Molecular Sciences, 22*(10), https://doi.org/10.3390/ijms22105069

Saravana Kumar, R. M., Wang, Y., Zhang, X., Cheng, H., Sun, L., He, S., & Hao, F. (2020). Redox components: Key regulators of epigenetic modifications in plants. Art. 4 *International Journal of Molecular Sciences, 21*(4), https://doi.org/10.3390/ijms21041419

Satour, P., Youssef, C., Châtelain, E., Vu, B. L., Teulat, B., Job, C., ... Montrichard, F. (2018). Patterns of protein carbonylation during Medicago truncatula seed maturation. *Plant, Cell & Environment, 41*(9), 2183–2194. https://doi.org/10.1111/pce.13194

Sattler, S. E., Gilliland, L. U., Magallanes-Lundback, M., Pollard, M., & DellaPenna, D. (2004). Vitamin E is essential for seed longevity and for preventing lipid peroxidation during germination. *The Plant Cell Online, 16*(6), 1419–1432. https://doi.org/10.1105/tpc.021360

Shacter, E. (2000). Quantification and significance of protein oxidation in biological samples. *Drug Metabolism Reviews, 32*(3–4), 307–326. https://doi.org/10.1081/DMR-100102336

Shen, Y., Issakidis-Bourguet, E., & Zhou, D.-X. (2016). Perspectives on the interactions between metabolism, redox, and epigenetics in plants. *Journal of Experimental Botany, 67*(18), 5291–5300. https://doi.org/10.1093/jxb/erw310

Smolikova, G., Leonova, T., Vashurina, N., Frolov, A., & Medvedev, S. (2020). Desiccation tolerance as the basis of long-term seed viability. *International Journal of Molecular Sciences, 22*(1), 101. https://doi.org/10.3390/ijms22010101

Steinbrecher, T., & Leubner-Metzger, G. (2018). Tissue and cellular mechanics of seeds. *Current Opinion in Genetics & Development, 51*, 1–10. https://doi.org/10.1016/j.gde.2018.03.001

Su, T., Li, W., Wang, P., & Ma, C. (2019). Dynamics of peroxisome homeostasis and its role in stress response and signaling in plants. *Frontiers in Plant Science, 10*. https://www.frontiersin.org/articles/10.3389/fpls.2019.00705

Sun, L., Wu, Q., & Mao, X. (2022). Effects of oxidation modification by malondialdehyde on the structure and functional properties of walnut protein. Art. 16 *Foods, 11*(16), https://doi.org/10.3390/foods11162432

Tola, A. J., Jaballi, A., & Missihoun, T. D. (2021). Protein carbonylation: Emerging roles in plant redox biology and future prospects. Art. 7 *Plants, 10*(7), https://doi.org/10.3390/plants10071451

van Zanten, M., Koini, M. A., Geyer, R., Liu, Y., Brambilla, V., Bartels, D., ... Soppe, W. J. J. (2011). Seed maturation in Arabidopsis thaliana is characterized by nuclear size reduction and increased chromatin condensation. *Proceedings of the National Academy of Sciences, 108*(50), 20219–20224. https://doi.org/10.1073/pnas.1117726108

van Zanten, M., Tessadori, F., Peeters, A. J. M., & Fransz, P. (2012). Shedding light on large-scale chromatin reorganization in Arabidopsis thaliana. *Molecular Plant, 5*(3), 583–590. https://doi.org/10.1093/mp/sss030

Vancostenoble, B., Blanchet, N., Langlade, N. B., & Bailly, C. (2022). Maternal drought stress induces abiotic stress tolerance to the progeny at the germination stage in sunflower. *Environmental and Experimental Botany, 201*, 104939. https://doi.org/10.1016/j.envexpbot.2022.104939

Wang, H., Tang, S., Wang, J., Shen, H., & Yang, L. (2021). Interaction between reactive oxygen species and hormones during the breaking of embryo dormancy in Sorbus pohuashanensis by exogenous nitric oxide. *Journal of Forestry Research*. https://doi.org/10.1007/s11676-021-01330-y

Waszczak, C., Carmody, M., & Kangasjärvi, J. (2018). Reactive oxygen species in plant signaling. *Annual Review of Plant Biology, 69*(1), 209–236. https://doi.org/10.1146/annurev-arplant-042817-040322

Wong, J. H., Cai, N., Balmer, Y., Tanaka, C. K., Vensel, W. H., Hurkman, W. J., & Buchanan, B. B. (2004). Thioredoxin targets of developing wheat seeds identified by complementary proteomic approaches. *Phytochemistry, 65*(11), 1629–1640. https://doi.org/10.1016/j.phytochem.2004.05.010

Xiang, Y., Nakabayashi, K., Ding, J., He, F., Bentsink, L., & Soppe, W. J. J. (2014). Reduced Dormancy5 encodes a protein phosphatase 2C that is required for seed dormancy in Arabidopsis. *The Plant Cell, 26*(11), 4362–4375. https://doi.org/10.1105/tpc.114.132811

Yang, X., Zhang, F., Yang, M., He, Y., Li, Z., Yang, J., & Wang, X. (2020). The NADPH-oxidase LsRbohC1 plays a role in lettuce (Lactuca sativa) seed germination. *Plant Physiology and Biochemistry, 154*, 751–757. https://doi.org/10.1016/j.plaphy.2020.05.042

Zhang, H., He, D., Li, M., & Yang, P. (2017). Carbonylated protein changes between active germinated embryos and quiescent embryos give insights into rice seed germination regulation. *Plant Growth Regulation, 83*(2), 335–350. https://doi.org/10.1007/s10725-017-0299-7

Zhang, K., Zhang, Y., Sun, J., Meng, J., & Tao, J. (2021). Deterioration of orthodox seeds during ageing: Influencing factors, physiological alterations and the role of reactive oxygen species. *Plant Physiology and Biochemistry, 158*, 475–485. https://doi.org/10.1016/j.plaphy.2020.11.031

Zhang, W., Cochet, F., Ponnaiah, M., Lebreton, S., Matheron, L., Pionneau, C., ... Baudouin, E. (2019). The MPK8-TCP14 pathway promotes seed germination in Arabidopsis. *The Plant Journal: For Cell and Molecular Biology*. https://doi.org/10.1111/tpj.14461

CHAPTER SEVEN

ROS metabolism and ripening of fleshy fruits

Francisco J. Corpas[a,*], Luciano Freschi[b], and José M. Palma[a]
[a]Group of Antioxidants, Free Radicals and Nitric Oxide in Biotechnology, Food and Agriculture, Department of Biochemistry, Cell and Molecular Biology of Plants, Estación Experimental del Zaidín, Spanish National Research Council (CSIC), Granada, Spain
[b]Department of Botany, Institute of Biosciences, University of São Paulo (USP), São Paulo, Brazil
*Corresponding author. e-mail address: javier.corpas@eez.csic.es

Contents

Abstract

The ripening of fleshy fruit is a complex physiological process that involves drastic changes at phenotypic, genomic, proteomic, and biochemical levels, where the different subcellular compartments undergo biochemical and metabolic reorganizations, the shift of chloroplasts to chromoplasts being perhaps the best example. In this process, the unripe fruits become an edible ripe product in which the metabolism of reactive oxygen species (ROS) has a relevant leading role. ROS are involved in the regulation of the ripening of climacteric and non-climacteric fruits due to their functions as signaling molecules in coordination with other compounds, including phytohormones (ethylene), nitric oxide (NO), hydrogen sulfide (H_2S), and melatonin. ROS metabolism is also very relevant during postharvest storage of fruits, where it can exert a dual effect since it can trigger oxidative damage contributing to the loss of fruit quality but, on the other hand, it can exert a protective role against pathogens. From an anthropic point of view, there are different categories of fleshy and edible fruits, with many representatives possessing high content of antioxidant properties. Their nutraceutical properties can improve human health, especially considering fruits are predominantly eaten fresh, consequently maintaining their antioxidant properties at consumption, which could be lost otherwise by cooking.

Advances in Botanical Research, Volume 105
ISSN 0065-2296, https://doi.org/10.1016/bs.abr.2022.08.024

1. Introduction

Fruit ripening implies drastic changes at the phenotypic level, flavor, and nutritional components that are the consequence of transcriptomic, biochemical, cellular, and structural modifications whose final physiological goal is the generation and dispersal of seeds allowing the species survival over generations (Dardick & Callahan, 2014). However, from an anthropic and nutritional point of view, fruit ripening requires that they have certain attractive organoleptic and visual characteristics for their consumption. Fleshy fruits are generally cataloged into two main groups: (i) climacteric fruits, which are characterized to have an increased respiration rate, followed by a burst of ethylene biosynthesis during ripening; and (ii) non-climacteric fruits, which do not have any peak of ethylene production or respiration during ripening. Fig. 1 depicts some representative fruits of both categories, most displaying a wide diversity of colors due to the pigments they accumulate both externally and internally. From the industrial perspective, this difference among fruits is very relevant because climacteric fruits can ripen after harvest, while non-climacteric fruits cannot. Another aspect of great relevance for the fruit industry concerns the storage after harvest to maintain the fruits under optimal conditions once they reach the consumers. Thus, fruit should display the best visual and organoleptic properties until distribution and consumption. For this reason, the fruit industry has developed different procedures for their conservation, such as storing at low temperatures, a low-oxygen atmosphere, or involving the treatment with tested and approved chemical products to avoid or delay the damage associated with postharvest storage. These protocols are also designed to prevent possible attacks by pathogens (viruses, bacteria or fungi) which, in most cases, are associated with an uncontrolled overproduction of reactive oxygen species (ROS) that can initiate oxidative damages and loss of the fruit quality.

The present work aims to provide a broad perspective of the relevance of ROS metabolism during the ripening of fleshy fruits, which usually involves physiological oxidative stress (phystress). Likewise, it will review how the application of exogenous molecules with signaling functions such as nitric oxide (NO), hydrogen sulfide (H_2S), or melatonin can contribute to regulating ROS metabolism and how these can exert beneficial effects during postharvest storage. Furthermore, the antioxidant composition of fleshy fruits, recognized as a nutraceutical trait that contributes to human health when regularly consumed, will also be addressed.

Fig. 1 Representatives of different climacteric and non-climacteric fleshy fruits.

2. General overview of reactive oxygen species (ROS) metabolism

ROS are a family of molecules that result from the partial reduction of oxygen and include radical and non-radical species. The principal ROS are superoxide ($O_2^{\bullet -}$), hydroxyl ($^{\bullet}OH$), alkoxyl ($RO^{\bullet}$), and peroxyl

($ROO^{\bullet}$) as radical molecules, and hydrogen peroxide (H_2O_2), singlet oxygen ($^1O_2$) and ozone (O_3) as non-radicals. Under physiological conditions, ROS are produced in plant cells by many metabolic pathways as normal by-products, being respiration, photosynthesis, photorespiration, and fatty acid oxidation, the most relevant (Asada, 2006; Corpas, Gupta, & Palma, 2015; Foyer, 2018; Kohli et al., 2019; Li & Kim, 2021; Moller, 2001). Therefore, mitochondria, chloroplasts, and peroxisomes are particularly relevant ROS sources at subcellular levels. But also, some enzymatic reactions involving polyamine oxidases (PAO) (Yang, Wang, Li, Liu, & Hu, 2022) or the membrane-bound respiratory burst oxidase homologs (RBOH) (Gémes et al., 2016; Marino, Dunand, Puppo, & Pauly, 2012; Orman-Ligeza et al., 2016; Otulak-Kozieł, Kozieł, & Valverde, 2019; Torres & Dangl, 2005) are also ROS sources in plants. Likewise, free metals, such as iron, or released from metalloprotein complexes can also contribute, via a Haber-Weiss reaction, to ROS production within the cells (Liochev & Fridovich, 2002).

Plants harbor a vast collection of enzymatic and non-enzymatic antioxidant systems, which usually are located nearby the ROS production site; this is an excellent mechanism to avoid the undesirable potential cascade of negative effects of ROS on biomolecules since oxidative damage usually causes dysfunction of the affected targets (oxidative stress). Among the most relevant enzymatic antioxidants are superoxide dismutase (SOD), catalase, and all components of the ascorbate-glutathione cycle that include ascorbate peroxidase (APX), monodehydroascorbate reductase (MDAR), dehydroascorbate reductase (DHAR) and glutathione reductase (GR). These enzymes also require the presence of the non-enzymatic antioxidants ascorbate (ASC) and glutathione (GSH) (Foyer & Noctor, 2011). All these enzyme systems are composed of a family of different isozymes with multiple subcellular localizations to control the ROS production in those compartments (Kohli et al., 2019). Additionally, there are other enzymatic antioxidants, such as glutathione peroxidase (Bela et al., 2015) or peroxiredoxins (Prxs), which are thiol-based peroxidase enzymes that can decompose various peroxide substrates, such as H_2O_2, alkyl hydroperoxide (ROOH), and peroxynitrite ($ONOO^-$) (Liebthal, Maynard, & Dietz, 2018).

All these antioxidant systems modulate the ROS content, which, at low levels, have signaling functions in some physiological processes such as seed and pollen germination (Bailly, 2019; Maksimov, Evmenyeva, Breygina, & Yermakov, 2018; Xie, Zheng, Zhou, Kanwar, & Zhou, 2022), root development (Dunand, Crèvecoeur, & Penel, 2007; Mabuchi et al., 2018; Tsukagoshi, Busch, & Benfey, 2010; Yamada, Han, & Benfey, 2020), plant growth and development (Airaki et al., 2015; Considine & Foyer, 2021), leaf senescence (Palma et al., 2006; Zentgraf et al., 2022), flowering (Zafra, Rejón, Hiscock, & Alché Jde, 2016) and fruit ripening (Corpas et al., 2022; González-Gordo et al., 2019; González-Gordo, Rodríguez-Ruiz, Palma, & Corpas, 2020; Huan et al., 2016; Vall-Llaura et al., 2022). Furthermore, ROS metabolism is also critical in the mechanism of response against abiotic and biotic environmental stresses (Signorelli et al., 2013; Berrios & Rentsch, 2022; Chen & Fluhr, 2018; Choudhury, Rivero, Blumwald, & Mittler, 2017; del Río, Corpas, López-Huertas, & Palma, 2018; Houmani et al., 2022). Thus, in Arabidopsis plants, ROS seems to work as a switch to trigger ethylene biosynthesis through the activation of mitogen-activated protein kinase 6 (MPK6), provoking the activation of ethylene response factors (ERFs) (ERF1 and ERF6), which in turn activate several genes encoding for antioxidant enzymes such as catalase 1, APX4, and DHAR1 to counteract the effects of both biotic and abiotic stresses (Müller & Munné-Bosch, 2015; Sewelam et al., 2013). One of the plant research targets is to decipher how ROS signals are transmitted from cells to cells and among organs. In this context, accumulating evidence indicates the existence of ROS waves that are transmitted from cell to cell under different stress conditions (Xiong et al., 2021; Zandalinas & Mittler, 2021; Zandalinas, Sengupta, Burks, Azad, & Mittler, 2019), with the participation of H_2O_2 receptors (Li, He, et al., 2021; Li, Zhao, et al., 2021; Singh, Mishra, Tripathi, Corpas, & Singh, 2022; Wu et al., 2022).

Besides the enzymatic antioxidant systems, plants and particularly fruits contain a wide group of compounds, above 200,000, with recognized antioxidant capacity. They are grouped into several large families that include phenolics, terpenes, or sulfur derivatives and vitamins. Fig. 2 shows the list of the most representative compounds with some examples of fruits that contain them. The presence of each family of compounds varies depending on the type of fruit as well as its ripening stage (Agati, Azzarello, Pollastri, & Tattini, 2012). In fact, compounds with antioxidant capacity have also been associated with pharmacological properties such as anticancer, antidiabetic, and anti-inflammatories, among others. Therefore, the regular consumption of antioxidant-rich fleshy fruits has significant potential benefits for human health (Raiola, Tenore, Barone, Frusciante, & Rigano, 2015; Wahyuni, Ballester, Sudarmonowati, Bino, & Bovy, 2011; Wahyuni, Ballester, Sudarmonowati, Bino, & Bovy, 2013; Zhao et al., 2022).

Ascorbate (vitamin C) is one of the most abundant antioxidants in plant cells and exerts multiple functions in plant development and key molecules against a wide range of stresses (Fenech, Amaya, Valpuesta, & Botella, 2019;

Main compounds with antioxidant activity

Phenolic derivatives

- **Flavonoids**
  - -Chalcones. E.g.: citruses, apples
  - - Flavones. E.g.: citrus family
  - - Flavonols
  - . Flavonoids (quecetin, catechins)
  - - Anthocyanins: E.g.: blueberries, black grapes, raisins, blackberries, plums.
  - -Tannins. E.g.: Pomegranate, persimmon, grapes, blueberries
  - - Aurones
- **Hydroxytyrosol and oleuropein. E.g.: olive fruit**
- **Phenolic acids. E.g.: carvacrol (oregano), ellagic acid, cinnamic acid derivatives (rosmarinic and lithospermic acid)**
- **Coumarins**
- **Quinone**

Terpene derivatives

- **Monoterpenes**
- **Diterpernes**
- **Carotenoids: α-carotene, β-carotene**
- **Lycopene**

Sulfur derivatives

- **Mercaptans**
- **Thiols**
- **S-allyl-L-cysteine (SAC), S-allyl-L-cysteine sulfoxide**

Vitamins

- **Vitamin C**
- **Vitamin E: α-, β-, γ- and δ tocopherols**

Fig. 2 Main groups of compounds with antioxidant capacity usually found in climacteric and non-climacteric fruits and other vegetables.

Foyer, Kyndt, & Hancock, 2020). However, from a consumer point of view, it is widely accepted that habitual consumption of fruits containing vitamin C, whose active forms are ascorbic acid (reduced form) and dehydroascorbic acid (oxidized form) (Fig. 3A) can reduce oxidative stress effectively and provide health benefits (Gan, Camarena, Mustafi, & Wang, 2019; Guaiquil, Vera, & Golde, 2001; Li, He, et al., 2021; Li, Zhao, et al., 2021). On the other hand, **vitamin E** is a non-enzymatic lipid-soluble antioxidant system that comprises eight compounds: four tocopherols, namely, α-, β-, γ-, and δ-tocopherols; and four tocotrienols, namely, α-, β-, γ, and δ-tocotrienols, being the α-tocopherol the most abundant in fruits and vegetables (Fig. 3B) (Chuna, Leeb, Yea, Exlerc, & Eitenmiller, 2006; Cortés-Herrera, Chacón, Artavia, & Granados-Chinchilla, 2019; Wahyuni et al., 2011; Zaaboul & Liu, 2022). Table 1 provides some of the fruits with the highest content of vitamin C and vitamin E per 100 g of fresh fruit weight.

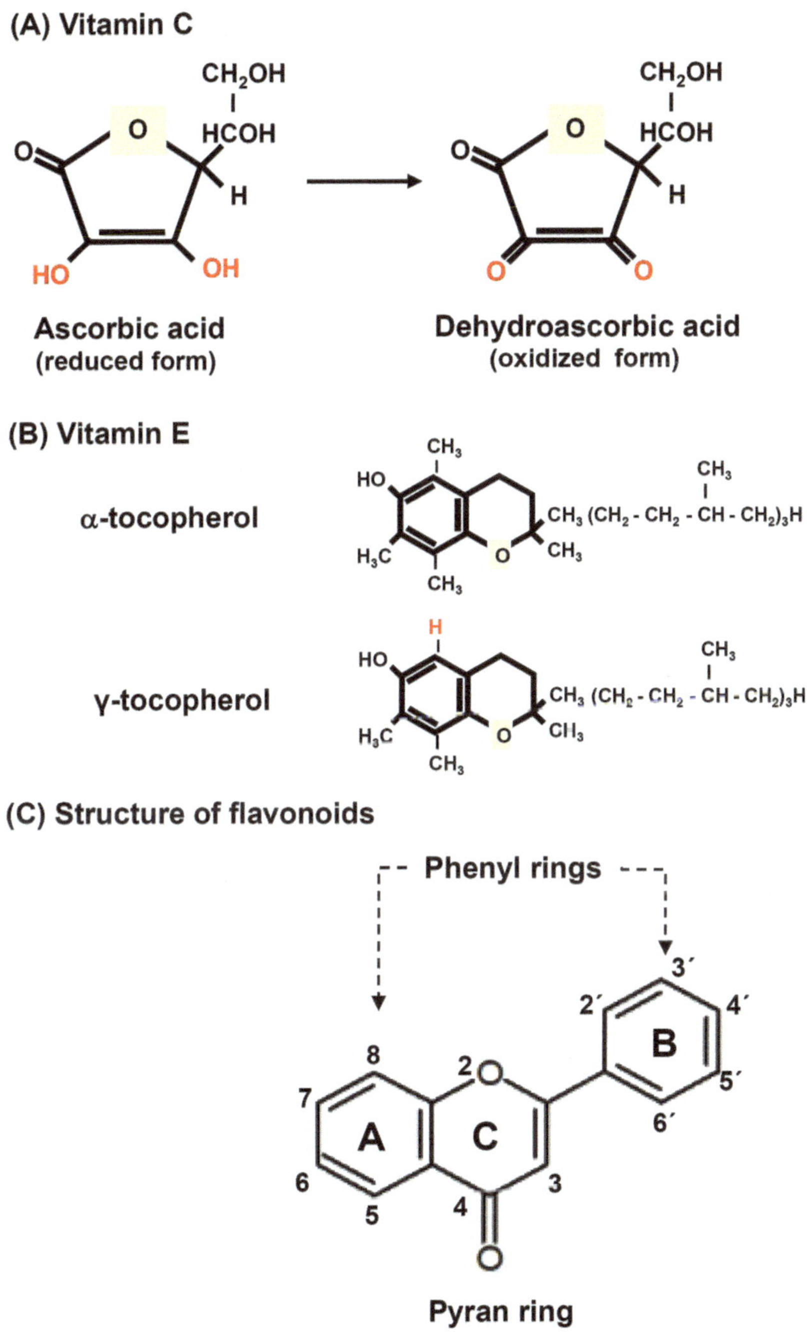

Fig. 3 (A) Structure of vitamin C, ascorbic acid (reduced form) and dehydroascorbic acid (oxidized form). (B) Vitamin E (α- and γ-tocopherol). (C) Basic chemical structure of flavonoids. It contains two phenyl rings (A and B) linked by a three-carbon chain that forms a closed pyran ring (heterocyclic ring containing oxygen, the C ring) with phenyl A ring.

Table 1 Content of vitamin C and vitamin E in different fleshy fruits.

Fruit specie	mg /100 g FW	References
Vitamin C		
Kakadu plum (*Terminalia ferdinandiana*)	3100	Sultanbawa, Chaliha, Cusack A, Edwards, and Williams (2018), Cozzolino, Phan, Netzel, Smyth, and Sultanbawa (2021), Mohanty and Cock (2012)
Camu camu (*Myrciaria dubia*)	2800	Sixto, Bravo, Sotero, and Oliva (2011)
Acerola cherry (*Malpighia glabra* L)	1600	Prakash and Baskaran (2018)
Jujube (*Ziziphus jujuba*)	200–600	Markovski and Velkoska-Markovska (2016)
Pepper (*Capsicum annuum* L.)	90–200	Wahyuni et al. (2013), Kantar et al. (2016)
Kiwifruit (*Actinidia deliciosa* cv. Hayward)	65–90	Nishiyama et al. (2004), Drummond (2013)
Strawberry (*Fragaria* × *ananassa*)	40–70	Sapei and Hwa (2014)
Orange (*Citrus* × *sinensis*)	59	Martí, Mena, Cánovas, Micol, and Saura (2009)
Grapefruit (*Citrus* × *paradisi*)	49	Najwa and Azlan (2017)
Lemon (*Citrus* × *limón*)	44	Najwa and Azlan (2017)
Lime (*Citrus aurantiifolia*)	28	Najwa and Azlan (2017)
Mango (*Mangifera indica* L.)	10–186	Manthey and Perkins-Veazie (2009)

(*continued on next page*)

Table 1 (*continued*)

Fruit specie	mg /100 g FW	References
Vitamin E		
Olive (*Olea europaea* L.)	1.3–9	Rocha, Borges, and Pinho (2020)
Peppers (*Capsicum annuum* L.)	5.4	Matsufuji, Ishikawa, Nunomura, Chino, and Takeda (2007)
Red gooseberry (*Ribes uva-crispa* L.)	4–6	Orsavová, Hlaváčová, Mlček, Snopek, and Mišurcová (2019)
Kiwifruit (*Actinidia deliciosa*)	1.5	Drummond (2013)
Black sapote (*Diospyros digyna* J.)	2.06	Corral-Aguayo, Yahia, Carrillo-Lopez, and González-Aguilar (2008)
Avocado (*Persea americana*)	1.7	Corral-Aguayo et al. (2008)
Mango (*Mangifera indica* L.)	0.8–1.02	Maldonado-Celis et al. (2019)
Strawberries (*Fragaria ananassa* L.)	0.3	Corral-Aguayo et al. (2008)
Papaya (*Carica papaya*)	0.2	Corral-Aguayo et al. (2008)

FW, Fresh weight. Note: The provided data are indicative since the content can differ depending on the fruit cultivar, ripening stage, and the analytical method used for quantification

Another group of compounds with recognized antioxidant properties is the **flavonoids** (from the Latin *flavus*, "yellow") which are considered the most abundant polyphenols and popularly called "vitamin P" (Agati et al., 2012; Al-Khayri et al., 2022). At present, it is difficult to determine their number accurately, although over 4000 have been identified. The general chemical structure of flavonoids contains one pyran ring and two phenyl rings (Fig. 3C), and the antioxidant capacity of flavonoids is due to (i) the presence of O-dihydroxy structure in ring B, which confers higher stability to the radical form and participates in the delocalization of electrons; (ii) presence of double bonds, in conjunction with the 4-oxo function of ring C; (iii) 3- and 5-OH groups with 4-oxo function in rings A and C, which are necessary to exert maximum antioxidant potential; (iv) ability to chelate transient metal ions, such as Fe^{2+}, Cu^{2+}, Zn^{2+}, involved in Fenton reactions, thus generating ROS; and, (v) inhibition of oxidases such as lipoxygenase, cyclooxygenase, myeloperoxidase, and xanthine oxidase, thus avoiding the formation of ROS and organic hydroperoxides.

On the other hand, certain compounds with antioxidant capacity are associated with certain fruits, where olive (*Olea europaea* L.) fruit is a good example. In addition to the main "classical" antioxidant enzymatic and non-enzymatic antioxidants, which are modulated during ripening (López-Huertas & del Río, 2014; López-Huertas & Palma, 2020), olive fruits contain some phenolics such as oleuropein, hydroxytyrosol, and tyrosol with an antioxidant capacity which are chiefly present in these fruits. Hydroxytyrosol and tyrosol appear by the hydrolysis of oleuropein during the olive fruit ripening (Conde, Delrot, & Gerós, 2008; Skodra et al., 2021).

3. Fruit ripening and ROS metabolism

Fruit ripening involves degradative reactions associated with membrane deterioration and complex biosynthetic reactions as well as the loss of cell structures. The relevance of ROS metabolism during fruit development and ripening has been increasingly associated with an oxidative syndrome. In fact, there are numerous examples to support this issue. In an early report, Brennan and Frenkel (1977) showed that endogenous H_2O_2 increases during the senescence of pear (*Pyrus communis*) fruits. Furthermore, the exogenous application of ethylene triggered an increase in H_2O_2 content, which accelerated fruit softening; however, the inhibition of the

H_2O_2-generating glycolate oxidase activity with α-hydroxy-2-pyridinemethanesulfonic acid decreased the content of this ROS and delayed fruit senescence. On the other hand, an increase in lipid peroxidation and protein oxidation was found during the ripening of tomato fruits which was associated with an increase in H_2O_2 (Jiménez et al., 2002). These data corroborate the relevance of H_2O_2 in the ripening of this climacteric fruit. Furthermore, in the cell walls of ripening fruits of diverse plants, including apple, avocado, banana (*Musa acuminata* Colla), mango, nectarine (*Prunus persica* L. Batsch), pear, plum (*Prunus domestica* L.), strawberry, and strawberry tree (*Arbutus unedo*), the hydroxyl radical ($^{\bullet}OH$), which seems generated by a Fenton-like reaction mediated by Cu^+, accelerated the disassembly of the polysaccharides allowing an increase in the sugar content (Airianah, Vreeburg, & Fry, 2016; Fry, Dumville, & Miller, 2001; Vreeburg, Airianah, & Fry, 2014). In tomato fruits, the increase of both Cu^+ and ascorbate in the apoplast triggered the generation of $^{\bullet}OH$ which promotes the non-enzymatic cleave of pectins and consequently the natural softening of the fruits (Dumville & Fry, 2003). Likewise, during nectarine fruit development and ripening, there is a wave of ROS content (Vall-Llaura et al., 2022). Thus, at the beginning of nectarine development, an increase in respiration rate and an oxidative burst has been associated with the activation of ethylene response factor 1 (ERF1) and a concomitant increase in the expression of genes encoding antioxidant enzymes and proteins involved in the biosynthesis of non-enzymatic antioxidants such as phenylpropanoids and their derivatives, such as flavonoids and glycosides (Korkina, 2007). After the nectarine fruit development occurs, both respiration rate and ERF1 expression are reduced, leading to low contents of non-enzymatic antioxidants and, consequently, an increase in H_2O_2 levels (Vall-Llaura et al., 2022). In grapevine and tomato, there is an accumulation of H_2O_2 and a general modulation of the ROS scavenging enzymes at the onset of ripening. Similarly, in pineapple (*Ananas comosus* L.), a non-climacteric tropical fruit, the fruit ripening is accompanied by oxidative processes and the concurrent activation of enzymatic and non-enzymatic ROS scavenging systems (Léchaudel, Darnaudery, Joët, Fournier, & Joas, 2018).

Fig. 4 shows the main changes of different components of ROS metabolism during ripening and senescence of sweet pepper as a model of non-climacteric fruit (Fig. 4A) and tomato as a model of climacteric fruit (Fig. 4B). During the ripening of sweet pepper fruit from green to ripe red, an increase in lipid oxidation (as a marker of oxidative damage) and the

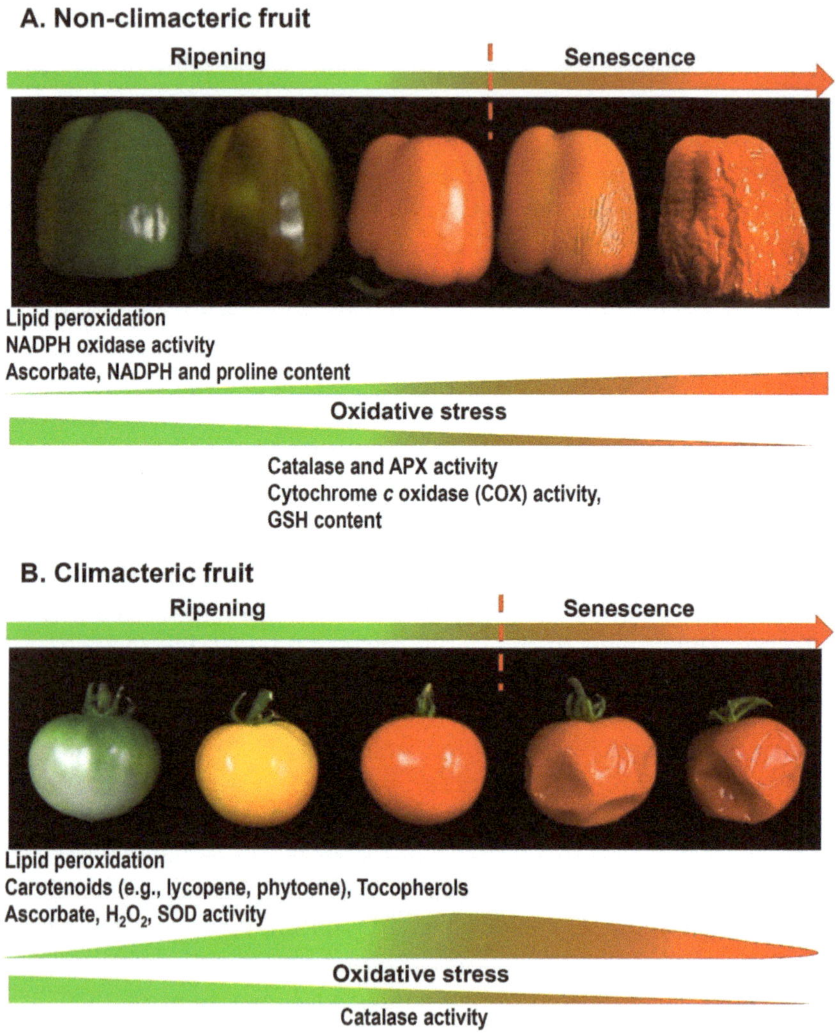

Fig. 4 Evolution of some parameters related to ROS metabolism in sweet pepper and tomato fruit ripening. *APX*, Ascorbate peroxidase; *GSH*, reduced glutathione; *SOD*, superoxide dismutase.

$O_2^{\bullet-}$-generating NADPH-oxidase activity, as well as a higher content of ascorbate, and NADPH were found (Palma, et al., 2015; Chu-Puga, González-Gordo, Rodríguez-Ruiz, Palma, & Corpas, 2019; González-Gordo et al., 2019, 2022; González-Gordo, Palma, & Corpas, 2020; González-Gordo, Rodríguez-Ruiz, et al., 2020; Mateos et al., 2013; Mateos, Bonilla-Valverde, del Río, Palma, & Corpas, 2009). On the other hand, the antioxidant activities of catalase and APX were downregulated during ripening, linked to a lower GSH content (Chaki et al., 2015; González-Gordo et al., 2019). However, the activity of SOD remained unaffected. Conversely, tomato fruits also undergo a drastic metabolic change during their ripening. Thus, it has been reported an increase in lipid peroxidation and H_2O_2 content as well as an increase in SOD enzyme activity and non-enzymatic ascorbate. As ripening progresses, tomato fruits produce and accumulate large amounts of compounds with antioxidant properties such as carotenoids (lycopene and phytoene) and tocopherols, which are an important source of vitamin E. On the contrary, catalase activity was downregulated during ripening (Liu, Shao, Zhang, & Wang, 2015; Zuccarelli et al., 2021).

Genetic approaches have also demonstrated the relevance of ROS metabolism in tomato ripening. The GDP-L-galactose phosphorylase (GGP) catalyzes the first step in the ascorbic acid biosynthesis, and *Slggp* defective tomato mutants, characterized to have a 30% lower content of ascorbic acid in comparison to wild-type, the fruit ripening was delayed. However, the exogenous application of 20 mM ascorbic acid to the *Slggp* mutant lines resulted in a ripening pattern similar to the wild-type. On the other hand, H_2O_2 treatment prolonged the ripening time not only in the *Slggp* mutant but also in the wild-type (Steelheart et al., 2020). Furthermore, *Slggp* mutant plants display reduced fruit yield (Alegre et al., 2020). On the contrary, ascorbate overproduction in the tomato mutant designated upstream open reading frame (uORF)-GGP1 negatively impacted flower development by impairing both anther and pollen development, thereby provoking seedless fruits (Deslous et al., 2021). A recent study also demonstrated that the master regulator of tomato ripening NON-RIPENING (NOR) is a target of sulfoxidation, which is a redox-mediated posttranslational modification (PTM) of methionine residues, and such PTM interferes with the DNA-binding capacity of NOR, consequently affecting NOR-dependent activation of gene expression in tomato fruits (Jiang et al., 2020).

Recently, Helmus Sies has proposed that oxidative stress has two sides designated "oxidative eustress" associated with physiology and healthy situations and "oxidative stress" which has associated injury to biomolecules with potentially deleterious consequences in pathophysiology and disease (Sies, 2020, 2021). Based on the available experimental information on pepper and tomato fruits (Chaki et al., 2015; González-Gordo et al., 2019; González-Gordo, Rodríguez-Ruiz, et al., 2020, 2022; Zuccarelli et al., 2021), it seems tempting to consider the highly coordinated ROS metabolism during ripening in both climacteric and non-climacteric as a developmentally-regulated physiological oxidative stress or "oxidative eustress".

Several studies using either traditional breeding or transgenic plants have also shown the importance of certain antioxidants in the ripening process, postharvest storage, and defense against pathogens. **Anthocyanins** are a type of flavonoid with recognized antioxidant capacity (Rice-Evans, Miller, & Paganga, 1996). For example, the delphinidin-3-(p-coumaroyl-rutinoside)-5-glucoside is the most prominent anthocyanin in pepper and eggplant fruits (Azuma et al., 2008; Lightbourn et al., 2008; Sadilova, Stintzing, & Carle, 2006) which has recognized antioxidant activity in the plant as well as nutrients for human (Husain et al., 2022; Liu et al., 2018; Yousuf, Gul, Wani, & Singh, 2016). Lately, it has been found that the biosynthesis of anthocyanin is triggered by light in pepper (*Capsicum annuum* L., HNUCA7454) fruits, particularly the delphinidin 3-O-glucoside being the major light-regulated anthocyanin (Zhou et al., 2022). The mechanism was explained since light stimulates the transcription of structural genes, transcription factor families (MYB4-like, MYB113-like, MYB308-like, and EGL1), as well as long non-coding RNAs (lncRNA), circular RNA (circRNA), and microRNA (miRNA).

Although most tomato cultivars do not produce high anthocyanin levels in fruit, a variety named "Indigo Rose", which contains the dominant *Aft* (*Anthocyanin fruit*) locus combined with the recessive *atv* (*atroviolacium*) locus from wild tomato species, over accumulates anthocyanin in the fruit skin, a response linked to the fact that *Aft* encodes for an anthocyanin activator (known as SlAN2-like), whose expression is promoted by light (Sun et al., 2020). The accumulation of anthocyanins in genetically modified tomatoes allowed to extend the fruit shelf life as well to reduce the susceptibility to the pathogen *Botrytis cinerea* (Bassolino et al., 2013; Butelli et al., 2008; Liu et al., 2021; Yan et al., 2020; Zhang et al., 2015). Alongside the antioxidant properties of anthocyanins in plant cells, these anthocyanin-enriched fruits have also great relevance from a nutritional and human health point of view (Gonzali & Perata, 2020; Kruger, Davies, Myburgh, & Lecour, 2014; Wang, Cao, & Prior, 1997; Yousuf et al., 2016).

Table 2 summarizes additional examples of climacteric and non-climacteric fleshy fruits where ROS metabolism is significantly modulated during ripening.

After the ripening onset, pepper and tomato fruits gradually acquire the optimal quality and good external appearance for human consumption.

Table 2 Representative examples of climacteric and non-climacteric fleshy fruits where changes in ROS metabolism associated with their ripening have been described.

Fruit type	ROS metabolism during ripening	References
Climacteric		
Tomato (*Solanum lycopersicum* L.)	Increase in MDA and H_2O_2 content and LOX activity, whereas there is a decrease in SOD, APX, GR, and POX	Mondal, Sharma, Malhotra, Dhawan, and Singh (2004)
Tomato (*S. lycopersicum* L.) 'Micro-Tom'	Increase of H_2O_2 content. Decreased APX and catalase activities. Increases lycopene, phytoene, phytofluene, total carotenoids and tocopherol levels.	Zuccarelli et al. (2021)
	High light stress induces H_2O_2 production and accelerates fruit ripening in tomato	Steelheart et al. (2022)
Banana (*Musa* spp., AAA group, cultivar 'Brazil')	ROS participate in the disassembly of the cellular wall	Cheng et al. (2008)
Guava (*Psidium guajava* L.)	Increase in MDA and H_2O_2 content and LOX activity, which was also accompanied by an increase in ascorbate content as well as catalase and APX activities	Mondal, Malhotra, Jain, and Singh (2009)
Sweet cherries (*Prunus avium* L)	Increases net CO_2 and ethylene production. Accumulation of H_2O_2 and MDA	Giné-Bordonaba et al. (2017)
Mango (*Mangifera indica* L. cv. 'Cogshall')	Increase in the content of H_2O_2 and MDA. Decrease of ascorbate content and decline in the activity of SOD, catalase, and APX.	Rosalie et al. (2018)
Avocado (*Persea americana* cv. Hass)	Total phenolics, hydrophilic and lipophilic antioxidant capacity increase	Campos et al. (2020)

(*continued on next page*)

Table 2 *(continued)*

Fruit type	ROS metabolism during ripening	References
Pears (*Pyrus communis* L.)	Increases MDA content	Lindo-García et al. (2020)
Non-climacteric		
Sweet orange (*Citrus sinensis* (L.) Osbeck)	Decreases the activities of SOD, catalase, guaiacol peroxidase, APX, DHAR, as well as the ratios of ascorbate/dehydroascorbate and GSH/GSSG.	Huang, Xia, Hu, and Wang (2007)
Grape berries (*Vitis vinifera* cv. Pinot Noir)	Accumulation of H_2O_2, $^1O_2$ (singlet oxygen) and lipid peroxidation. Increases LOX protein expression. Enhances catalase activity	Pilati et al. (2014)
Pineapple (*Ananas comosus* L.)	Increases the expression of genes associated with phenolic compound synthesis (phenylalanine ammonia lyase and 4-coumarate:CoA ligase), and oxidation (polyphenol oxidase). Increases APX3, GR and MDAR transcripts.	Léchaudel et al. (2018)
Strawberry (*Fragaria* × *ananassa*)	Increases SOD, catalase and POX activities. Also increases SOD and glutathione transferase protein abundance while decreases DHAR, 2-Cys peroxiredoxin, catalase, 1-Cys peroxiredoxin and APX levels.	López, Gochicoa, and Franco (2010), Song et al. (2020)
Olive (*Olea europaea* cv Picual)	Increased the content of GSH and ascorbate as well as hydroxytyrosol and luteolin. Reduction of catalase activity and lower protein content of APX and DHAR	López-Huertas and Palma (2020)
Sweet pepper (*Capsicum annuum* L.)	Increases MDA and proline contents, and NADPH oxidase activity. Decreases catalase, APX and GSH content.	Chu-Puga et al. (2019), González-Gordo et al. (2019), González-Gordo, Rodríguez-Ruiz, et al., (2020), González-Gordo, Palma, et al. (2020)

APX, Ascorbate peroxidase; *DHAR*, dehydroascorbate reductase; *GR*, glutathione reductase; *GSH*, reduced glutathione; *GSSG*, oxidized glutathione; *LOX*, lipoxygenase; *MDA*, malondialdehyde; *MDAR*, monodehydroascorbate reductase; *POX*, peroxidase; *SOD*, superoxide dismutase.

Soon after, fruit senescence is initiated, characterized by a progressive loss of water content, turgidity, and protein content, among other parameters. However, it should be mentioned that, under controlled conditions of humidity, light, and temperature, fruit senescence, which involves its dehydration either naturally or by an industrial process, can be considered a mechanism to preserve fruit material for later uses as a seasoning product (Degwale et al., 2022; Famurewa, Oluwamukomi, & Adenuga, 2006; McSweeney & Seetharaman, 2015; Zhang et al., 2017).

4. ROS metabolism and postharvest storage

During postharvest storage, the fruits usually undergo an increase in the production of ROS, which leads to oxidative damage, and this provokes fruit deterioration (Rubio-Melgarejo et al., 2021). This may be associated with the proliferation of pathogens that can accelerate the deterioration process, thus involving significant economic losses in the fruit industry (Wang et al., 2019). However, this ROS production could also have beneficial effects on combating biotic stresses (Meitha, Pramesti, & Suhandono, 2020). Different strategies have been sought to alleviate oxidative damage, usually associated with the type of fruit, either climacteric or non-climacteric, including storage at low temperature (Rosalie, Léchaudel, Dhuique-Mayer, Dufossé, & Joas, 2018), moderate heat shock treatment (Gómez et al., 2008), or the exogenous addition of different compounds such as nitric oxide (NO), hydrogen sulfide (H_2S), or melatonin (Aghdam et al., 2021; Corpas & Palma, 2018; Molinett et al., 2021). These situations either delay the ripening process or induce the production of enzymatic and non-enzymatic antioxidant systems.

Nitric oxide ($^{\bullet}NO$) is a free radical that has a family of derived molecules designated reactive nitrogen species (RNS) such as peroxynitrite ($ONOO^-$), nitrogen dioxide ($^{\bullet}NO_2$), or *S*-nitrosoglutathione (GSNO). These molecules, directly or indirectly, can modulate protein functions through PTMs such as *S*-nitrosation and tyrosine nitration. *S*-nitrosation (previously designated as *S*-nitrosylation) is a reversible modification that affects the thiol group (-SH) of certain cysteine present in proteins that can modify its function either positively or negatively. Furthermore, NO can interact with peptides such as the antioxidant GSH to generate GSNO. Protein nitration is an irreversible process that can affect aromatic residues such as tyrosine and tryptophan, and usually provokes a loss of function of the target proteins (Corpas, González-Gordo, & Palma, 2021). Accumulating data have demonstrated that some key antioxidant enzymes that regulate H_2O_2 content, such as APX and the peroxisomal catalase are modulated by NO. Thus, whereas APX activity is positively affected by *S*-nitrosation (Begara-Morales et al., 2014; Yang et al., 2015), Tyr-nitration triggers its inhibition (Begara-Morales et al., 2014). In the case of catalase, the activity is downregulated by both *S*-nitrosation and Tyr-nitration (Palma et al., 2020). **Hydrogen sulfide** (H_2S) is another gasotransmitter endogenously generated in plant cells as part of the cysteine metabolism (González-Gordo, Palma, et al., 2020). This molecule is also known to modify protein function through a PTM designated persulfidation (previously called sulfhydration), which involves the non-enzymatic interaction between H_2S in the form of hydrosulfide (HS^-) with the thiol group of cysteine proteins to generate the persulfidated proteins (Aroca, Benito, Gotor, & Romero, 2017; Aroca, Gotor, & Romero, 2018; Aroca, Serna, Gotor, & Romero, 2015; Corpas et al., 2022; Corpas, Barroso, González-Gordo, Muñoz-Vargas, & Palma, 2019; Corpas, González-Gordo, Cañas, & Palma, 2019). Similar to NO, experimental data have shown that APX and catalase are also targets of persulfidation. Thus, it has been shown that whereas APX is positively modulated (Aroca et al., 2015), catalase is inhibited (Corpas, Barroso, et al., 2019; Corpas, González-Gordo, et al., 2019; Palma et al., 2020). **Melatonin** is a master regulator generated in plant cells (Arnao & Hernández-Ruiz, 2019a) with recognized antioxidant properties to combat oxidative stress due to its interactions with ROS and RNS to form hydroxymelatonin, N-nitromelatonin and nitrosomelatonin (Arnao & Hernandez-Ruiz, 2019b; Corpas et al., 2022; Galano & Reiter, 2018; Reiter, Tan, Manchester, & Qi, 2001). However, melatonin has also signaling properties in a wide range of plant functions, from seed germination, root development, plant growth, leaf senescence, flowering and fruit ripening to environmental stress response mechanisms with the capacity to induce antioxidant systems (Aghdam et al., 2021; Arnao & Hernández-Ruiz, 2018, 2021). Table 3 provides representative examples of the beneficial effects triggered by the exogenous application of either NO, H_2S or melatonin in fruits during ripening or postharvest storage to palliate oxidative damages.

Taken together these data, it seems evident that, independently of whether these molecules (NO, H_2S and melatonin) are endogenously generated during the fruit ripening process, their exogenous application under controlled conditions could be a relevant biotechnological tool to

Table 3 Beneficial effects triggered by exogenous application of nitric oxide, hydrogen sulfide, or melatonin on ROS metabolism in fruits during ripening or postharvest storage, also a mechanism to palliate damages associated with chilling injury and pathogens.

Fruits	Main effects	References
Nitric oxide		
Kiwifruit (*Actinidia chinensis* L. Planch cv. Xuxiang)	Extends postharvest life and enhances antioxidant activity	Zhu, Sun, and Zhou (2010)
Bayberry (*Myrica rubra* Sieb. & Zucc. cv. Dongkui)	Decrease in firmness and increases the activity of SOD, catalase and APX	Wu et al. (2012)
Pepper (*Capisum annuum* L)	Delays ripening and increases vitamin C content	Rodríguez-Ruiz, Mateos, Codesido, Corpas, and Palma (2017)
Tomato (*Solanum lycopersicum*)	Overaccumulation of ascorbate (25%) and flavonoids (60%),	Zuccarelli et al. (2021)
Hydrogen sulfide		
Kiwifruit (*Actinidia deliciosa*),	Triggers ROS generation and increases antifungal effects against *Aspergillus niger* and *Penicillium italicum*	Fu et al. (2014)
Kiwifruit (*Actinidia chinensis*)	Delays ripening and senescence. Inhibits ethylene production. Increases antioxidant activities. Regulates the cell wall degrading enzyme gene	Lin et al. (2020)
Banana (*Musa* spp.)	Lower accumulation of ROS (superoxide and peroxide ions), elevated phenol content and antioxidant capacity. The ripening inhibitory effects of H_2S delay chlorophyll loss and reduces ethylene and CO_2 production.	Siddiqui et al. (2022)

(*continued on next page*)

Table 3 *(continued)*

Fruits	Main effects	References
Peach (*Prunus persica* Batsch cv. 'Jinhuangjin')	Induces the accumulation of ascorbic acid, glutathione and the enhancement of antioxidant enzyme activities. Suppresses the cell-wall degradation and alleviates chilling injury	Wang et al. (2022)
Melatonin		
Peach (*Prunus persica* Batsch)	Delays postharvest senescence. Reduces ROS accumulation and diminishes membrane lipid peroxidation. Induces antioxidant enzyme activities and leads to decreased oxidative stress.	Gao et al. (2016)
Sweet cherry (*Prunus avium*)	Delays water and weight loss, improving cuticle integrity. Lowers respiration and ethylene production. Lowers pedicle browning, Increases SOD, CAT, APX, DHAR and GR gene expression, Lower MDA accumulation. Enhances anthocyanins accumulation.	Miranda et al. (2020)
Pomegranate (*Punica granatum* cv. Malas Saveh)	Confers chilling tolerance because it triggers higher SOD, APX, CAT and GR activities. Lowers H_2O_2 accumulation. Higher phenols accumulation and DPPH scavenging capacity. Higher PAL/PPO enzymes activity. Higher membrane integrity. Lower electrolyte leakage and MDA accumulation.	Jannatizadeh (2019)
Strawberry (*Fragaria* × *ananassa* Duch.)	Higher fruit firmness. Higher phenols and vitamin C content. Higher DPPH scavenging capacity	Zahedi, Hosseini, Abadía, and Marjani (2020)

DPPH, 2,2-Diphenyl-1-picrylhydrazyl; *PAL*, phenylalanine ammonia-lyase; *PPO*, polyphenol oxidase.

control the damage associated with the fruits postharvest storage since they participate in the modulation of ROS metabolism.

5. Antioxidants, fruit color and associated nutraceutical properties

Fruits contain a wide diversity of metabolites with antioxidant properties whose physiological relevance is to preserve the fruits against adverse environmental and biotic (e.g., pathogens, insects, birds) conditions to ensure optimal seed dispersal for the next generation. Although the vitamin C is probably the most powerful antioxidant and is widely known by most consumers (Szeto, Tomlinson, & Benzie, 2002), fruits contain a large number of molecules derived from the secondary metabolism that have antioxidant properties (Barros, Andrade, Denadai, Nunes, & Narain, 2017; Basu & Maier, 2016; Brabo de Sousa, de Andrade Mattietto, Campos Chisté, & Carvalho, 2018; Crozier et al., 2000; Jideani et al., 2021; Kumari, Parida, Rangani, & Panda, 2017; Lanfer-Marquez, Barros, & Sinnecker, 2005; Lenucci, Tornese, Mita, & Durante, 2022; Mueller & Boehm, 2011; Orlikova, Tasdemir, Golais, Dicato, & Diederich, 2011). Therefore, fruits are of great importance for their benefits to human health, mainly because this type of food is predominantly consumed raw and fresh, so most of these compounds preserve their antioxidant properties (Fig. 2). These natural compounds with antioxidant capacities can exert additive and synergistic effects to endow protective functions against diverse chronic diseases such as cancer, diabetes, cardio- and cerebrovascular pathologies, ocular and neurological diseases, stroke, hypertension, and blood-related diseases (Blumberg et al., 2013; Fresco, Borges, Diniz, & Marques, 2006; Gomes-Rochette et al., 2016; Nowak, Gośliński, & Kłębukowska, 2022).

Fruits are organs with a great color palette due to their chemical content, and there is a correlation between color and the antioxidant capacity of fruits (Cömert, Mogol, & Gökmen, 2020). Photosynthetic pigments (chlorophylls and carotenoids) are relevant for the color displayed by fruit. Carotenoids, which are isoprene derivatives, can be classified into two groups: carotenes, which are strictly hydrocarbons; and xanthophylls, with oxidized substitutions in the extreme rings. These pigments are associated with the red, yellow, or orange color present in tomatoes, pumpkins, or peppers (Xavier & Pérez-Gálvez, 2016), and chlorophyll provides the green color shown by fruits like avocado, kiwifruits, pears, apples, lime, or green grapes (Lanfer-Marquez et al., 2005). Furthermore, anthocyanins are one of the main phenolic groups responsible for a wide range of diverse colors, including red, blue, black, and purple displayed by blueberries, black grapes, raisins, blackberries, pepper, and plums (Mattoo et al., 2022) as well as tomato (Gonzali & Perata, 2020).

On the other hand, the exogenous ethylene treatment of the climacteric Japanese plum (*Prunus salicina* Lindl. cv Santa Rosa) and its non-climacteric mutant Sweet Miriam triggered an accumulation of anthocyanin but affected negatively the biosynthesis of flavonoids (flavonol and flavan-3-ol). Furthermore, it was found that the content of sucrose and galactose was positively associated with the content of anthocyanin, whereas sorbitol, fructose, glucose and other minor sugars were linked with flavonoid metabolism. These data suggest that the ethylene treatment could be a biotechnological tool to improve the plum fruit coloration as well as its health properties (Farcuh, Tajima, Lerno, & Blumwald, 2022).

6. Concluding remarks

The ripening of both climacteric and non-climacteric fruits involves a very active ROS metabolism, affecting both ROS production and the different enzymatic and non-enzymatic antioxidant systems that entail physiological oxidative stress. In many cases, the ripening process is associated with the accumulation of different types of antioxidants (e.g., vitamin C, polyphenols, flavonoids) depending on the type of fruit what, from an anthropological point of view, is of great interest to consumers as it brings beneficial, healthy properties when fruits are consumed regularly, especially as raw and fresh food. At the commercial level of the fruit industry, the target is to preserve the fruits with all their organoleptic properties from their harvesting, storage, and distribution until they reach the consumer. For this reason, different strategies are being developed that, through the exogenous application of different signaling molecules (NO, H_2S, melatonin, or phytohormones), are capable of stimulating the ROS metabolism, thus being able to avoid possible damage, either mechanical or pathogenic, that may affect the fruit quality. Likewise, some of these treatments of fruits can be of great interest since they can favor the accumulation of certain antioxidants, so fruits can be used as vehicles with great nutraceutical potentiality.

Acknowledgments

FJC and JMP research work is supported by European Regional Development Fund cofinanced grants from the Ministry of Science and Innovation (PID2019-103924GB-I00), and the Plan Andaluz de Investigación, Desarrollo e Innovación (P18-FR-1359), Spain. LF research work is supported by São Paulo Research Foundation (FAPESP) (grant 2021/05714-8).

References

Agati, G., Azzarello, E., Pollastri, S., & Tattini, M. (2012). Flavonoids as antioxidants in plants: Location and functional significance. *Plant Science, 196*, 67–76.

Aghdam, M. S., Mukherjee, S., Flores, F. B., Arnao, M. B., Luo, Z., & Corpas, F. J. (2021). Functions of melatonin during postharvest of horticultural crops. *Plant Cell Physiology*. https://doi.org/10.1093/pcp/pcab175

Airaki, M., Leterrier, M., Valderrama, R., Chaki, M., Begara-Morales, J. C., Barroso, J. B., ... Corpas, F. J. (2015). Spatial and temporal regulation of the metabolism of reactive oxygen and nitrogen species during the early development of pepper (*Capsicum annuum*) seedlings. *Annals of Botany, 116*(4), 679–693.

Airianah, O. B., Vreeburg, R. A., & Fry, S. C. (2016). Pectic polysaccharides are attacked by hydroxyl radicals in ripening fruit: Evidence from a fluorescent fingerprinting method. *Annals of Botany, 117*(3), 441–455.

Alegre, M. L., Steelheart, C., Baldet, P., Rothan, C., Just, D., Okabe, Y., ... Bartoli, C. G. (2020). Deficiency of GDP-L-galactose phosphorylase, an enzyme required for ascorbic acid synthesis, reduces tomato fruit yield. *Planta, 251*(2), 54.

Al-Khayri, J. M., Sahana, G. R., Nagella, P., Joseph, B. V., Alessa, F. M., & Al-Mssallem, M. Q. (2022). Flavonoids as potential anti-inflammatory molecules: A review. *Molecules (Basel, Switzerland), 27*(9), 2901.

Arnao, M. B., & Hernández-Ruiz, J. (2018). Melatonin and its relationship to plant hormones. *Annals of Botany, 121*, 195–207.

Arnao, M. B., & Hernández-Ruiz, J. (2019a). Melatonin: A new plant hormone and/or a plant master regulator? *Trends in Plant Science, 24*, 38–48.

Arnao, M. B., & Hernandez-Ruiz, J. (2019b). Melatonin and reactive oxygen and nitrogen species: A model for the plant redox network. *Melatonin Research, 2*, 152–168.

Arnao, M. B., & Hernández-Ruiz, J. (2021). Melatonin against environmental plant stressors: A review. *Current Protein & Peptide Science, 22*, 413–429.

Aroca, A., Benito, J. M., Gotor, C., & Romero, L. C. (2017). Persulfidation proteome reveals the regulation of protein function by hydrogen sulfide in diverse biological processes in Arabidopsis. *Journal of Experimental Botany, 68*, 4915–4927.

Aroca, A., Gotor, C., & Romero, L. C. (2018). Hydrogen sulfide signaling in plants: Emerging roles of protein persulfidation. *Frontiers in Plant Science, 9*, 1369.

Aroca, Á., Serna, A., Gotor, C., & Romero, L. C. (2015). S-sulfhydration: A cysteine posttranslational modification in plant systems. *Plant Physiology, 168*, 334–342.

Asada, K. (2006). Production and scavenging of reactive oxygen species in chloroplasts and their functions. *Plant Physiology, 141*, 391–396.

Azuma, K., Ohyama, A., Ippoushi, K., Ichiyanagi, T., Takeuchi, A., Saito, T., & Fukuoka, H. (2008). Structures and antioxidant activity of anthocyanins in many accessions of eggplant and its related species. *Journal of Agricultural and Food Chemistry, 56*(21), 10154–10159.

Bailly, C. (2019). The signalling role of ROS in the regulation of seed germination and dormancy. *Biochemical Journal, 476*(20), 3019–3032.

Barros, R. G. C., Andrade, J. K. S., Denadai, M., Nunes, M. L., & Narain, N. (2017). Evaluation of bioactive compounds potential and antioxidant activity in some Brazilian exotic fruit residues. *Food Research International, 102*, 84–92.

Bassolino, L., Zhang, Y., Schoonbeek, H. J., Kiferle, C., Perata, P., & Martin, C. (2013). Accumulation of anthocyanins in tomato skin extends shelf life. *The New Phytologist, 200*(3), 650–655.

Basu, P., & Maier, C. (2016). In vitro antioxidant activities and polyphenol contents of seven commercially available fruits. *Pharmacognosy Research, 8*(4), 258–264.

Bela, K., Horváth, E., Gallé, Á., Szabados, L., Tari, I., & Csiszár, J. (2015). Plant glutathione peroxidases: Emerging role of the antioxidant enzymes in plant development and stress responses. *Journal of Plant Physiology, 176*, 192–201.

Berrios, L., & Rentsch, J. D. (2022). Linking reactive oxygen species (ROS) to abiotic and biotic feedbacks in plant microbiomes: The dose makes the poison. *International Journal of Molecular Science, 23*(8), 4402.

Blumberg, J. B., Camesano, T. A., Cassidy, A., Kris-Etherton, P., Howell, A., Manach, C., ... Vita, J. A. (2013). Cranberries and their bioactive constituents in human health. *Advances in Nutrition, 4*(6), 618–632.

Brabo de Sousa, S. H., de Andrade Mattietto, R., Campos Chisté, R., & Carvalho, A. V. (2018). Phenolic compounds are highly correlated to the antioxidant capacity of genotypes of *Oenocarpus distichus* Mart. fruits. *Food Research International, 108*, 405–412.

Brennan, T., & Frenkel, C. (1977). Involvement of hydrogen peroxide in the regulation of senescence in pear. *Plant Physiology, 59*, 411–416.

Butelli, E., Titta, L., Giorgio, M., Mock, H. P., Matros, A., Peterek, S., ... Martin, C. (2008). Enrichment of tomato fruit with health-promoting anthocyanins by expression of select transcription factors. *Nature Biotechnology, 26*(11), 1301–1308.

Campos, D., Teran-Hilares, F., Chirinos, R., Aguilar-Galvez, A., García-Ríos, D., Pacheco-Avalos, A., & Pedreschi, R. (2020). Bioactive compounds and antioxidant activity from harvest to edible ripeness of avocado cv. Hass (*Persea americana*) throughout the harvest seasons. *International Journal of Food Science and Technology, 55*, 2208–2218.

Chaki, M., Álvarez de Morales, P., Ruiz, C., Begara-Morales, J. C., Barroso, J. B., Corpas, F. J., & Palma, J. M. (2015). Ripening of pepper (*Capsicum annuum*) fruit is characterized by an enhancement of protein tyrosine nitration. *Annals of Botany, 116*, 637–647.

Chen, T., & Fluhr, R. (2018). Singlet oxygen plays an essential role in the root's response to osmotic stress. *Plant Physiology, 177*(4), 1717–1727.

Cheng, G., Duan, X., Shi, J., Lu, W., Luo, Y., Jiang, W., & Jiang, Y. (2008). Effects of reactive oxygen species on cellular wall disassembly of banana fruit during ripening. *Food Chemistry, 109*, 319–324.

Choudhury, F. K., Rivero, R. M., Blumwald, E., & Mittler, R. (2017). Reactive oxygen species, abiotic stress and stress combination. *Plant Journal, 90*(5), 856–867.

Chuna, J., Leeb, J., Yea, L., Exlerc, J., & Eitenmiller, R. R. (2006). Tocopherol and tocotrienol contents of raw and processed fruits and vegetables in the United States diet. *Journal of Food Composition and Analysis, 19*, 196–204.

Chu-Puga, Á., González-Gordo, S., Rodríguez-Ruiz, M., Palma, J. M., & Corpas, F. J. (2019). NADPH Oxidase (Rboh) activity is up regulated during sweet pepper (*Capsicum annuum* L.) fruit ripening. *Antioxidants (Basel), 8*(1), 9.

Cömert, E. D., Mogol, B. A., & Gökmen, V. (2020). Relationship between color and antioxidant capacity of fruits and vegetables. *Current Research in Food Science, 2*, 1–10.

Conde, C., Delrot, S., & Gerós, H. (2008). Physiological, biochemical and molecular changes occurring during olive development and ripening. *Journal of Plant Physiology, 165*(15), 1545–1562.

Considine, M. J., & Foyer, C. H. (2021). Oxygen and reactive oxygen species-dependent regulation of plant growth and development. *Plant Physiology, 186*(1), 79–92.

Corpas, F. J., Barroso, J. B., González-Gordo, S., Muñoz-Vargas, M. A., & Palma, J. M. (2019). Hydrogen sulfide: A novel component in Arabidopsis peroxisomes which triggers catalase inhibition. *Journal of Integrative Plant Biology, 61*(7), 871–883.

Corpas, F. J., González-Gordo, S., & Palma, J. M. (2021). Protein nitration: A connecting bridge between nitric oxide (NO) and plant stress. *Plant Stress, 2*, 100026.

Corpas, F. J., Gupta, D. K., & Palma, J. M. (2015). Production sites of reactive oxygen species (ROS) in organelles from plant cells. In Gupta, (Ed.). *Reactive oxygen species and oxidative damage in plants under stress* (pp. 1–22). Switzerland: Springer International Publishing. https://doi.org/10.1007/978-3-319-20421-5_1

Corpas, F. J., & Palma, J. M. (2018). Nitric oxide on/off in fruit ripening. *Plant Biology, 20*, 805–807.

Corpas, F. J., Rodríguez-Ruiz, M., Muñoz-Vargas, M. A., González-Gordo, S., Reiter, R. J., & Palma, J. M. (2022). Interactions of melatonin, ROS and NO during fruit ripening: An update and prospective view. *Journal of Experimental Botany*. https://doi.org/10.1093/jxb/erac128

Corpas, F. J., González-Gordo, S., Cañas, A., & Palma, J. M. (2019). Nitric oxide and hydrogen sulfide in plants: Which comes first? *Journal of Experimental Botany, 70*, 4391–4404.

Corral-Aguayo, R. D., Yahia, E. M., Carrillo-Lopez, A., & González-Aguilar, G. (2008). Correlation between some nutritional components and the total antioxidant capacity measured with six different assays in eight horticultural crops. *Journal of Agricultural and Food Chemistry, 56*(22), 10498–10504.

Cortés-Herrera, C., Chacón, A., Artavia, G., & Granados-Chinchilla, F. (2019). Simultaneous LC/MS analysis of carotenoids and fat-soluble vitamins in costa rican avocados (Persea americana Mill.). *Molecules (Basel, Switzerland), 24*(24), 4517.

Cozzolino, D., Phan, A. D. T., Netzel, M. E., Smyth, H., & Sultanbawa, Y. (2021). The use of vibrational spectroscopy to predict vitamin C in Kakadu plum powders (*Terminalia ferdinandiana* Exell, Combretaceae). *Journal of the Science of Food and Agriculture, 101*(8), 3208–3213.

Crozier, A., Burns, J., Aziz, A. A., Stewart, A. J., Rabiasz, H. S., Jenkins, G. I., ... Lean, M. E. J. (2000). Antioxidant flavonols from fruits, vegetables and beverages: Measurements and bioavailability. *Biological Research, 33*(2), 79–88.

Dardick, C., & Callahan, A. M. (2014). Evolution of the fruit endocarp: Molecular mechanisms underlying adaptations in seed protection and dispersal strategies. *Frontiers in Plant Science, 5*, 284.

Degwale, A., Asrat, F., Eniyew, K., Asres, D., Tesfa, T., & Avalew, A. (2022). Influence of dehydration temperature and time on physicochemical properties of tomato (*Solanum lycopersicum* L.) powder. *Frontiers in Sustainable Food Systems*. https://doi.org/10.3389/fsufs.2022.839385

del Río, L. A., Corpas, F. J., López-Huertas, E., & Palma, M. (2018). *Plant superoxide dismutases: Function under abiotic stress conditions. Antioxidants and Antioxidant Enzymes in Higher Plants*. Cham, Switzerland: Springer International Publishing,1–26. ⟨10.1007/978-3-319-75088-0_1⟩.

Deslous, P., Bournonville, C., Decros, G., Okabe, Y., Mauxion, J. P., Jorly, J., ... Baldet, P. (2021). Overproduction of ascorbic acid impairs pollen fertility in tomato. *Journal of Experimental Botany, 72*(8), 3091–3107.

Drummond, L. (2013). The composition and nutritional value of kiwifruit. *Advances in Food and Nutrition Research, 68*, 33–57.

Dumville, J. C., & Fry, S. C. (2003). Solubilisation of tomato fruit pectins by ascorbate: A possible non-enzymic mechanism of fruit softening. *Planta, 217*(6), 951–961.

Dunand, C., Crèvecoeur, M., & Penel, C. (2007). Distribution of superoxide and hydrogen peroxide in Arabidopsis root and their influence on root development: Possible interaction with peroxidases. *New Phytology, 174*(2), 332–341.

Famurewa, J. A. V., Oluwamukomi, M. O., & Adenuga, A. L. (2006). Dehydration of osmosised red bell pepper (*Capsicum annuum*). *Journal of Food Technology, 4*, 249–252.

Farcuh, M., Tajima, H., Lerno, L. A., & Blumwald, E. (2022). Changes in ethylene and sugar metabolism regulate flavonoid composition in climacteric and non-climacteric plums during postharvest storage. *Food Chemistry (Oxford), 4*, 100075.

Fenech, M., Amaya, I., Valpuesta, V., & Botella, M. A. (2019). Vitamin C content in fruits: Biosynthesis and regulation. *Frontiers in Plant Science, 9*, 2006.

Foyer, C. H. (2018). Reactive oxygen species, oxidative signaling and the regulation of photosynthesis. *Environmental and Experimental Botany, 154*, 134–142.

Foyer, C. H., Kyndt, T., & Hancock, R. D. (2020). Vitamin C in plants: Novel concepts, new perspectives, and outstanding issues. *Antioxidant Redox Signaling, 32*(7), 463–485.

Foyer, C. H., & Noctor, G. (2011). Ascorbate and glutathione: The heart of the redox hub. *Plant Physiology, 155*(1), 2–18.

Fresco, P., Borges, F., Diniz, C., & Marques, M. P. (2006). New insights on the anticancer properties of dietary polyphenols. *Medicinal Research Reviews, 26*(6), 747–766.

Fry, S. C., Dumville, J. C., & Miller, J. G. (2001). Fingerprinting of polysaccharides attacked by hydroxyl radicals in vitro and in the cell walls of ripening pear fruit. *Biochemical Journal, 357*(Pt 3), 729–737.

Fu, L. H., Hu, K. D., Hu, L. Y., Li, Y. H., Hu, L. B., Yan, H., ... Zhang, H. (2014). An antifungal role of hydrogen sulfide on the postharvest pathogens *Aspergillus niger* and *Penicillium italicum*. *PLoS One, 9*(8), e104206.

Galano, A., & Reiter, R. J. (2018). Melatonin and its metabolites vs oxidative stress: From individual actions to collective protection. *Journal Pineal Research, 65*, e12514.

Gan, L., Camarena, V., Mustafi, S., & Wang, G. (2019). Vitamin C inhibits triple-negative breast cancer metastasis by affecting the expression of YAP1 and synaptopodin 2. *Nutrients, 11*(12), 2997.

Gao, H., Zhang, Z. K., Chai, H. K., Cheng, N., Yang, Y., & Wang, D. N. (2016). Melatonin treatment delays postharvest senescence and regulates reactive oxygen species metabolism in peach fruit. *Postharvest Biology and Technology, 118*, 103–110.

Gémes, K., Kim, Y. J., Park, K. Y., Moschou, P. N., Andronis, E., Valassaki, C., ... Roubelakis-Angelakis, K. A. (2016). An NADPH-oxidase/polyamine oxidase feedback loop controls oxidative burst under salinity. *Plant Physiology, 172*(3), 1418–1431.

Giné-Bordonaba, J., Echeverria, G., Ubach, D., Aguiló-Aguayo, I., López, M. L., & Larrigaudière, C. (2017). Biochemical and physiological changes during fruit development and ripening of two sweet cherry varieties with different levels of cracking tolerance. *Plant Physiology and Biochemistry, 111*, 216–225.

Gomes-Rochette, N. F., Da Silveira Vasconcelos, M., Nabavi, S. M., Mota, E. F., Nunes-Pinheiro, D. C., Daglia, M., & De Melo, D. F. (2016). Fruit as potent natural antioxidants and their biological effects. *Current Pharmaceutical Biotechnology, 17*, 986–993.

Gómez, F., Fernández, L., Gergoff, G., Guiamet, J. J., Chaves, A., & Bartoli, C. G. (2008). Heat shock increases mitochondrial H_2O_2 production and extends postharvest life of spinach leaves. *Postharvest Biology and Technology, 49*, 229–234.

González-Gordo, S., Palma, J. M., & Corpas, F. J. (2020). Appraisal of H_2S metabolism in *Arabidopsis thaliana*: In silico analysis at the subcellular level. *Plant Physiology and Biochemistry, 155*, 579–588.

González-Gordo, S., Rodríguez-Ruiz, M., López-Jaramillo, J., Muñoz-Vargas, M. A., Palma, J. M., & Corpas, F. J. (2022). Nitric oxide (NO) differentially modulates the ascorbate peroxidase (APX) isozymes of sweet pepper (*Capsicum annuum* L.) fruits. *Antioxidants (Basel), 11*(4), 765.

González-Gordo, S., Bautista, R., Claros, M. G., Cañas, A., Palma, J. M., & Corpas, F. J. (2019). Nitric oxide-dependent regulation of sweet pepper fruit ripening. *Journal of Experimetal Botany, 70*, 4557–4570.

González-Gordo, S., Rodríguez-Ruiz, M., Palma, J. M., & Corpas, F. J. (2020). Superoxide radical metabolism in sweet pepper (*Capsicum annuum* L.) fruits is regulated by ripening and by a NO-enriched environment. *Frontiers in Plant Science, 11*, 485.

Gonzali, S., & Perata, P. (2020). Anthocyanins from purple tomatoes as novel antioxidants to promote human health. *Antioxidants (Basel), 9*(10), 1017.

Guaiquil, V. H., Vera, J. C., & Golde, D. W. (2001). Mechanism of vitamin C inhibition of cell death induced by oxidative stress in glutathione-depleted HL-60 cells. *Journal of Biological Chemistry, 276*(44), 40955–40961.

Houmani, H., Debez, A., Freitas-Silva, L., Abdelly, C., Palma, J. M., & Corpas, F. J. (2022). Potassium (K^+) starvation-induced oxidative stress triggers a general boost of antioxidant and NADPH-generating systems in the halophyte *Cakile maritima. Antioxidants (Basel), 11*(2), 401.

Huan, C., Jiang, L., An, X., Yu, M., Xu, Y., Ma, R., & Yu, Z. (2016). Potential role of reactive oxygen species and antioxidant genes in the regulation of peach fruit development and ripening. *Plant Physiology and Biochemisty, 104*, 294–303.

Huang, R., Xia, R., Hu, L., & Wang, M. (2007). Antioxidant activity and oxygen scavenging system in orange pulp during fruit ripening and maturation. *Scientia Horticulturae, 113*, 166–172.

Husain, A., Chanana, H., Khan, S. A., Dhanalekshmi, U. M., Ali, M., Alghamdi, A. A., & Ahmad, A. (2022). Chemistry and pharmacological actions of delphinidin, a dietary purple pigment in anthocyanidin and anthocyanin forms. *Frontiers in Nutrition, 9*, 746881.

Jannatizadeh, A. (2019). Exogenous melatonin applying confers chilling tolerance in pomegranate fruit during cold storage. *Scientia Horticulturae, 246*, 544–549.

Jiang, G., Zeng, J., Li, Z., Song, Y., Yan, H., He, J., ... Duan, X. (2020). Redox regulation of the NOR transcription factor is involved in the regulation of fruit ripening in tomato. *Plant Physiology, 183*, 671–685.

Jideani, A. I. O., Silungwe, H., Takalani, T., Omolola, A. O., Udeh, H. O., & Anyasi, T. A. (2021). Antioxidant-rich natural fruit and vegetable products and human health. *International Journal of Food Properties, 24*, 41–67.

Jiménez, A., Creissen, G., Kular, B., Firmin, J., Robinson, S., Verhoeyen, M., & Mullineaux, P. (2002). Changes in oxidative processes and components of the antioxidant system during tomato fruit ripening. *Planta, 214*, 751–758.

Kantar, M. B., Anderson, J. E., Lucht, S. A., Mercer, K., Bernau, V., Case, K. A., ... Baumler, D. J. (2016). Vitamin variation in Capsicum Spp. provides opportunities to improve nutritional value of human diets. *PLoS One, 11*(8), e0161464.

Kohli, S. K., Khanna, K., Bhardwaj, R., Abd Allah, E. F., Ahmad, P., & Corpas, F. J. (2019). Assessment of subcellular ROS and NO metabolism in higher plants: Multifunctional signaling molecules. *Antioxidants (Basel), 8*(12), 641.

Korkina, L. G. (2007). Phenylpropanoids as naturally occurring antioxidants: From plant defense to human health. *Cellullar and Molecular Biology, 53*(1), 15–25.

Kruger, M. J., Davies, N., Myburgh, K. H., & Lecour, S. (2014). Proanthocyanidins, anthocyanins and cardiovascular diseases. *Food Research International, 59*, 41–52.

Kumari, A., Parida, A. K., Rangani, J., & Panda, A. (2017). Antioxidant activities, metabolic profiling, proximate analysis, mineral nutrient composition of *Salvadora persica* fruit unravel a potential functional food and a natural source of pharmaceuticals. *Frontiers in Pharmacology, 8*, 61.

Lanfer-Marquez, U. M., Barros, R. M. C., & Sinnecker, P. (2005). Antioxidant activity of chlorophylls and their derivatives. *Food Research International, 38*, 885–891.

Léchaudel, M., Darnaudery, M., Joët, T., Fournier, P., & Joas, J. (2018). Genotypic and environmental effects on the level of ascorbic acid, phenolic compounds and related gene expression during pineapple fruit development and ripening. *Plant Physiology and Biochemisty, 130*, 127–138.

Lenucci, M. S., Tornese, R., Mita, G., & Durante, M. (2022). Bioactive compounds and antioxidant activities in different fractions of mango fruits (*Mangifera indica* L., cultivar Tommy Atkins and Keitt). *Antioxidants (Basel), 11*(3), 484.

Li, M., & Kim, C. (2021). Chloroplast ROS and stress signaling. *Plant Communication, 3*(1), 100264.

Li, P., Zhao, L., Qi, F., Htwe, N. M. P. S., Li, Q., Zhang, D., ... Liang, Y. (2021). The receptor-like cytoplasmic kinase RIPK regulates broad-spectrum ROS signaling in multiple layers of plant immune system. *Molecular Plant, 14*(10), 1652–1667.

Li, Z., He, P., Long, Y., Yuan, G., Shen, W., Chen, Z., ... Zhang, X. (2021). Drug repurposing of pantoprazole and vitamin c targeting tumor microenvironment conditions improves anticancer effect in metastatic castration-resistant prostate cancer. *Frontiers in Oncology, 11*, 660320.

Liebthal, M., Maynard, D., & Dietz, K. J. (2018). Peroxiredoxins and redox signaling in plants. *Antioxidant Redox Signaling, 28*(7), 609–624.

Lightbourn, G. J., Griesbach, R. J., Novotny, J. A., Clevidence, B. A., Rao, D. D., & Stommel, J. R. (2008). Effects of anthocyanin and carotenoid combinations on foliage and immature fruit color of *Capsicum annuum* L. *The Journal of Heredity, 99*(2), 105–111.

Lin, X., Yang, R., Dou, Y., Zhang, W., Du, H., Zhu, L., & Chen, J. (2020). Transcriptome analysis reveals delaying of the ripening and cell-wall degradation of kiwifruit by hydrogen sulfide. *Journal of the Science of Food and Agriculture, 100*(5), 2280–2287.

Lindo-García, V., Larrigaudière, C., Duaigües, E., López, M. L., Echeverria, G., & Giné-Bordonaba, J. (2020). Elucidating the involvement of ethylene and oxidative stress during on- and off-tree ripening of two pear cultivars with different ripening patterns. *Plant Physiology and Biochemistry, 155*, 842–850.

Liochev, S. I., & Fridovich, I. (2002). The Haber-Weiss cycle – 70 years later: An alternative view. *Redox Report, 7*(1), 55–57.

Liu, L., Shao, Z., Zhang, M., & Wang, Q. (2015). Regulation of carotenoid metabolism in tomato. *Molecular Plant, 8*(1), 28–39.

Liu, M., Zhang, Z., Xu, Z., Wang, L., Chen, C., & Ren, Z. (2021). Overexpression of SlMYB75 enhances resistance to Botrytis cinerea and prolongs fruit storage life in tomato. *Plant Cell Reports, 40*(1), 43–58 2021 Jan.

Liu, Y., Tikunov, Y., Schouten, R. E., Marcelis, L. F. M., Visser, R. G. F., & Bovy, A. (2018). Anthocyanin biosynthesis and degradation mechanisms in solanaceous vegetables: A review. *Frontiers in Chemistry, 6*, 52.

López, A. P., Gochicoa, M. T. N., & Franco, A. R. (2010). Activities of antioxidant enzymes during strawberry fruit development and ripening. *Biology Plantarum, 54*, 349–352.

López-Huertas, E., & del Río, L. A. (2014). Characterization of antioxidant enzymes and peroxisomes of olive (*Olea europaea* L.) fruits. *Journal of Plant Physiology, 171*(16), 1463–1471.

López-Huertas, E., & Palma, J. M. (2020). Changes in glutathione, ascorbate, and antioxidant enzymes during olive fruit ripening. *Journal of Agricultural Food Chemistry, 68*(44), 12221–12228.

Mabuchi, K., Maki, H., Itaya, T., Suzuki, T., Nomoto, M., Sakaoka, S., ... Tsukagoshi, H. (2018). MYB30 links ROS signaling, root cell elongation, and plant immune responses. *Proceedings of the National Academy of Sciences of the United States of America, 115*(20), E4710–E4719.

Maksimov, N., Evmenyeva, A., Breygina, M., & Yermakov, I. (2018). The role of reactive oxygen species in pollen germination in Picea pungens (blue spruce). *Plant Reproduction, 31*(4), 357–365.

Maldonado-Celis, M. E., Yahia, E. M., Bedoya, R., Landázuri, P., Loango, N., Aguillón, J., ... Guerrero Ospina, J. C. (2019). Chemical composition of mango (*Mangifera indica* L.) fruit: Nutritional and phytochemical compounds. *Front in Plant Science, 10*, 1073.

Manthey, J. A., & Perkins-Veazie, P. (2009). Influences of harvest date and location on the levels of beta-carotene, ascorbic acid, total phenols, in vitro antioxidant capacity, and phenolic profiles of five commercial varieties of mango (*Mangifera indica* L.). *Journal of Agricultural and Food Chemistry, 57*, 10825–10830.

Marino, D., Dunand, C., Puppo, A., & Pauly, N. (2012). A burst of plant NADPH oxidases. *Trends in Plant Science, 17*(1), 9–15.
Markovski, A., & Velkoska-Markovska, L. (2016). Content of vitamin C in the fruits of jujube (*Ziziphus jujube* Mill.) varieties and their progenies. *Agriculture & Forestry, 62*(2), 217–225.
Martí, N., Mena, P., Cánovas, J. A., Micol, V., & Saura, D. (2009). Vitamin C and the role of citrus juices as functional food. *Natural Product Communications, 4*(5), 677–700.
Mateos, R. M., Bonilla-Valverde, D., del Río, L. A., Palma, J. M., & Corpas, F. J. (2009). NADP-dehydrogenases from pepper fruits: Effect of maturation. *Physiologia Plantarum, 135*(2), 130–139.
Mateos, R. M., Jiménez, A., Román, P., Romojaro, F., Bacarizo, S., Leterrier, M., ... Palma, J. M. (2013). Antioxidant systems from pepper (*Capsicum annuum* L.): Involvement in the response to temperature changes in ripe fruits. *International Journal of Molecular Science, 14*(5), 9556–9580.
Matsufuji, H., Ishikawa, K., Nunomura, O., Chino, M., & Takeda, M. (2007). Antioxidant content of different coloured sweet peppers, white, green, yellow, orange and red (*Capsicum annuum* L.). *International Journal of Food Science and Technology, 42*(12), 1482–1488.
Mattoo, A. K., Dwivedi, S. L., Dutt, S., Singh, B., Garg, M., & Ortiz, R. (2022). Anthocyanin-rich vegetables for human consumption-focus on potato, sweetpotato and tomato. *International Journal of Molecular Science, 23*(5), 2634.
McSweeney, M., & Seetharaman, K. (2015). State of polyphenols in the drying process of fruits and vegetables. *Critical Reviews in Food Science and Nutrition, 55*(5), 660–669.
Meitha, K., Pramesti, Y., & Suhandono, S. (2020). Reactive oxygen species and antioxidants in postharvest vegetables and fruits. *International Journal of Food Science, 2020*, 8817778.
Miranda, S., Vilches, P., Suazo, M., Pavez, L., García, K., Méndez, M. A., ... Del Pozo, T. (2020). Melatonin triggers metabolic and gene expression changes leading to improved quality traits of two sweet cherry cultivars during cold storage. *Food Chemistry, 319*, 126360.
Mohanty, S., & Cock, I. E. (2012). The chemotherapeutic potential of *Terminalia ferdinandiana*: Phytochemistry and bioactivity. *Pharmacognosy Reviews, 6*(11), 29–36.
Molinett, S. A., Alfaro, J. F., Sáez, F. A., Elgueta, S., Moya-León, M. A., & Figueroa, C. R. (2021). Postharvest treatment of hydrogen sulfide delays the softening of chilean strawberry fruit by downregulating the expression of key genes involved in pectin catabolism. *International Journal of Molecular Science, 22*(18), 10008.
Moller, I. M. (2001). Plant mitochondria and oxidative stress: Electron transport, NADPH turnover, and metabolism of reactive oxygen species. *Annual Review of Plant Physiology and Plant Molecular Biology, 52*, 561–591.
Mondal, K., Malhotra, S. P., Jain, V., & Singh, R. (2009). Oxidative stress and antioxidant systems in Guava (*Psidium guajava* L.) fruits during ripening. *Physiology and Molecular Biology of Plants, 15*(4), 327–334.
Mondal, K., Sharma, N. S., Malhotra, S. P., Dhawan, K., & Singh, R. (2004). Antioxidant systems in ripening tomato fruits. *Biologia Plantarum, 48*, 49–53.
Mueller, L., & Boehm, V. (2011). Antioxidant activity of β-carotene compounds in different in vitro assays. *Molecules (Basel, Switzerland), 16*(2), 1055–1069.
Müller, M., & Munné-Bosch, S. (2015). Ethylene response factors: A key regulatory hub in hormone and stress signaling. *Plant Physiology, 169*(1), 32–41.
Najwa, R., & Azlan, A. (2017). Comparison of vitamin C content in citrus fruits by titration and high performance liquid chromatography (HPLC) methods. *International Food Research Journal, 24*(2), 726–733.

Nishiyama, I., Yamashita, Y., Yamanaka, M., Shimohashi, A., Fukuda, T., & Oota, T. (2004). Varietal difference in vitamin C content in the fruit of kiwifruit and other actinidia species. *Journal of Agricultural and Food Chemistry, 52*(17), 5472–5475.

Nowak, D., Gośliński, M., & Kłębukowska, L. (2022). Antioxidant and antimicrobial properties of selected fruit juices. *Plant Foods for Human Nutrition.* https://doi.org/10.1007/s11130-022-00983-2

Orlikova, B., Tasdemir, D., Golais, F., Dicato, M., & Diederich, M. (2011). Dietary chalcones with chemopreventive and chemotherapeutic potential. *Genes & Nutrition, 6*(2), 125–147.

Orman-Ligeza, B., Parizot, B., de Rycke, R., Fernandez, A., Himschoot, E., Van Breusegem, F., ... Draye, X. (2016). RBOH-mediated ROS production facilitates lateral root emergence in Arabidopsis. *Development (Cambridge, England), 143*(18), 3328–3339.

Orsavová, J., Hlaváčová, I., Mlček, J., Snopek, L., & Mišurcová, L. (2019). Contribution of phenolic compounds, ascorbic acid and vitamin E to antioxidant activity of currant (Ribes L.) and gooseberry (Ribes uva-crispa L.) fruits. *Food Chemistry, 284*, 323–333 Jun 30.

Otulak-Kozieł, K., Kozieł, E., & Valverde, R. A. (2019). The respiratory burst oxidase homolog D (RbohD) cell and tissue distribution in potato-potato virus Y (PVYNTN) hypersensitive and susceptible reactions. *International Journal of Molecular Science, 20*(11), 2741.

Palma, J. M., Jiménez, A., Sandalio, L. M., Corpas, F. J., Lundqvist, M., Gómez, M., ... del Río, L. A. (2006). Antioxidative enzymes from chloroplasts, mitochondria, and peroxisomes during leaf senescence of nodulated pea plants. *Journal of Experimental Botany, 57*(8), 1747–1758 2006.

Palma, J. M., Mateos, R. M., López-Jaramillo, J., Rodríguez-Ruiz, M., González-Gordo, S., Lechuga-Sancho, A. M., & Corpas, F. J. (2020). Plant catalases as NO and H_2S targets. *Redox Biology, 34*, 101525.

Pilati, S., Brazzale, D., Guella, G., Milli, A., Ruberti, C., Biasioli, F., ... Moser, C. (2014). The onset of grapevine berry ripening is characterized by ROS accumulation and lipoxygenase-mediated membrane peroxidation in the skin. *BMC Plant Biology, 14*, 87.

Prakash, A., & Baskaran, R. (2018). Acerola, an untapped functional superfruit: A review on latest frontiers. *The Journal of Food Science and Technology, 55*(9), 3373–3384.

Raiola, A., Tenore, G. C., Barone, A., Frusciante, L., & Rigano, M. M. (2015). Vitamin E content and composition in tomato fruits: Beneficial roles and bio-fortification. *International Journal of Molecular Science, 16*(12), 29250–29264 8.

Reiter, R. J., Tan, D. X., Manchester, L. C., & Qi, W. (2001). Biochemical reactivity of melatonin with reactive oxygen and nitrogen species: A review of the evidence. *Cell Biochemistry and Biophysics, 34*, 237–256.

Rice-Evans, C. A., Miller, J. A., & Paganga, G. (1996). Structure-antioxidant activity relationships of flavonoids and phenolic acids. *Free Radical Biology & Medicine, 20*, 933–956.

Rocha, J., Borges, N., & Pinho, O. (2020). Table olives and health: A review. *Journal of Nutritional Science, 9*, e57.

Rodríguez-Ruiz, M., Mateos, R. M., Codesido, V., Corpas, F. J., & Palma, J. M. (2017). Characterization of the galactono-1,4-lactone dehydrogenase from pepper fruits and its modulation in the ascorbate biosynthesis. Role of nitric oxide. *Redox Biology, 12*, 171–181.

Rosalie, R., Léchaudel, M., Dhuique-Mayer, C., Dufossé, L., & Joas, J. (2018). Antioxidant and enzymatic responses to oxidative stress induced by cold temperature storage and ripening in mango (*Mangifera indica* L. cv. 'Cogshall') in relation to carotenoid content. *Journal Plant Physiology, 224–225*, 75–85.

Rubio-Melgarejo, A., Balois-Morales, R., Ochoa-Jiménez, V. A., Casas-Junco, P. P., Jiménez-Zurita, J. O., Bautista-Rosales, P. U., & Berumen-Varela, G. (2021). Differential responses of antioxidative system during the interaction of soursop fruits (*Annona muricata* L.) and *Nectria haematococca* at postharvest storage. *Plants (Basel), 10*, 1432.

Sadilova, E., Stintzing, F. C., & Carle, R. (2006). Anthocyanins, colour and antioxidant properties of eggplant (*Solanum melongena* L.) and violet pepper (*Capsicum annuum* L.) peel extracts. *Zeitschrift für Naturforschung. C, A Journal of Biosciences, 61*(7–8), 527–535.

Sapei, L., & Hwa, L. (2014). Study on the kinetics of vitamin C degradation in fresh strawberry juices. *Procedia Chemistry, 9*, 62–68.

Sewelam, N., Kazan, K., Thomas-Hall, S. R., Kidd, B. N., Manners, J. M., & Schenk, P. M. (2013). Ethylene response factor 6 is a regulator of reactive oxygen species signaling in Arabidopsis. *PLoS One, 8*, e70289.

Siddiqui, M. W., Homa, F., Lata, D., Mir, H., Aftab, T., & Mishra, P. (2022). Hydrogen sulphide infiltration downregulates oxidative metabolism and extends postharvest life of banana. *Plant Biology (Stuttg), 24*(4), 697–703.

Sies, H. (2020). Findings in redox biology: From H_2O_2 to oxidative stress. *Journal of Biological Chemistry, 295*(39), 13458–13473.

Sies, H. (2021). Oxidative eustress: On constant alert for redox homeostasis. *Redox Biology, 41*, 101867.

Singh, P., Mishra, V., Tripathi, D. K., Corpas, F. J., & Singh, V. P. (2022). RIPK: A crucial ROS signaling component in plants. *Trends in Plant Science, 27*(3), 214–216.

Sixto, I., Bravo, L., Sotero, V., & Oliva, C. (2011). Vitamin C content in fruits of camu camu Myrciaria dubia (H.B.K) Mc Vaugh, in four states of maturation, coming from the Collection of Germoplasma of the INIA Loreto, Peru. *Scientia Agropecuaria, 2*(3), 123–130.

Skodra, C., Titeli, V. S., Michailidis, M., Bazakos, C., Ganopoulos, I., Molassiotis, A., & Tanou, G. (2021). Olive fruit development and ripening: Break on through to the "-omics" side. *International Journal of Molecular Science, 22*(11), 5806.

Song, J., CampbellPalmer, L., Vinqvist-Tymchuk, M., Fillmore, S., Forney, C., Luo, H., & Zhang, Z. (2020). Proteomic changes in antioxidant system in strawberry during ripening. *Frontiers in Plant Science, 11*, 594156.

Steelheart, C., Alegre, M. L., Baldet, P., Rothan, C., Bres, C., Just, D., ... Bartoli, C. G. (2020). The effect of low ascorbic acid content on tomato fruit ripening. *Planta, 252*(3), 36.

Steelheart, C., Alegre, M. L., Baldet, P., Rothan, C., Bres, C., Just, D., ... Bartoli, C. G. (2022). High light stress induces H_2O_2 production and accelerates fruit ripening in tomato. *Plant Science, 322*, 111348.

Sultanbawa, Y., Chaliha, M., Cusack, A. D. Edwards, D., & Williams, D. (2018). Monitoring quality and bioactivity of Kakadu plum in the Northern territory. In *Agri futures Australia* (pp. 1–50). ISBN 978-1-74254-997-2.

Sun, C., Deng, L., Du, M., Zhao, J., Chen, Q., Huang, T., ... Li, C. A. (2020). Transcriptional network promotes anthocyanin biosynthesis in tomato flesh. *Molecular Plant, 13*(1), 42–58 6.

Szeto, Y. T., Tomlinson, B., & Benzie, I. F. (2002). Total antioxidant and ascorbic acid content of fresh fruits and vegetables: Implications for dietary planning and food preservation. *The British Journal of Nutrition, 87*(1), 55–59.

Torres, M. A., & Dangl, J. L. (2005). Functions of the respiratory burst oxidase in biotic interactions, abiotic stress and development. *Current Opinion in Plant Biology, 8*(4), 397–403.

Tsukagoshi, H., Busch, W., & Benfey, P. N. (2010). Transcriptional regulation of ROS controls transition from proliferation to differentiation in the root. *Cell, 143*(4), 606–616.

Vall-Llaura, N., Fernández-Cancelo, P., Nativitas-Lima, I., Echeverria, G., Teixidó, N., Larrigaudière, C., ... Giné-Bordonaba, J. (2022). ROS-scavenging-associated transcriptional and biochemical shifts during nectarine fruit development and ripening. *Plant Physiology and Biochemisty, 171*, 38–48.

Vreeburg, R. A., Airianah, O. B., & Fry, S. C. (2014). Fingerprinting of hydroxyl radical-attacked polysaccharides by N-isopropyl-2-aminoacridone labelling. *Biochemical Journal, 463*(2), 225–237.

Wahyuni, Y., Ballester, A. R., Sudarmonowati, E., Bino, R. J., & Bovy, A. G. (2011). Metabolite biodiversity in pepper (Capsicum) fruits of thirty-two diverse accessions: Variation in health-related compounds and implications for breeding. *Phytochemistry, 72*(11–12), 1358–1370.

Wahyuni, Y., Ballester, A. R., Sudarmonowati, E., Bino, R. J., & Bovy, A. G. (2013). Secondary metabolites of Capsicum species and their importance in the human diet. *Journal of Natural Products, 76*(4), 783–793.

Wang, H., Cao, G., & Prior, R. L. (1997). Oxygen radical absorbing capacity of anthocyanins. *Journal of Agricultural and Food Chemistry, 45*, 304–309.

Wang, L., Chen, S., Shao, J., Zhang, C., Mei, L., Wang, K., ... Zheng, Y. (2022). Hydrogen sulfide alleviates chilling injury in peach fruit by maintaining cell structure integrity via regulating endogenous H_2S, antioxidant and cell wall metabolisms. *Food Chemistry, 391*, 133283.

Wang, Y., Ji, D., Chen, T., Li, B., Zhang, Z., Qin, G., & Tian, S. (2019). Production, signaling, and scavenging mechanisms of reactive oxygen species in fruit-pathogen interactions. *International Journal of Molecular Science, 20*, 2994.

Wu, B., Li, P., Hong, X., Xu, C., Wang, R., & Liang, Y. (2022). The receptor-like cytosolic kinase RIPK activates NADP-malic enzyme 2 to generate NADPH for fueling ROS production. *Molecular Plant, 15*(5), 887–903.

Wu, F., Yang, H., Chang, Y., Cheng, J., Bai, S., & Yin, J. (2012). Effects of nitric oxide on reactive oxygen species and antioxidant capacity in Chinese Bayberry during storage. *Scientia Horticulturae, 135*, 106–111.

Xavier, A. A., & Pérez-Gálvez, A. (2016). Carotenoids as a source of antioxidants in the diet. *Sub-Cellular Biochemistry, 79*, 359–375 2016.

Xie, D. L., Zheng, X. L., Zhou, C. Y., Kanwar, M. K., & Zhou, J. (2022). Functions of redox signaling in pollen development and stress response. *Antioxidants (Basel), 11*(2), 287.

Xiong, H., Hua, L., Reyna-Llorens, I., Shi, Y., Chen, K. M., Smirnoff, N., ... Hibberd, J. M. (2021). Photosynthesis-independent production of reactive oxygen species in the rice bundle sheath during high light is mediated by NADPH oxidase. *Proceedings of the National Academy of Sciences of the United States of America, 118*(25), e2022702118.

Yamada, M., Han, X., & Benfey, P. N. (2020). RGF1 controls root meristem size through ROS signalling. *Nature, 577*(7788), 85–88.

Yan, S., Chen, N., Huang, Z., Li, D., Zhi, J., Yu, B., ... Qiu, Z. (2020). Anthocyanin Fruit encodes an R2R3-MYB transcription factor, SlAN2-like, activating the transcription of SlMYBATV to fine-tune anthocyanin content in tomato fruit. *The New Phytologist, 225*(5), 2048–2063.

Yang, H., Mu, J., Chen, L., Feng, J., Hu, J., Li, L., ... Zuo, J. (2015). S-nitrosylation positively regulates ascorbate peroxidase activity during plant stress responses. *Plant Physiology, 167*, 1604–1615.

Yang, J., Wang, P., Li, S., Liu, T., & Hu, X. (2022). Polyamine oxidase triggers H_2O_2-mediated spermidine improved oxidative stress tolerance of tomato seedlings subjected to saline-alkaline stress. *International Journal of Molecular Science, 23*(3), 1625.

Yousuf, B., Gul, K., Wani, A. A., & Singh, P. (2016). Health benefits of anthocyanins and their encapsulation for potential use in food systems. *Critical Reviews in Food Science and Nutrition, 56*, 2223–2230.

Zaaboul, F., & Liu, Y. (2022). Vitamin E in foodstuff: Nutritional, analytical, and food technology aspects. *Comprehensive Reviews in Food Science and Food Safety, 21*(2), 964–998.

Zafra, A., Rejón, J. D., Hiscock, S. J., & Alché Jde, D. (2016). Patterns of ROS accumulation in the stigmas of angiosperms and visions into their multi-functionality in plant reproduction. *Front in Plant Science,* 7, 1112.

Zahedi, S. M., Hosseini, M. S., Abadía, J., & Marjani, M. (2020). Melatonin foliar sprays elicit salinity stress tolerance and enhance fruit yield and quality in strawberry (*Fragaria* × *ananassa* Duch.). *Plant Physiology and Biochemistry: PPB/Societe Francaise de Physiologie Vegetale, 149,* 313–323.

Zandalinas, S. I., & Mittler, R. (2021). Vascular and nonvascular transmission of systemic reactive oxygen signals during wounding and heat stress. *Plant Physiology, 186*(3), 1721–1733.

Zandalinas, S. I., Sengupta, S., Burks, D., Azad, R. K., & Mittler, R. (2019). Identification and characterization of a core set of ROS wave-associated transcripts involved in the systemic acquired acclimation response of Arabidopsis to excess light. *Plant Journal, 98*(1), 126–141.

Zhang, M., Chen, H., Mujumdar, A. S., Tang, J., Miao, S., & Wang, Y. (2017). Recent developments in high-quality drying of vegetables, fruits, and aquatic products. *Critical Reviews in Food Science and Nutrition, 57*(6), 1239–1255.

Zhang, Y., De Stefano, R., Robine, M., Butelli, E., Bulling, K., Hill, L., ... Schoonbeek, H. J. (2015). Different reactive oxygen species scavenging properties of flavonoids determine their abilities to extend the shelf life of tomato. *Plant Physiology, 169*(3), 1568–1583.

Zhao, Z., Chen, J., Ci, F., Pang, H., Cheng, N., & Xing, A. (2022). α-Carotene: A valuable carotenoid in biological and medical research. *Journal of the Science of Food and Agriculture.* https://doi.org/10.1002/jsfa.11966

Zhou, Y., Mumtaz, M. A., Zhang, Y., Yang, Z., Hao, Y., Shu, H., ... Wang, Z. (2022). Response of anthocyanin biosynthesis to light by strand-specific transcriptome and miRNA analysis in *Capsicum annuum. BMC Plant Biology, 22*(1), 79.

Zhu, S., Sun, L., & Zhou, J. (2010). Effects of different nitric oxide application on quality of kiwifruit during 20°C storage. *International Journal of Food Science and Technology, 45,* 245–251.

Zuccarelli, R., Rodríguez-Ruiz, M., Lopes-Oliveira, P. J., Pascoal, G. B., Andrade, S. C. S., Furlan, C. M., ... Freschi, L. (2021). Multifaceted roles of nitric oxide in tomato fruit ripening: NO-induced metabolic rewiring and consequences for fruit quality traits. *Journal of Experimetal Botany, 72*(3), 941–958.

Further reading

Andrade-Galan, A. G., & Bieker, S. (2022). Specificity of H_2O_2 signaling in leaf senescence: Is the ratio of H_2O_2 contents in different cellular compartments sensed in Arabidopsis plants? *Cellular & Molecular Biology Letters, 27*(1), 4.

Begara-Morales, J. C., Sánchez-Calvo, B., Chaki, M., Valderrama, R., Mata-Pérez, C., López-Jaramillo, J., Padilla, M. N., Carreras, A., Corpas, F. J., & Barroso, J. B. (2014). Dual regulation of cytosolic ascorbate peroxidase (APX) by tyrosine nitration and S-nitrosylation. *Journal of Experimental Botany, 65*(2), 38 527.

Bouthour, D., Kalai, T., Chaffei, H. C., Gouia, H., & Corpas, F. J. (2015). Differential response of NADP-dehydrogenases and carbon metabolism in leaves and roots of two durum wheat (*Triticum durum* Desf.) cultivars (Karim and Azizi) with different sensitivities to salt stress. *Journal Plant Physiology, 179,* 56–63.

Corpas, F. J., & Palma, J. M. (2020). H_2S signaling in plants and applications in agriculture. *Journal of Advanced Research, 24,* 131–137.

Corpas, F. J., Freschi, L., Rodríguez-Ruiz, M., Mioto, P. T., González-Gordom, S., & Palma, J. M. (2018). Nitro-oxidative metabolism during fruit ripening. *Journal of Experimetal Botany, 69*, 3449–3463.

Gramegna, G., Rosado, D., Sánchez Carranza, A. P., Cruz, A. B., Simon-Moya, M., Llorente, B., ... Rossi, M. (2019). PHYTOCHROME-INTERACTING FACTOR 3 mediates light-dependent induction of tocopherol biosynthesis during tomato fruit ripening. *Plant Cell Environment, 42*(4), 1328–1339.

Palma, J. M., Corpas, F. J., & del Río, L. A. (2011). Proteomics as an approach to the understanding of the molecular physiology of fruit development and ripening. *Journal of Proteomics, 74*, 1230–1243.

Palma, J. M., Sevilla, F., Jiménez, A., del Río, L. A., Corpas, F. J., Álvarez de Morales, P., & Camejo, D. M. (2015). Physiology of pepper fruit and the metabolism of antioxidants: Chloroplasts, mitochondria and peroxisomes. *Annals of Botany, 116*, 627–636.

Rodríguez-Ruiz, M., Aparicio-Chacón, M. V., Palma, J. M., & Corpas, F. J. (2019a). Arsenate disrupts ion balance, sulfur and nitric oxide metabolisms in roots and leaves of pea (*Pisum sativum* L.) plants. *Environmental and Experimental Botany, 161*, 143–156.

Rodríguez-Ruiz, M., González-Gordo, S., Cañas, A., Campos, M. J., Paradela, A., Corpas, F. J., & Palma, J. M. (2019b). Sweet pepper (*Capsicum annuum* L.) fruits contain an atypical peroxisomal catalase that is modulated by reactive oxygen and nitrogen species. *Antioxidants (Basel), 8*, 374.

Signorelli, S., Corpas, F. J., Borsani, O., Barroso, J. B., & Monza, J. (2013). Water stress induces a differential and spatially distributed nitro-oxidative stress response in roots and leaves of *Lotus japonicus*. *Plant Science, 201–202*, 137–146.

Smirnoff, N. (2018). Ascorbic acid metabolism and functions: A comparison of plants and mammals. *Free Radical Biology and Medicine, 122*, 116–129.

Zentgraf, U., Andrade-Galan, A. G., & Bieker, S. (2022). Specificity of H2O2 signaling in leaf senescence: is the ratio of H2O2 contents in different cellular compartments sensed in Arabidopsis plants? *Cellular & Molecular Biology Letters, 27*(1), 4.

CHAPTER EIGHT

Reactive oxygen species (ROS) in mycorrhizal fungi and symbiotic interactions with plants

Maaria Rosenkranz[a], Huili Shi[b], Johannes Ballauff[b], Jörg-Peter Schnitzler[a], and Andrea Polle[b,*]

[a]Research Unit Environmental Simulation, Institute of Biochemical Plant Pathology, Helmholtz Munich, Neuherberg, Germany

[b]Forest Botany and Tree Physiology, Georg-August University of Goettingen, Göttingen, Germany

*Corresponding author. e-mail address: apolle@gwdg.de

Contents

Abstract

The interaction of plant roots with mycorrhizal fungi leads to the formation of a novel structure, the ectomycorrhiza. Ectomycorrhizal symbioses benefit both organisms by mutual nutrient exchange and improve the host´s stress resilience. Here, we review

ISSN 0065-2296, https://doi.org/10.1016/bs.abr.2022.11.001

the roles of reactive oxygen species and antioxidative systems in the fungal partner and the host plant for the establishment of the symbiosis and for the induction of local and systemic mycorrhizal resistance. Ectomycorrhizal symbioses can be formed with different fungal species each equipped with a distinct set of extracellular oxidative enzymes. While one function of the extracellular enzymes is the access to recalcitrant nutrient sources, some examples suggest neofunctionalization to enable intraradical colonization. The colonization also involves small secreted proteins, NADPH oxidases, H_2O_2 production, and affects the plant catalase, superoxide dismutase and peroxidase-based defenses. The fungal antioxidative defense is mainly based on glutathione and thioredoxin-based systems but its role in host colonization is largely unexplored. Among signals likely involved in systemic responses are jasmonates and salicylic acid in the plant, and volatile organic compounds released by the fungus. An emerging player with a role in plant-fungal interactions is nitric oxide. Higher stress tolerance of the mycorrhizal fungal species generally confers higher stress tolerance to the host and can vary among different isolates of the same fungal species. This review emphasizes that many puzzle parts are present, but that a complete picture has not yet emerged. We discuss further studies required to increase our understanding of the inter-kingdom dialog of both organisms.

1. Introduction

Fungi occur globally in all biomes. They have versatile life-styles, colonizing soil, plants and animals with many astonishing shapes and properties (Bunyard, 2022). In terrestrial forest ecosystems, their most important functions are the biotrophic or necrotrophic interactions with plants as mutualists or pathogens and their saprotrophic activities leading to the degradation of organic matter and solubilization of minerals. Fungi occur as unicellular (yeasts) or multicellular (hyphae) organisms and have been used by humans as resource for baking, brewing, and food, for example, truffles (*Tuber melanosporum*, Pezizales), chanterelle (*Cantharellus cibarius*, Canthrellales), cep (*Boletus edulis*, Boletales), Shiitake (*Lentiluna edodes*, Agaricales), etc. Fungi are also a source for an enormous range of chemical compounds such as the well-known drugs and poisons of fly agaric (*Amanita muscaria*) and death cap (*Amanita phalloides*, Agaricales). An outstanding example involving reactive oxygen species (ROS) in fungal metabolism is bioluminescence. Several saprotrophic fungi in the genus *Mycena* (Agaricales) glow in darkness. They contain a luciferase-enzyme complex (Ke et al., 2020), which can use molecular oxygen to oxidize luciferin to a high-energy endoperoxide, which then releases CO_2 and chemiluminescence (Garcia-Iriepa, Marazzi, & Navizet, 2020). It is obvious that different fungal species harbor highly diverse metabolism and biochemistry to engage in a wide array of ecological functions and trophic modes (Li et al., 2012). Recent advances in genome sequencing provided novel insights into their metabolic potentials (Kohler et al., 2015; Martin et al., 2008; Miyauchi et al., 2020) but many fungal species still await discovery (Martin et al., 2018). It is, therefore, important to note that even within a selected group of fungi, we can only scratch the surface of the many roles and functions of fungal ROS metabolism and their interaction with plants.

Here, we focus on ectomycorrhizal fungi (EMF). EMF form mutualistic associations with roots of woody plant species. The interaction of soil fungi with plant roots was initially observed by Hartig (1840) and later functionally described by Frank (1885) as "fungus root" (mýkēs (fungus), rhiza (root)); Frank postulated that the mycorrhiza was the sole organ for water and mineral supply to the tree and sustained by provision of carbon compounds from the host (cited after Trappe, 2005). A typical ectomycorrhiza consists of a hyphal mantle, which ensheathes the root tip and an extramatrical mycelium which emanates into the soil; inside the mantle facing the root surface, hyphae invade the walls between adjacent rhizodermal cells, forming a novel interface (Hartig net) for nutrient exchange between the fungus and the plant (Smith and Read, 2010) (Fig. 1). In addition to ectomycorrhizas, various other forms of mycorrhizal associations with woody and non-woody plants exist such as arbuscular mycorrhizal fungi (AMF), ericoid mycorrhizas, orchid mycorrhizas, etc. AMF (phylum Mucoromycota) (Bonfante & Venice, 2020) are the most ancient symbionts, whereas EMF are relatively recent evolutionary inventions of polyphyletic origin (Basidiomycota, Ascomycota) (Lebreton et al., 2021; Strullu-Derrien, Selosse, Kenrick, & Martin, 2018). Their ancestors are wood-degrading brown and white rot fungi as well as litter decomposers (Kohler et al., 2015). Wood degrading fungi rely on extracellular production of ROS, mainly hydroxyl radicals generated by Fenton chemistry, to access recalcitrant nutrient sources from lignin or cellulose (Castaño, Zhang, Anderson, & Schilling, 2018). EMF have retained a range of these molecular tools (Miyauchi et al., 2020), enabling many species to thrive as free-living hyphae without a host plant, whereas AMF are obligate biotroph (Bago & Bécard, 2002).

In this review, we compile an overview on ROS metabolism in EMF and ROS functions for nutrient acquisition and plant defense responses. Examples from other mycorrhizal symbioses have been interspersed occasionally to illustrate communalities or differences. We discuss the question of how extra- and intracellular ROS-forming and detoxifying

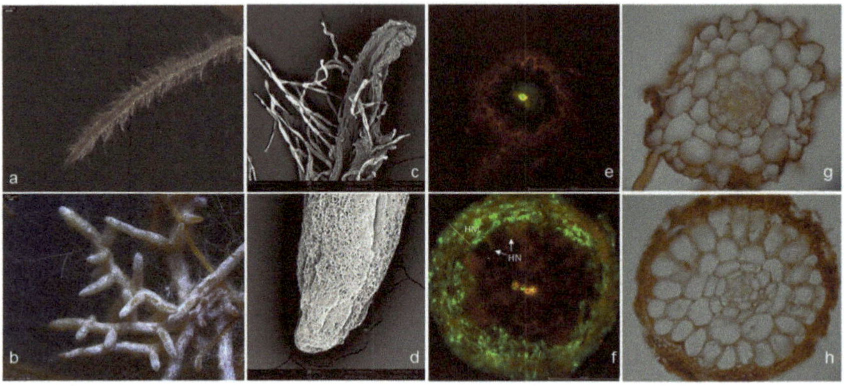

Fig. 1 Morphology and anatomical characteristics of non-mycorrhizal (A, C, E, G) and ectomycorrhizal (B, D, F, H) poplar root tips in association with *Paxillus involutus*. Photos of poplar root tips were taken under a compound microscope (A: non-mycorrhizal root tip, B: ectomycorrhizal root tips) and a scanning electron microscope (C: non-mycorrhizal root tip, D: ectomycorrhizal root tips). Plates A and C show abundant root hairs emanating from non-mycorrhizal roots and plates B and D the colonized root tips with very fine emanating hyphae (B) and the net-like structure formed around the root tip by the fungus (D). Cross sections of non-mycorrhizal (E) and ectomycorrhizal (F) poplar root tips were stained with propidium iodide to view plant cell walls (*red*) and with WGA-Alexa488 to view fungal hyphae (*green*). HM: hyphal mantle, HN: Hartig net.·H_2O_2 accumulation in cross-section of non-mycorrhizal (G) and ectomycorrhizal (H) poplar root tips is visualized by staining with diaminobenzidine.

systems of EMF fungi are controlled. Since ROS and antioxidative systems also play important roles in plant development (Dat et al., 2003; Huang, Ullah, Zhou, Yi, & Zhao, 2019), we address the general role of ROS in the establishment of host-fungal interactions and their impact on host defenses.

2. ROS production and antioxidative systems in ectomycorrhizal fungi

2.1 Reactive oxygen species in ectomycorrhizal fungi

The production and detoxification of ROS is a hallmark of aerobic life. The main forms of ROS occurring in cellular metabolism are superoxide radicals ($O_2^{\bullet -}$) and hydrogen peroxide (H_2O_2). In the presence of transition metals (Mn, Fe, Cr), H_2O_2 and $O_2^{\bullet -}$ can drive the production of highly reactive hydroxyl radicals (*OH) (Fig. 2). This reaction is

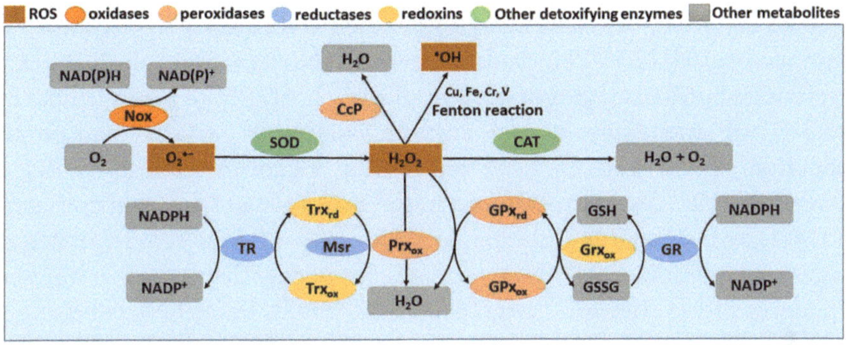

Fig. 2 Schematic representation of the intracellular ROS-producing and antioxidative system in EMF. NAD(P)H oxidases produce $O_2^{\bullet-}$. SODs convert $O_2^{\bullet-}$ to O_2 and H_2O_2. CATs catalyze H_2O_2 degradation to H_2O and O_2. EMF control intracellular ROS mainly by glutathione and thiol-related redox systems. GPx remove cellular H_2O_2 by oxidation of GSH to GSSG; the regeneration of GSH is achieved by GR with NADPH as the reductant. In thioredoxin systems, Prx reduce H_2O_2 to H_2O with Trx as donor; oxidized Trx is reduced by TR. Grx and Msr participate in antioxidant systems indirectly. Oxidized Grx is reduced by GSH non-enzymatically, whereas Msr reduces methionine sulfoxide to methionine and is regenerated by Trxs. Besides, H_2O_2 can be removed by CcP or utilized in the Fenton reaction. SOD, Superoxide dismutases; CAT, catalases; GPx, glutathione peroxidase; GR, glutathione reductase; GSH, glutathione; GSSG, glutathione disulfide; Grx, glutaredoxin; Prx, thioredoxin-dependent peroxiredoxin; Trx, thioredoxin; TR, thioredoxin reductases; Msr, methionine sulfoxide reductases; CcP, cytochrome c peroxidases.

particularly relevant in some clades of EMF, as described below. ROS have functions as signaling compounds and moreover, in the extracellular compartment, they help the fungus to mine the surroundings for nutrients. ROS levels increase in response to abiotic and biotic stress in free living EMF as well as in plant symbioses and induce antioxidative systems (Bellion, Courbot, Jacob, Blaudez, & Chalot, 2006; Chen, Hawighorst, Sun, & Polle, 2014; Chot & Reddy, 2022; Luo et al., 2014; Schützendübel & Polle, 2002). Unattended, excessive ROS production causes oxidation of membranes, DNA, and proteins, leading to lipid peroxides, DNA strand breaks and deregulation of the metabolic redox state (as described elsewhere in this volume). Under these conditions, plants activate programmed cell death (Gaspar & Pawlowska, 2022). Whether regulated ROS-induced cell death also occurs in fungi is unknown but ferroptosis-induced lipid peroxidation (the release of iron and ROS production by NADPH oxidases) has been found in fungi under stress (Gaspar & Pawlowska, 2022).

Like in other organisms, controlled $O_2^{\bullet-}$ formation is mediated by ectomycorrhizal NADPH oxidases (termed Nox in fungi or RBOH = respiratory burst oxidase homolog in plants). NADPH oxidases (Nox, EC 1.6.3.1) are membrane-bound enzyme complexes, which catalyze the reduction of O_2 into $O_2^{\cdot-}$ by transferring electrons from NADPH to oxygen (Fig. 2). $O_2^{\cdot-}$ are produced in the extracellular space and converted to H_2O_2 by dismutation, superoxide dismutases or by the activity of certain classes of peroxidases. NADPH oxidases are widely distributed in different kingdoms of life (Bedard, Lardy, & Krause, 2007). Fungal Noxs belong to three subfamilies NoxA, NoxB, and NoxC, in addition to NoxR, which encodes a regulatory subunit (Takemoto, Tanaka, & Scott, 2007). Mining for homologous genes encoding Nox enzymes indicated that there are at least two Noxs (NoxA or B and NoxR) in ectomycorrhizal fungal genomes (Mattila, Österman-Udd, Mali, & Lundell, 2022).

Noxs play major roles in plant immune responses, mediating biotic interactions (Wrzaczek, Brosché, & Kangasjärvi, 2013). In the beneficial *Tricholoma*-plant interaction, the expression of plant jasmonate-related defense was enhanced in the presence of fungal strains with deletions of Nox genes (Villalobos-Escobedo et al., 2020). However, the vitality of the fungal deletion strains was reduced, supporting that Nox activities are important for normal life and for plant-fungal interactions. A further indication for a role for ROS in plant-EMF interaction is a strong accumulation of H_2O_2 in the hyphal mantle of *Paxillus involutus* ectomycorrhizas compared with non-mycorrhizal poplar roots (Fig. 1G and H; Gafur, Schützendübel, Langenfeld-Heyser, Fritz, & Polle, 2004). Whether this ROS production is due to ectomycorrhizal Nox activities is unknown. It has been speculated that H_2O_2 in the ectomycorrhizal mantle may have antimicrobial properties, preventing overgrowth by other microbes and stimulating the host´s defense systems (Gafur et al., 2004; Langenfeld-Heyser et al., 2007). It has also been shown that growth of EMF is sensitive to H_2O_2 (Liu et al., 2022; Mucha, Napierała-Filipiak, Gabała, Pawłowski, & Zadworny, 2019) and therefore, ROS may be a host-controlled mechanism to attenuate fungal proliferation at the root surface.

2.2 Intracellular ROS control

2.2.1 Superoxide dismutases and catalases

As in all known aerobic organisms, superoxide dismutases (SOD, EC 1.15.1.1) are major antioxidant enzymes in EMF protecting cells from damage by $O_2^{\bullet-}$ (Fig. 2). SODs convert $O_2^{\bullet-}$ to O_2 and H_2O_2

(McCord & Fridovich, 1969). Recent genomic analyses of potential SOD genes predicted the existence of different types of SODs in wide range of EMF species from different phyla (Bellion et al., 2006; Bolchi et al., 2011; Mattila et al., 2022). Different SODs can be distinguished by their reactive center and were predicted to be predominantly localized to the cytoplasm and mitochondria or to other organelles like peroxisomes (SOD1 = Cu/Zn-SOD = generally cytosolic enzymes, SOD2/3 = Fe- or Mn-SOD = generally mitochondrial or peroxisomal enzymes). In addition, some SODs, present in the genomes of EMF were predicted to be localized in the extracellular space of the fungal cell wall, for example, SOD1 from *Laccaria bicolor* and SOD4–6 from *Tulasnella calospora* (Mattila et al., 2022).

The stress responses of SODs differ among EMF. For example, *Pisolithus tinctorius* showed a massive enhancement of SOD activities under very high Mn concentrations, whereas *Cenococcum geophilum* and *Suillus granulatus* lost their SOD activities entirely and showed strong growth impairment under excessive Mn (Qi, Zhao, Liu, & Huang, 2016). A comparison of ten *C. geophilum* isolates under salt stress showed differences in SOD responses, which were however, unrelated to the salt tolerance or susceptibility of the isolates (Li et al., 2022). Only a few studies established causal links between ectomycorrhizal SODs and stress. A Mn-SOD was cloned from *Paxillus involutus* (Jacob et al., 2001). Complementation of an *E.coli* null sod-mutant with the *P. involutus* Mn-SOD rescued the bacteria from oxidative stress imposed by paraquat (a $O_2^{\bullet-}$ producing chemical) or cadmium (Cd) (Jacob et al., 2001). Likewise, Cd and Zn stress enhanced Cu/Zn-SOD activities (Chiapello, Martino, & Perotto, 2015), while disruption of a Cu/Zn-SOD gene in the ericoid mycorrhizal fungus *Oidiodendron maius* enhanced fungal stress susceptibility to Cd, Zn and menadione (a radical producing agent) (Abba, Khouja, Martino, Archer, & Perotto, 2009). Although the *O. maius* strain with the loss-of-Cu/Zn-SOD still contained another SOD, i.e., SOD2, its vitality, conidia production and ability for mycorrhization was decreased compared with the wildtype (Abba et al., 2009). These studies underpin the importance of SODs for fungal fitness. Jacob et al., (2001) found that oxidative stress hardly affected the transcript abundance of SOD in *P. involutus* and speculated that the enzyme was post-transcriptionally regulated. It is now known that yeast SOD can be regulated by phosphorylation or other protein modifications (Tsang, Liu, Thomas, Zhang, & Zheng, 2014). However, regulatory details are unknown for SODs in EMF.

Catalases (CAT, EC 1.11.1.6), together with SODs function as the first line of antioxidant defense (Fig. 2). CATs convert two molecules of H_2O_2 into H_2O and O_2. Despite their importance for detoxification of H_2O_2 (Vandenabeele et al., 2004), to the best of our knowledge, there is currently no evidence for the purification and biochemical characterization of a CAT enzyme from an EMF species. Putative CAT encoding genes were found in all genomes of EMF species inspected to date (Bellion et al., 2006; Bolchi et al., 2011; Mattila et al., 2022) and CAT activities were measured in ectomycorrhizal mycelia (Kothamasi et al., 2019; Ott, Fritz, Polle, & Schützendübel, 2002). For instance, exposure to radioactive gamma radiation of pure cultures of *Suillus luteus*, *S. bovinus,* and *Rhizopon luteolus* resulted in increased enzymatic activities of CAT and SOD (Kothamasi et al., 2019). Cu and Cd also caused significant increases in CAT, SOD and other antioxidant enzyme activities in mycelia of *Lepista sodaria* (Dachuan & Jinyu, 2021). Ott et al. (2002) examined the time courses of SOD and CAT activities under different Cd levels in *P. involutus* cultures. They showed that CAT activities were relatively stress susceptible, declining with increasing Cd accumulation, whereas SOD activities showed time-dependent activity peaks. Therefore, Ott et al. (2002) suggested that other defense mechanisms than CATs might have taken over H_2O_2 removal under stress.

2.2.2 Glutathione and thiol-related redox systems

EMF contain multiple enzymatic and non-enzymatic systems for the control of H_2O_2, mainly based on thiol- and NADPH-dependent redox reactions. The NADPH-dependent glutathione system is composed of glutathione peroxidase (GPx, EC 1.11.1.9) and glutathione reductase (GR, EC 1.8.1.7), and requires glutathione (GSH) and NADPH as reductants (Fig. 2). GPxs remove cellular H_2O_2 or lipid peroxides by oxidation of GSH to glutathione disulfide (GSSG) and concomitant reduction of the peroxide to H_2O. The regeneration of GSH from GSSG is achieved by GR with NADPH as the reductant. Some studies also support that fungal GPxs use thioredoxin instead of GSH for the reduction of H_2O_2 (Adriani et al., 2021; Tanaka, Izawa, & Inoue, 2005). In vitro, GPx activity of *Paxillus involutus* can be measured with GSH in a coupled reaction consuming NADPH (Ott et al., 2002). However, whether GSH is replaced by thioredoxin in the cellular metabolism of an ectomycorrhizal fungus is not clear. Thioredoxins are also reduced by NADPH in a TR mediated reaction (Fernandez & Wilson, 2014). Since $NAD(P)^+$ reduction is achieved by glucose degradation, surveillance and keeping adequate H_2O_2 levels is directly linked to the cellular energy metabolism of the fungus.

When the first genome of an ectomycorrhizal fungus was sequenced, Morel et al. (2008) conducted a comprehensive screening of the genes encoding GPxs and the thioredoxin system in *Laccaria bicolor* (Morel, Kohler, Martin, Gelhaye, & Rouhier, 2008). For *L. bicolor*, the following gene families (with the number of homolog genes) were reported: thioredoxin reductase (EC 1.8.1.9, TR: 1), thioredoxin (EC 1.8.1.9, Trx: 4), glutathione reductase (EC 1.8.1.7, GR:1), glutaredoxins (EC 1.20.4, Grx: 5), thioredoxin-dependent peroxiredoxin [Prx, EC 1.11.1.24,Tpx: 6, including 2-Cys Prx (1), 1-Cys Prx (1), PrxQ (2), class II peroxidases EC1.11.1.14, Prx II (2)], glutathione peroxidase (EC 1.11.1.9, Gpx: 1), methionine sulfoxide reductases (MsrA, EC 1.8.4.13 and MsrB, EC 1.8.4.14, Msr: 2) but no sulfiredoxins (Srx, EC 1.8.98.2, Srx: 0) (Morel, Kohler, Martin, Gelhaye, & Rouhier, 2008). Sulfiredoxins are oxidoreductases, which have been identified in yeast (*Saccharomyces cerevisiae*) (Biteau, Labarre, & Toledano, 2003) and which are involved in re-activating peroxiredoxins (2-Cys Prx) by reducing sulfinic acids formed on the peroxidatic cysteines under oxidizing conditions. Methionine sulfoxide reductases reduce methionine sulfoxide to methionine and are regenerated by thioredoxins (Hage, Rosso, & Tarrago, 2021). The expression of thiol-related antioxidative genes was supported by microarray analyses (Morel et al., 2008). To date, many fungal genomes have been sequenced and the homologs to genes described by Morel et al. (2008) were annotated in a wide array of EMF species (Bolchi et al., 2011; Mattila et al., 2022). EMF in the phylum of Basidiomycota (Mattila et al., 2022) generally contained numbers of homolog genes similar to those reported by Morel et al. (2008).

2.2.3 Other ROS handling systems

Fungi contain class I and class II peroxidases (discussed below) but not class III peroxidases, which are typical for plants (Maruta, Sawa, Shigeoka, & Ishikawa, 2016). Accordingly, guaiacol peroxidase activities (class III Prxs) could not be measured in mycorrhizal hyphal mantle tissue of *Laccaria amethysthea* collected from larch (*Larix decidua*) and spruce (*Picea abies*) roots (Münzenberger, Otter, Polle, & Wüstrich, 1997).

EMF taxa contain cytochrome c peroxidases (CcP, EC 1.11.1.5), which are heme-containing enzymes removing H_2O_2 produced by the cell respiration. CcP was isolated and extensively characterized from aerobically grown yeast and is related to plant ascorbate peroxidases (APX1.11.1.11) since they shared the PF00141 (peroxidase) PFAM domain (Lyall, Nikoloski, & Gechev, 2020; Zámocký, Gasselhuber, Furtmüller, & Obinger, 2014). While CcPs are present in photosynthetic and non-photosynthetic eukaryotes, APXs are almost exclusively found in plastid-containing photosynthetic organisms (Maruta et al., 2016). The genomes of several EMF taxa contain one CcP gene per species (Mattila et al., 2022), while APXs were not present. Furthermore, APX requires ascorbate as a reductant, which was not discovered in *Paxillus involutus* (Ott et al., 2002).

In animals and plants, the final step of ascorbate biosynthesis is catalyzed by L-gulono-1,4-lactone oxidase (GGLO) and L-galactone-1,4-lactone dehydrogenase (GLDH), respectively (Smirnoff, 2018). Fungi produce an analog, D-erythroascorbate, employing D-arabino-1,4-lactone oxidase (ALO) for the final biosynthetic step (Smirnoff, 2018). We blasted the sequences of the *Arabidopsis thaliana* GLDH, the mammalian (*Mus mus*) GGLO and the yeast (*Saccharomyces cerevisiae*) ALO against several EMF genomes using blastp v.2.9.0 (Altschul, Gish, Miller, Myers, & Lipman, 1990) (Supplementary Table S1). The basidomycetes *Laccaria bicolor, Amanita muscaria, Lactarius quietus, Russula ochroleuca,* and *Paxillus involutus* show high homology to the mammalian enzyme (e-values $< E^{-80}$, high bit score and overlap with query length), while the tested ascomycetes produced two hits under these conditions, one for the mammalian GGLO (e-values $< E^{-80}$) and one for the yeast ALO enzyme (e-values close to E^{-100}) (Supplementary Table S1). High similarities with the plant enzyme GLDH were not detected (Supplementary Table S1). Morel et al. (2008) did not discover genes for dehydroascorbate reductase and monodehydroascorbate radical reductase in *L. bicolor*. Thus, the potential role of erythroascorbate and its reduction for ROS detoxification in EMF is elusive.

Some reports indicated that heavy metals (Cd, Cu) or drought stress treatment induced ascorbate activities in EMF (Alvarez et al., 2009; Dachuan & Jinyu, 2021). In the absence of ascorbate peroxidase genes and most likely also of ascorbate, confounding non-enzymatic reaction might have mimicked erroneously the detected activities (Smirnoff, 2018). Reis et al. (2011) reported the presence of ascorbate in a range of EMF grown in pure culture. However, their analysis was based on a relatively unspecific method, which measures spectrophotometrically absorbance changes due to the reduction of 2,6-dichloroindophenol by fungal extracts. Therefore, antioxidative reactions of compounds other than ascorbate in these extracts cannot be excluded. Tocopherols, another important group of antioxidant compounds known from the plant metabolism, have been discovered by

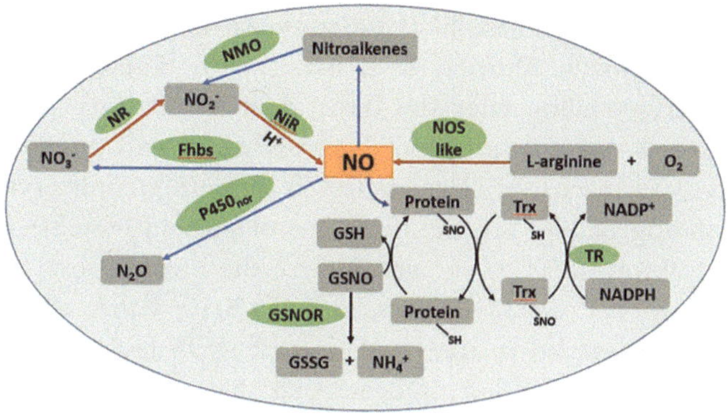

Fig. 3 Schematic, partly hypothetical overview of NO production and removal in fungi. NO may be produced by oxidative and reductive pathways. Oxidative pathway: L-arginine and O_2 catalyze NO formation by NOS-like proteins. Reductive pathways: NO_3^- is reduced by NR and NiR to NO_2^- and NO. The removal of NO is achieved by multiple reactions. Fhbs convert NO to NO_3^-, $P450_{nor}$ reduce NO to N_2O in an NADPH-dependent reaction. Higher levels of NO are converted to nitroalkenes during nitro-oxidative conditions, then further metabolized by NMO to NO_2^-. A specific NO-inducible nitrosothionein is capable of directly scavenging NO in coordination with the Trx/TR redox system. GSNOR is involved in the NO-dependent nitrosative signaling. Red arrows indicate NO production and blue arrows NO removal. NO_3^-, nitrate; NO_2^-, nitrite; NO, nitric oxide; NR, nitrate reductase; NiR, nitrite reductase; NOS, nitric oxide synthases; Fhbs, flavohemoglobins; P450nor, Cytochrome P450 nitric oxide reductase; NMO, nitronate monooxygenase; GSH, glutathione; GSNO, *S*-nitrosoglutathione; GSNOR, S-nitrosoglutathione reductase; GSSG, glutathione disulfide; Protein-SH, proteins with free thiols in cysteine; Protein-SNO, *S*-nitrosylated proteins; Trx, thioredoxin; TR, thioredoxin reductase.

high pressure liquid chromatographic and comparison with authentic standards in EMF species from various genera (Amanita, Chroogomphus, Cortinarius, Lactarius, Russula Suillus, and Tricholoma) (Reis et al., 2011). However, physiological studies on the role of tocopherols in EMF are lacking. Since antioxidative systems in mycorrhizal fungi are often investigated with the aim to select stress tolerant EMF species for phytoremediation, there is an urgent need to characterize the metabolism of mycorrhizal tocopherols, ascorbate-like compounds and related enzyme activities to better understand fungal ROS handling.

A further important reactive compound for redox regulation is nitric oxide NO* (Fig. 3). NO* participates in the redox cycle by accepting an electron (NO^-, nitroxyl anion) or donating an electron (NO^+, nitrosonium cation), while the radical NO* is uncharged (Correa-Aragunde, Foresi, & Lamattina, 2015). In plants, NO* can be produced by nitrate reductase or NO synthase-like proteins through different pathways (Moreau, Lindermayr, Durner, & Klessig, 2010; Rekhter et al., 2019). NO* exerts its signaling effect by direct post-translational protein modifications, i.e., *S*-nitrosylation, metal nitrosylation, and tyrosine nitration (Lindermayr, Saalbach, & Durner, 2005) (Fig. 3). *S*-nitrosylation, the covalent binding of NO* to the thiol side of protein cysteine residues to form nitrosothiols (SNOs), is considered the major post translational modification of NO* signaling (Moreau et al., 2010; Vanzo et al., 2016). In plants, NO* is involved in the regulation of ROS levels (Lindermayr & Durner, 2015) by enhancing or decreasing the activities of ROS-metabolizing enzymes such as NADPH oxidase, CAT, APX, and GR (Correa-Aragunde et al., 2015; Dat et al., 2003; Davletova, Schlauch, Coutu, & Mittler, 2005; Lindermayr et al., 2005; Vandenabeele et al., 2004; Vanzo et al., 2016), thereby adjusting the ROS pool. Recently, Martínez-Medina et al. (2019) demonstrated that NO-dependent regulation of phytoglobins plays a key role in establishing AMF-plant interactions. Phytoglobins (Pgb), along with S-nitrosoglutathione reductase (GSNOR), are the major scavenging pathways for NO in plants. Phytoglobins comprise a group of non-symbiotic hemoglobins that exhibit high affinity for both oxygen and NO under certain conditions such as hypoxia (Berger et al., 2018). In this process, NO is converted to nitrate by the oxygen-enriched ferrous phytoglobin (Pgb(Fe^{2+})), which is thereby transformed to the metPgb form (Pgb(Fe^{3+})). metPgb is reduced by a NAD(P)H-dependent reductase and subsequently re-oxygenated (Kuruthukulangarakoola et al., 2017). Nitrate can serve as a substrate for NR to produce nitrite and thus, eventually also NO, driving again Pgb reactions. This cycle is referred to as the "Pgb-NO cycle" (Kumari, Pathak, Loake, & Gupta, 2019). To what extent the regulation of NO content by the Pgb-NO cycle also plays a role in the interaction of EMF species with plants during the formation of ectomycorrhizas is unknown.

Although fungal NO* production may be ecologically very important with regard to nitrogen cycling, we have little information on its role in EMF. Environmental contamination with Cr(IV) enhances nitrate and nitrite reductase gene expression and enzymes activities in *Pisolithus* species (Shi et al., 2020). Under these conditions, the production of NO* and ROS detoxification by GSH were stimulated (Shi et al., 2022). Suppression of NO* production by supplementation of the fungus with the NO* inhibitor PTIO (2-phenyl-4,4,5,5,-tetramethylimidazoline-1-oxyl 3-oxide) increased ROS production and antioxidative activities under stress (Shi et al., 2022).

When the NO* donor, sodium nitroprusside (SNP) was used for soil amendment, root colonization with *Tuber indicum* increased (Zhang et al., 2019). Thus, the current studies hint towards a role of NO⋆ in redox regulation and root colonization. In natural EMF communities, considerable transcript abundances of nitrate reductase are present, although uptake of reduced nitrogen (NH_4^+, amino acids) is preferred (Rivera Pérez, Janz, Schneider, Daniel, & Polle, 2022). When nitrate reductase (NR, EC 1.7.1.1) was suppressed by genetic transformation (RNAi) in *L. bicolor*, viable ectomycorrhizas could not be formed anymore (Kemppainen, Duplessis, Martin, & Pardo, 2009). Therefore, either a direct or indirect nitrate signal via NO* radicals is important for the symbiosis. It is obvious that a better understanding of NO* in plant-fungal interactions is necessary.

2.3 Extracellular ROS producing and oxidative enzymes

Fungi produce an enormous variety of specialized extracellular enzymes, which reflect their abilities to thrive on a wide number of substrates and is due to different phylogenetic origins (Kohler et al., 2015; Miyauchi et al., 2020). Many of these enzymes are oxidases, resulting in ROS production when attacking recalcitrant potential nutrient sources. Since EMF have direct access to plant-derived simple carbohydrates when engaged in mutualistic symbiosis, the presence of multiple carbohydrate-active enzymes was surprising. However, comparative analyses of wood-degrading fungi and EMF in culture showed that the overall activities of EMF for degradation of recalcitrant compounds are lower than those of typical saprotrophs (Gramss, Kirsche, Voigt, Günther, & Fritsche, 1999). Nevertheless, substantial activities of cellobiose oxidase (cellobiose dehydrogenase, EC 1.1.99.18) and glucose oxidase (beta-D-glucose:oxygen 1-oxidoreductase, EC 1.1.3.4) along with H_2O_2 production were found when the EMF *Suillus variegatus, Pisolithus tinctorius,* and a *Cortinarius* sp. were grown on the corresponding substrates (Burke & Cairney, 1998). The ectomycorrhizal fungus *Paxillus involutus* is able to decompose organic matter by Fenton chemistry (Shah et al., 2015). In the Fenton reaction Fe^{2+} is oxidized by H_2O_2 to yield Fe^{3+}, OH* and HO^-. The re-reduction of Fe^{3+} to Fe^{2+} is driven by $O_2^{\cdot -}$ radicals and special metabolites, in the case of *P. involutus* probably by the diarylcyclopentenone involutin (Shah et al., 2015). Involutin production occurs only on organic but not on mineral substrates, thereby, enhancing nutrient accessibility to the host (Shah et al., 2015). Field studies on mycorrhizal root tips from forest trees also show environmental variation of carbohydrate-active enzyme activities and variation among different EMF species, pointing to flexible adaptation to changing environmental conditions (Agerer, Schloter, & Hahn, 2000; Courty et al., 2010; Pritsch & Garbaye, 2011). How plant roots cope with varying enzyme activities and exposure to different types of ROS is not clear. It could be that ROS also have functions in nutrient signaling but this area is unexplored for mycorrhizal fungi, despite the importance of EMF for tree nutrition (Becquer, Guerreo-Galan, Eibensteiner Houdinet Bücking, Zimmermann & Garcia, 2019).

Progress has been made in the identification of carbohydrate-active enzymes in EMF by systematic screening of fungal genomes and classification of the enzymes in the Cazy data base (www.cazy.org; Drula et al., 2022). We used published information (modified from Supplementary Table S6 in Miyauchi et al., 2020) to generate an overview on AA enzymes, i.e., enzymes with "axillary activities" (AA1 to AA14) in 62 ectomycorrhizal fungi (Fig. 4). Among these categories AA1, AA2, AA3, AA9, and AA11 are of special interest because they explain differences among different phylogenetic and ecological mycorrhizal forms (Fig. 4).

High gene counts in the AA1 family are mainly characteristic of AMF but some basidiomycetes (*Hydnum rufescens, Piloderma croceum, Lactarius quietus, Suillus brevipes and Tricholoma matsutake*) were also present in this category. AA1 enzymes are multicopper oxidases (EC 1.10.3.-), e.g., laccases (EC 1.10.3.2) that use diphenols with oxygen as the acceptor. Furthermore, genes encoding AA3 enzymes, which comprise functionally diverse glucose–methanol–choline oxidoreductases (GMC oxidoreductase) were more enriched in AMF than in most other mycorrhizal life forms. GMC oxidases are a source of H_2O_2. They include aryl-alcohol oxidoreductase (AAO, EC 1.1.3.7), alcohol oxidase (AOx, EC 1.1.3.13), cellobiose dehydrogenase (CDH, EC 1.1.99.18), glucose dehydrogenase (GDH, EC 1.1.5.9), pyranose dehydrogenase (PDH, EC 1.1.99.29), and pyranose oxidase (Pox, EC 1.1.3.10).

The AA2 family contains class-II peroxidases such as manganese peroxidase (EC 1.11.1.13), versatile peroxidase (EC 1.11.1.16), lignin peroxidase (EC 1.11.1.14), peroxidase (EC 1.11.1.-), cytochrome-c peroxidase (EC 1.11.1.5), and ascorbate peroxidase (EC 1.11.1.11). These PODs are heme-containing enzymes, which use H_2O_2 as the electron acceptor, occurring in many EMF species (except ascorbate peroxidase) (Bödeker, Nygren, Taylor, Olson, & Lindahl, 2009; Floudas et al., 2012). For example, Chen et al. (2001) showed the presence of lignin peroxidase (LiP) and manganese peroxidase (MnP) genes in a broad taxonomic range of EMF by PCR with a

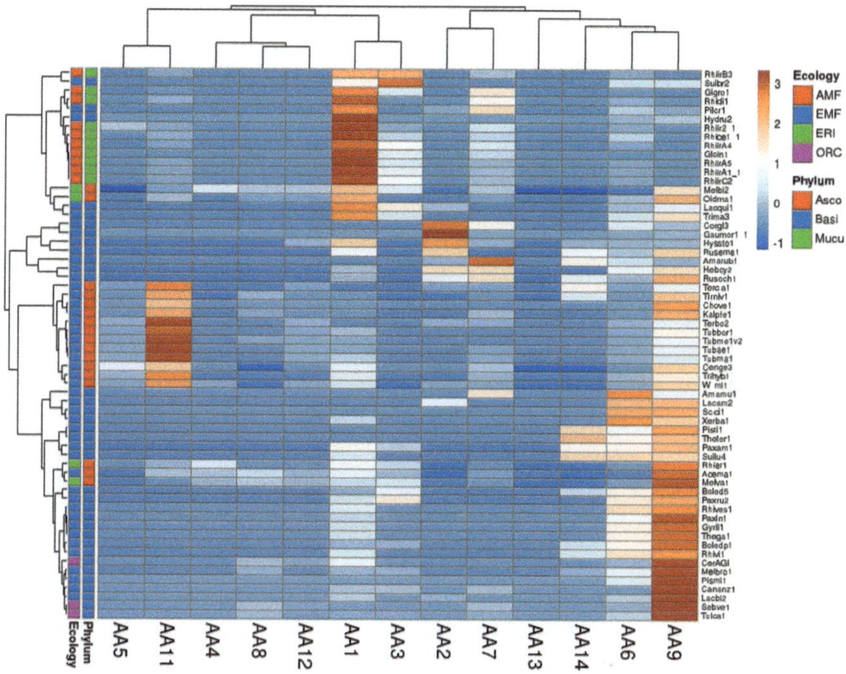

Fig. 4 Hierarchical clustering of gene abundances of carbohydrate-activing enzyme with axillary activities (AA). The heatmap was generated with the numbers of genes per genome and per AA family provided by Miyauchi et al. (2020). The AA families were taken from the Cazy classification system. The function of different families are described in the main text. Abbreviations and taxonomic information for the fungi is found in Supplementary Table S2.

specific primer and confirmed the identity of the amplified fragments by sequencing (Chen, Taylor, Burke, & Cairney, 2001). In our analysis, *Gautieria morchelliformis, Cortinarius glaucopus* and several Russulaceae stood out with regard to their AA2 gene inventory (Fig. 4).

The ectomycorrhizal Ascomycota were enriched with AA11 family members (Fig. 4). AA11 proteins are copper-dependent lytic polysaccharide monooxygenases (LPMOs), which have been detected just a decade ago (Courtade & Aachmann, 2019). They can cleave crystalline cellulose and chitin chains, thereby, complementing glycoside hydrolases, which act on amorphous cellulose (Courtade & Aachmann, 2019). Although the A11 gene counts clearly separated ascomycetes EMF from basidiomycetes EMF, we noted an exception: the AA11 gene inventory of Ascomycota forming ericoid mycorrhiza resembled more closely that of Basidiomycota EMF (Fig. 4).

The AA9 family also contains copper-dependent lytic polysaccharide monooxygenases but they were reported to catalyze oxidative cellulose degradation [lytic cellulose monooxygenase (C1-hydroxylating) (EC 1.14.99.54), lytic cellulose monooxygenase (C4-dehydrogenating) (EC 1.14.99.56)] (Drula et al., 2022). However, usage of other substrates cannot be excluded (Vandhana et al., 2022). Very high counts of genes encoding AA9 family members were present in *Pisolithus microcarpus, Melanogaster broomeianus, Cantharellus anzutake, Laccaria bicolor* and in the three orchid EMF (Basidiomycota) available in the list. Veneault-Fourrey et al. (2014) found that three AA9 LPMO genes in *Laccaria bicolor* were expressed during first root contact and during mycorrhizal maturation when the Hartig net is formed (Veneault-Fourrey et al., 2014). LPMO enzymes may be involved in remodeling of root cell walls by loosening the wall together with endoglucanases (Zhang et al., 2018) since they are localized at the interface between fungal hyphae and root cortex cells (Labourel et al., 2020). In as much LPMO activities from the AA9 and AA11 families functionally distinguish asco- and basidiomycetes should be studied in the future. The new insights open interesting avenues for further characterization of plant fungal interaction and their ecological consequences for carbon turnover.

3. Local and systemic defenses in plants induced by ectomycorrhizal fungi

3.1 ROS, volatile organic compounds and other signaling components involved in initiation of EMF-symbiosis

The establishment of a symbiotic association between EMF and plant root requires several coordinated events that are not yet completely understood. The initiation of an EMF symbiosis includes at least the secretion of small secreted proteins (SSPs) (Plett et al., 2011; Pellegrin, Morin, Martin, & Veneault-Fourrey, 2015), lipochitooligosaccharides (LCOs) (Cope et al., 2019) and other chitin-related compounds on the fungal side and exudation of flavonoids and phytohormones and chitinase activity (Garcia, Delaux, Cope, & Ané, 2015) on the plant side (Fig. 5).

The exudation of flavonoids and phytohormones is considered to initiate the chemical dialogue between the roots and the fungus. Plant chitinases release fungus-typical chitin-related compounds and are involved in plant response to pathogenic as well as beneficial fungi (Salzer, Hebe, & Hager, 1997; Schickler & Chet, 1997). Chitin-related compounds elicit

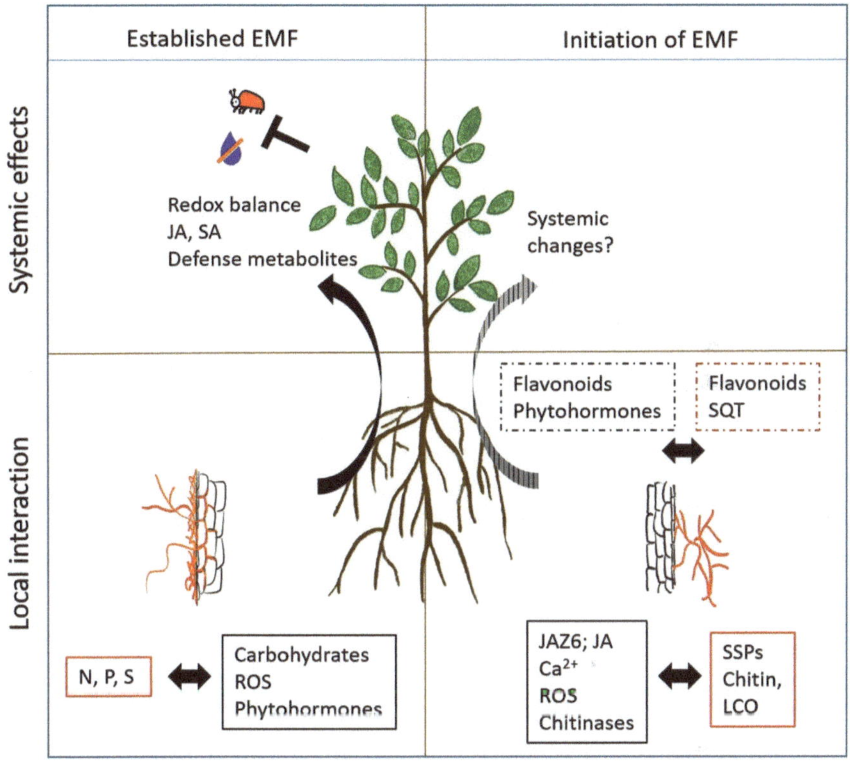

Fig. 5 Potential interaction signals involved in initiating and maintaining the ectomycorrhizal relationship. Systemic effects that are associated to the initiation stage and to established mycorrhiza are also shown. The dashed boxes include cues that are potentially involved in interaction through soil matrix, before the contact. JA: jasmonates; LCO: lipochitooligosaccharides; SA: salicylic acid; SQT: sesquiterpenes; SSPs: small secreted proteins; JAZ6: JASMONATE-ZIM-DOMAIN 6; N, P, S: nitrogen, phosphorus, sulfur.

further adjustments in plant signaling to prepare for the interaction (Punja & Zhang, 1993; Salzer et al., 1997; Sauter & Hager, 1989). Here, the extracellular oxidative enzymes could play important roles but general, overarching principles have not yet been clarified. The signal exchange between the partners leads to altered jasmonic acid (JA)-dependent signaling in roots and a locally reduced defense response (Garcia et al., 2015; Plett, Daguerre, et al., 2014; Plett, Khachane, et al., 2014). For *L. bicolor* it is known that fungal SSPs (specifically the mycorrhiza-induced SSP7 (MiSSP7)) function as effectors that interact with the poplar JASMONATE-ZIM-DOMAIN 6 (JAZ6)-protein. JAZ6 is a repressor of JA-signaling and its binding to MiSSP7 prevents induction of local JA-related defense responses, thus, probably allowing the symbiotic relationship to develop (Fig. 5; for detailed review see Garcia et al., 2015). Also other SSPs are able to modulate the plant phytohormone balance similar to various further fungal effectors that can manipulate plant hormonal pathways and regulate host defenses (Dreischhoff, Das, Jakobi, Kasper, & Polle, 2020). In later stages of symbiosis, ethylene- and JA-pathways are again induced probably to limit fungal growth within the roots (Plett, Khachane, et al., 2014). In addition to altered JA signaling, the early interaction mechanisms between plant and EMF include Ca^{2+} spiking (Cope et al., 2019; Garcia et al., 2015) and local adjustments in the root's redox balance (Baptista et al., 2007). Together these initial adjustments might be involved in initiating cell-to-cell signaling and further systemic responses. JA signaling is considered as antagonistic to salicylic acid (SA) signaling and involved in cross-talk with other plant signaling pathways (Erb, 2018; Vlot et al., 2021). Therefore, already the initial changes necessary to allow mycorrhizal interactions can cause comprehensive adjustments in plant performance. In support of this idea, circumstantial evidence shows that contact to ecto-mycorrhizal cell wall fragments is sufficient to induce systemic immunity against a biotrophic leaf pathogen (*Pseudomonas syringae*) in the non-mycorrhizal plant *Arabidopsis thaliana* (Vishwanathan et al., 2020). Obviously, a functional mycorrhiza is not required to induce systemic effects in plants. Moreover, EMF-induced systemic effects may have an evolutionary signature that is conserved irrespective of the trophic mode of the interacting fungus.

3.1.1 Involvement of ROS in initiation of an EMF symbiosis

Though all the mechanisms in "the interaction dance" behind the initiation of EMF symbiosis are not elucidated, ROS (H_2O_2, O_2^-) are known to play a role in this process. An EMF-related ROS burst was detected when the initial contact between *Picea abies* and two different EMF fungi, *Amanita muscaria* and *Hebeloma crustuliniforme* was studied (Salzer, Hubner, Sirrenberg, & Hager, 1997; Salzer et al., 1996; Schwacke & Hager, 1992). Early contact between *Castanea sativa* roots and *Pisolithus tinctorius* triggered three sequentially occurring H_2O_2 bursts, which were additionally accompanied by two O_2^- bursts (Baptista et al., 2007). Interestingly, adjustments in ROS scavenging enzymes SOD and CAT were observed in association with ROS bursts, suggesting that during the early stages of EMF establishment the ROS burst results from an inhibition of these enzymes, thus, underlining the tight regulation of the ROS signaling events. In *C. sativa,* the EMF-induced ROS bursts were followed by increased root exudation and subsequently induced hyphal growth of the EMF-fungus (Baptista, Martins, Pais, Tavares, & Lino-Neto, 2007). These transient adjustments in ROS pools upon EMF contact resemble those that were found for AMF during mycorrhization events. Dynamic ROS levels are known to play a central role in initial AMF colonization, subsequent symbiosis formation and eventually, in the degradation of arbuscules. The initial, transient ROS burst is overcome by accumulation of ROS scavenging enzymes (e.g., increased activity of SOD, CAT, APX) as well as by non-enzymatic antioxidants (carotenoids, α-tocopherol, proline, ascorbate and glutathione) in the plant (Zou, Wu, & Kuča, 2021). Also various antioxidative enzymes of AMF are known to contribute to regulation of the initial, dynamic oxidative burst, which is mainly restricted to AMF-containing cortical root cells (Zou et al., 2021). The exact function of ROS bursts and the possibility of an induction of ROS waves for communication between root cells and ectomycorrhizal fungus remain to be elucidated. Many different types of abiotic and biotic stresses can trigger ROS waves, which are considered as an essential signal to alert and prepare the plant for environmental changes (Fichman & Mittler, 2020). Likewise, ROS waves could also have a function in the initiation of a symbiotic relationship between plant and fungus.

Moreover, it is known that similar to rhizobia, AMF exude so-called Nod-factors, such as LCOs that are involved in the initiation process to allow plant roots to prepare for symbiosis. After perceiving LCOs from AMF, mevalonate biosynthesis is induced followed by activation of nuclear ion channels and Ca^{2+} flow. As Ca^{2+} is released from the nucleus and pumped again back by a calcium ATPase, LCOs trigger Ca^{2+} spiking (Bertoni, 2019). Further down-stream steps lead to an activation of transcription factors essential for the regulation of the "common symbiosis signaling pathway" (Bertoni, 2019). Recent transcriptomic analyses of EMF-colonized oak tree roots suggest that the "common symbiosis signaling pathway" of AMF is involved also in the initiation of ectomycorrhizal symbiosis (Bouffaud et al., 2020). Cope and colleagues revealed that *L. bicolor* produce LCOs that can trigger the "common symbiosis signaling pathway" and thus allow colonization of plant roots (Cope et al., 2019). *Laccaria bicolor* LCOs were shown to cause Ca^{2+} spiking in nuclei of *Populus* roots and enhance lateral root development and EMF colonization (Cope et al., 2019). While symbiotic organisms elicit the nuclear Ca^{2+} oscillations, pathogens trigger cytoplasmic spiking, a difference which is considered as a key to specify plant responses (Tian, Wang, Gao, Li, & Luan, 2020). In *Phaseolus vulgaris*-rhizobia interactions, the transient increase of Ca^{2+} correlated spatially and temporarily with changing ROS concentrations (Cárdenas, Martínez, Sánchez, & Quinto, 2008; Cárdenas & Quinto, 2008). The function of ROS in initiation of different beneficial relationships might, however, strongly depend on the interacting species. For example, the *Phaseolus vulgaris* NADPH-oxidase functions as positive or negative regulator of biotic interactions depending on the interacting partner: enhanced plant ROS levels induced enhanced nodule formation by rhizobia, whereas the opposite, reduced colonization was observed for AMF (Arthikala et al., 2014). On the other hand, in arbusculated *Medicago truncatula* cells NADPH oxidase (MtRbohE) gene expression was enhanced compared to control roots and knock-down of the *MtRbohE*-gene lead to altered root cortex colonization pattern of *Medicago* (Belmondo et al., 2016). Together, these results suggest important roles of NADPH-oxidase and ROS concentrations in AMF mycorrhization. Whether exposure to EMF-released LCOs alter ROS in roots and what might be their final role in initiation of EMF symbiosis remains to be elucidated. The redox balance may however be a key regulating step for EMF colonization of roots. At least production of ROS has been linked to the specificity of the EMF interactions: plants produce ROS when exposed to EMF strains that are capable of forming ectomycorrhizas (Fig. 1), whereas no increase in ROS levels was observed upon exposure to a *Paxillus* strain that cannot develop a complete ectomycorrhiza (Gafur et al., 2004). The host specialization of the EMF *Suillus* spp. is also linked to various oxidative stress-associated domains such as thioredoxin reductase, glutathione, SOD, CATs, pyridine nucleotide-disulfide oxidoreductase and multiple aldehyde dehydrogenases in the ectomycorrhiza (Lofgren et al., 2021). Targeted ROS deactivation was suggested as a possible mechanism behind host specificity of this EMF-host combination (Lofgren et al., 2021).

3.1.2 Potential volatile signals exchanged before a contact

Volatile organic compound (VOC)-mediated interactions may prepare for symbiosis initiation already before the interacting partners are in wall-to-wall contact with each other. *Laccaria bicolor* VOCs were shown to alter the root redox balance already before the contact. ROS accumulated in poplar and *Arabidopsis* roots exposed to a volatile sesquiterpene from *L. bicolor* in an NADPH-oxidase dependent manner (Fig. 5; Ditengou et al.,

2015). The altered ROS concentration was associated to enhanced lateral root and root hair development. Lateral roots are the preferred target for mycorrhizal fungi to colonize roots (Sun, Bonfante, & Tang, 2015) and their stimulation can facilitate the establishment of symbiosis. Several other studies reported altered root architecture and root growth upon exposure to microbial VOCs (Garnica-Vergara et al., 2016; Moisan et al., 2021; Sun et al., 2015; Werner, Polle, & Brinkmann, 2016), even though none of these studies tested explicitly the involvement of ROS signaling in the plants` response. Some years ago Matsui (2016) suggested that perceiving and sensing specific external VOCs in plant cells might generally be associated with plant redox adjustments (Matsui, 2016). This hypothesis is supported by a few studies, which detected increased ROS levels in conjunction with altered plant performance upon perceiving an external VOC cue (Ameye et al., 2020; Riedlmeier et al., 2017). Further, isoprene, a root endogenous volatile hemiterpene, is involved in an NADPH-oxidase dependent accumulation of ROS and altered lateral root growth of poplar (*P.* x *canescens*) (Miloradovic van Doorn et al., 2020). Thus, collectively recent studies support a ROS-related function of VOCs in roots. An intriguing question to be tackled in the future is if specific VOCs released by EMF lead to adjustments in root performance in natural conditions and what their importance is in initiation of a symbiotic relationship. Since transient ROS bursts are involved in early events of mycorrhization (Salzer et al., 1996; Schwacke & Hager, 1992) and since H_2O_2 is present in the mature ectomycorrhizal mantle (Gafur et al., 2004), in vivo imaging of H_2O_2 and redox compounds by the recently developed whole-plant ROS live imaging (Fichman et al., 2019) or by Hyper-lines (Belousov et al., 2006) may help to view the dynamics of ROS upon VOC and EMF exposure. Thereby, essential information about the role of VOCs and ROS in early and later events of symbiosis establishment and functioning could be disentangled. Recent comparative genomics further supports the potential role of terpenes in EMF-host interaction. Terpene biosynthesis related gene clusters are enriched in the specialist EMF-genus *Suillus*, suggesting that terpenes might have a function in initiating interaction between a specialist fungus and the host. As specificity was associated with overrepresentation of oxidative stress related domains (Lofgren et al., 2021), it seems evident that the link between specificity, terpene profile and redox balance deserves more attention when exploring the mechanisms of ectomycorrhizal establishment in the future.

3.2 Established ectomycorrhiza alters plant internal signaling and alleviate stress

Signaling is not only crucial in initiation of EMF symbiosis, but also in the established mycorrhizal state and in alleviating plant stress responses. For example, *Paxillus involutus* colonization was related to increased concentrations of SA and ABA in roots, whereas JA and auxin concentrations decreased (Luo et al., 2009), suggesting alterations in internal signaling of EMF-poplar roots that were exposed to salinity. The EMF-host interaction does not only alleviate acute plant stress tolerance but seems to adjust plant internal signaling already before stress exposure. In a recent study, the effects of three different EMFs were investigated by analyzing the transcriptomes of EMF colonized roots, systemic non-colonized roots and leaves of oak trees (*Quercus robur*) (Bouffaud et al., 2020). In general, more common differentially expressed genes induced by all the three EMF species, *L. bicolor*, *Paxillus involutus* and *Pisolithus microcarpus* were detected in the colonized roots than in any of the systemic tissues. The data suggest that the local responses to EMF are more conserved than the EMF-induced systemic responses, which varied depending on the interacting species. All three fungal species altered expression levels of various disease resistance-related genes in the systemic roots and SOD copper chaperones and genes encoding ABA degrading enzymes in leaves. The symbiosis with *P. microcarpus* affected redox process-related genes in the systemic leaves (Bouffaud et al., 2020). Together the results suggest that redox balance can be altered in systemic tissues of EMF-colonized trees already before experiencing a stress. Although not all the mechanisms and prevailing factors leading to altered plant signaling in mycorrhizal stage are elucidated, several studies linked EMF-based improved stress tolerance with alterations in ROS, ROS scavenging enzymes and phytohormones (Alvarez et al., 2009; Bai, Hao, Hu, & Leng, 2021; Kaling et al., 2018; Pfabel et al., 2012; Vishwanathan et al., 2020). In order to alert the plant for stress adaptation, ROS can act also as long distance signals conferring information within plant tissues and activate defense related signaling pathways (Baxter, Mittler, & Suzuki, 2014; Gilroy et al., 2014; Mittler & Blumwald, 2015). To efficiently utilize ROS as internal signaling molecules and to mitigate toxic effects of ROS, plants need to sustain a delicate balance between ROS generation and ROS scavenging (Baxter et al., 2014). Such a balance is a central component of various within-plant signaling events, including local and systemic responses to EMF (Bai et al., 2021; Sun et al., 2022).

3.2.1 The interplay between ROS and ROS regulating enzymes in plants is adjusted by EMF

The balance between different antioxidative enzyme activities, such as SOD, APX and CAT, is essential for suppressing toxic ROS levels in a cell (Mittler, 2017). Furthermore, EMF-related attenuation of oxidative stress is regulated by ROS scavenging enzymes and metabolites. The link between EMF-based stress tolerance, ROS and ROS scavenging enzymes has often been studied in association with abiotic stresses, which enhance the production of ROS, such as drought. For example, Alvarez et al. (2009) found that the combination of effective ROS prevention and ROS detoxification in EMF-roots resulted in reduced cellular damage and increased *Nothofagus dombeyi* growth under drought. Different EMF species can trigger different protective mechanisms to reduce oxidative stress: *Descolea antartica* mycorrhization helped to maintain SOD activity at a steady-state level, whereas *Pisolithus tinctorius*-colonized roots showed high activities of several ROS scavenging enzymes. However, different defense strategies were differentially effective since only ROS detoxification through the synchronized action of antioxidative enzymes led to reduced cellular damage in the host (Alvarez et al., 2009). In accordance with these results, AMF-plants are well known to have lower levels of H_2O_2, O_2^- and less lipid peroxidation as the consequence of enhanced antioxidant enzyme activities and mitigation of the oxidative burst especially under abiotic stress conditions, such as drought (Nath et al., 2016; Zou, Wu, & Kuča, 2021). Reduced malondialdehyde (a marker for lipid peroxidation), H_2O_2 and O_2^- levels were detected in drought stressed plants colonized with the AMF *Funneliformis mosseae* (Huang, Zou, & Wu, 2017). Interestingly, using the non-invasive micro-test technique, the authors detected increased efflux of H_2O_2 from roots, which could be a cause of the alleviated oxidative stress (Huang, Zou, & Wu, 2017).

3.2.2 EMF, ROS and plant systemic responses to abiotic stresses

The colonization by EMF can alleviate the plant abiotic stress response in systemic tissues by inducing a systemic acquired acclimation (SAA)-like state. SAA is distinguished from systemic acquired resistance (SAR) because SAA refers to the systemic acclimation of the plant to abiotic stresses, while SAR is associated with induced immunity against pathogen infection. Several rapid whole-plant systemic signals including Ca^{2+}, ROS and hydraulic and electric waves are associated with mediating SAA (Fichman & Mittler, 2020). Furthermore, EMF induced systemic plant acclimation is associated with redox regulation (Bai, Hao, Hu, & Leng, 2021; Sun et al., 2022). For example, Bai et al. (2021) showed higher CAT and peroxidase activities, that were associated with the salt stress tolerance in leaves of EMF-colonized oak trees (*Quercus mongolica*) (Bai et al., 2021). The authors found more pronounced effects in the trees that were colonized by *Gomphidius viscidus* than by *Suillus luteus*, suggesting, like Bouffaud and colleagues (Bouffaud et al., 2020), that different EMF species induce different systemic responses (Bai et al., 2021). *Suillus luteus* colonization has also been shown to modulate ROS and the oxidative stress response of oak (*Quercus acutissima*) and pine trees (*Pinus massoniana*) (Liu, Chen, Ding, Li, & Ren, 2020; Sun et al., 2022). In leaves of EMF oak trees, SOD and GR were enhanced (Sun et al., 2022), whereas in pine trees the mycorrhization was connected with improved growth performance, enhanced antioxidant activities (SOD and POD) and reduced malondialdehyde accumulation (Liu et al., 2020). Recent transcriptomic analyses revealed, furthermore, an activated ROS scavenging machinery in ectomycorrhizal pine prior to the stress. Such EMF-induced primed state could allow rapid responses to abiotic and biotic stresses.

EMF can improve plant performance under salinity and enhance tolerance against heavy metals (Chen et al., 2014; Langenfeld-Heyser et al., 2007; Schützendübel & Polle, 2002; Zhang et al., 2017). The capability of EMF to exclude Na^+ (Chen et al., 2014) and enhance the uptake of nutrients (Fig. 5) can enhance the tolerance against high salinity. The EMF-induced SAA-like response has, for example, been associated with decreased unsaturated-to-saturated fatty acid ratios, which together with altered nutrient status and osmo-regulation reduced the salinity induced damage in the leaves of mycorrhizal poplars (Luo et al., 2011). In addition, the plant phytoremediation potential can be facilitated by ectomycorrhizas in a ROS-dependent manner. The colonization by the EMF-species *Paxillus involutus* was shown to enhance H_2O_2 production and a H^+-pumping activity in poplar that grew under high Cd^{2+} or salt concentrations (Deng et al., 2021; Li et al., 2012). This, in turn, activated plasma membrane Ca^{2+} channels that were employed for the enhanced uptake of Cd^{2+} (Zhang et al., 2017) and thus, improved the phytoremediation.

Together these studies reveal adjustments in ROS, several signaling pathways and plant metabolome and, thereby, provide a basis to better understand the EMF-induced improvements in abiotic stress tolerance. In several studies, however, the final cause of improved performance due to EMF has remained unexplained. For example, the mechanism behind the improved growth and increased unsaturation level of membrane lipids in the *P. tinctorius* mycorrhizal oak trees under drought was not discovered (Sebastiana et al., 2018). The transient nature of the ROS signals and the concentration-dependent action of some signaling compounds make it challenging to detect the components involved in plant signaling and mounting an improved resistance.

3.2.3 ROS in mycorrhiza induced resistance (MIR) against biotic stresses

Colonization by EMF does not only improve plant tolerance against abiotic stresses, but it can also improve the resistance against aboveground biotic threats. Altered response of EMF-trees to leaf herbivores has been shown for poplar (*P.* x *canescens*), eucalyptus (*Eucalyptus urophylla*), American chestnut (*Castanea dentata*) and birch (*Betula pubescens*) (Gange, Gane, Chen, & Gong, 2005; Kaling et al., 2018; Oddsdottir, Eilenberg, Sen, Harding, & Halldorsson, 2010; Rieske, Rhoades, & Miller, 2003). EMF-caused a state of induced systemic resistance (ISR) is commonly referred to as "mycorrhiza induced resistance" (MIR) and is considered to share characteristics of ISR and systemic acquired resistance (SAR) (Cameron, Neal, van Wees, & Ton, 2013). Both, ISR and SAR induction in systemic tissues include adjustments in phytohormone-signaling and redox balance (El-Shetehy et al., 2015; Vlot et al., 2021; Yu et al., 2022). There are hints that induction of MIR requires similar adjustments. Recent studies on poplar – *L. bicolor* interactions, for example, revealed a MIR-like state of EMF poplar-trees and improved resistance towards the specialist poplar leaf feeding herbivore, *Chrysomela populi* (Kaling et al., 2018; Sivaprakasam Padmanaban et al., 2022). The improved resistance was associated with prominent changes in plant secondary metabolites and enzymes such as VOCs, nitriles, aldoximes and chitinases. Transcriptomic analyses of poplar leaves have shown that EMF can also trigger SA- and JA-signaling related changes in systemic tissues: *Laccaria bicolor* colonization altered the gene expression of the SA-signaling essential protein, NONEXPRESSOR OF PATHOGENESIS-RELATED PROTEINS1 (NPR1) and various transcription factors related to JA and JAZ1 (Kaling et al., 2018). SA concentrations were, moreover, altered in poplars that were colonized by *Hebeloma mesophaeum* (Pfabel et al., 2012). For AMF, it is known that adjustments in different phytohormones, including SA and JA, are involved in such systemic responses (Bedini, Mercy, Schneider, Franken, & Lucic-Mercy, 2018; Benjamin, Pandharikar, & Frendo, 2022). EMF-based induced immunity and altered phytohormone levels have been observed even in non-host plants: in *Arabidopsis* heat-killed *L. bicolor* fragments triggered SA- and JA-signaling and immunity against the aboveground pest *Trichoplusia ni* (Vishwanathan et al., 2020). In general, ROS levels are interconnected to phytohormone adjustments, making it thus likely that ROS can also be essential in conferring immunity to systemic tissues. Leastwise, ROS is a central element of SAR induction in systemic tissues: The accumulation of pipecolic acid (Pip) depends on the interplay between ROS and NO* in addition to other signaling compounds such as azelaic acid and glycerol-3-phosphate (Rekhter et al., 2019; Vlot et al., 2021). In addition, adjustments in ROS levels are known to be involved in ISR (Yu et al., 2022) and even in plant responses to fungal volatile sequiterpenes (Ditengou et al., 2015). For example, *Laccaria bicolor* interacts with *Arabidopsis* through the air phase by volatile compounds in controlled laboratory set-ups. VOCs from *Laccaria bicolor* elicit ROS accumulation and alter the lateral root growth of both, the host plant poplar and the non-host *Arabidopsis* (Ditengou et al., 2015).

3.2.4 Potential signals in transmitting information to systemic tissues

How the information is transmitted from local to systemic tissues remains unknown at the moment. As MIR resembles both, ISR and SAR, further analyses of the mobile signals associated to these two routes of induced resistance to shed light on EMF-based resistance are needed. Various mobile signals have been suggested to transmit SAR-related information, e.g., SA and Pip having verified functions in phloem-based SAR fortification (Vlot et al., 2021). Pip and NHP have also been detected along the transport route from root to leaves in xylem sap (Kasper et al., 2022). Other suggested compounds that are related to propagation of SAR include methyl salicylate (MeSA), azelaic acid and glycerol-3-phosphate (Dreischhoff et al., 2020; Vlot et al., 2021). Regarding ISR, the involved signaling routes are less clear. In the last decade, evidence accumulated that not only JA/ethylene signaling related cues but also SA signaling are involved in inducing ISR (Yu et al., 2022). Both ISR and SAR require adjustments of ROS levels as discussed above (El-Shetehy et al., 2015; Vlot et al., 2021; Yu et al., 2022). ROS are considered essential in cell-to-cell propagation of SAR and ISR signals. The defense responses that are activated by SA and Pip to propagate SAR, are dependent on activation of CALCIUM-DEPENDENT PROTEIN KINASE 5 (CPK5). The activation of CPK5 is associated to altered Ca^{2+} levels and activation of NADPH-oxidases. The phosphorylation of NADPH-oxidase is followed by elevated ROS levels which, in turn, activate the same mechanisms in the recipient cells to further propagate the signal (Vlot et al., 2021; Yu et al., 2022). The enhanced Ca^{2+} influx and related strong production of NADPH-mediated ROS is also essential for transmission of the ISR signal to systemic tissues (Yu et al., 2022). The role of ROS in transmitting EMF-based induced resistance is, however, unknown and even their final role in propagating SAR or ISR needs still to be validated.

Considering the speed of the movement of soluble signals and the rapidness of the response in the systemic tissues, volatile cues might also contribute to eliciting induced resistance. To date, there is no direct evidence of the involvement of VOCs in inducing EMF-based resistance in systemic tissues. However, it is possible that VOCs contribute to conferring systemic immunity to non-colonized roots or to aboveground tissues. Interestingly, also microbial VOCs have functions in inducing systemic resistance. Perception of distinct microbial VOCs can elicit ISR in *Arabidopsis* either in JA- or SA-dependent manner without wall-to-wall contact between the organisms (Naznin et al., 2014). Specific VOCs can also propagate SAR-related information in aboveground interactions (Frank et al., 2021; Wenig et al., 2019). It remains to be elucidated if EMF-based induced resistance might be transmitted to systemic tissues through the air or through the soil matrix.

4. Conclusions and outlook

This review highlights the composition of the antioxidative system of ectomycorrhizal fungi and identifies important gaps, such as a possible involvement of erythroascorbate in the defense system. Although several relevant metabolites and proteins required to mediate contact and establish a functional mycorrhiza have been identified, a clear mechanistic picture is yet to emerge. Genetic suppression of fungal SOD and NR established their roles in ectomycorrhizal formation and fitness, underpinning the crucial roles of reactive oxygen species in this interaction. These enzymes may control plant ROS burst and NO* signaling, which occur in response to biotic invasion. In the extracellular space, secreted oxidative enzymes drive not only oxidative degradation of potential nutrient sources but are also critical for intra-radical fungal growth. In mature mycorrhizas, plant-fungal interactions are characterized by mutual growth control, involving H_2O_2. Local and systemic responses of the plant to mycorrhizal colonization involve ROS, phytohormones, and Ca^{2+} signaling. Less is known about the fungal metabolism in the interaction. It will be important to understand how mycorrhizal fungal species, able to produce *OH radicals via Fenton reaction, control these activities depending on host and nutrient source. A picture is now emerging that the genetic inventory of different mycorrhizal fungal species varies with their phylogenetic origin. How this affects fungal-plant interactions is not yet well understood but evidence is accumulating that different fungi evoke different systemic effects in their host. Mechanistic molecular understanding of mycorrhizal functions and their metabolic range in different host-fungal species interactions is an outstanding task for future research. These studies are of fundamental interest. They have also an important applied perspective since soil amended with ectomycorrhizal fungi has often been recommended to improve growth and yield. In climate-stressed habitats and on polluted soils, the selection of suitable mycorrhizal species to increase the establishment and stress tolerance of saplings is of high practical relevance. Deeper insights into the genetic make-up and the underlying processes that control the host-plant interaction and resilience can open new avenues for plant improvement.

Acknowledgments

A.P., M.R., and J.P.S. thank the DFG (Deutsche Forschungsgemeinschaft) for continuous support, especially for financial support to the projects PO362/22-2, SCHN653/7-2, RO5311/4-1, and IRTG 2172-PRoTECT M2.2. H.S. gratefully acknowledges a PhD scholarship provided by the Chinese Scholarship Counsel (CSC, P.R. China).

Appendix A. Supporting information

Supplementary data associated with this article can be found in the online version at doi:10.1016/bs.abr.2022.11.001.

References

Abba, S., Khouja, H., Martino, E., Archer, D., & Perotto, S. (2009). SOD1-targeted gene disruption in the ericoid mycorrhizal fungus *Oidiodendron maius* reduces conidiation and the capacity for mycorrhization. *Molecular Plant Microbe Interactions, 22*(11), 1412–1421.

Adriani, P. P., de Paiva, F. C. R., de Oliveira, G. S., Leite, A. C., Sanches, A. S., Lopes, A. R., & Chambergo, F. S. (2021). Structural and functional characterization of the glutathione peroxidase-like thioredoxin peroxidase from the fungus *Trichoderma reesei*. *International Journal of Biological Macromolecules, 167*, 93–100.

Agerer, R., Schloter, M., & Hahn, C. (2000). Fungal enzymatic activity in fruitbodies. *Nova Hedwigia, 71*, 315–336.

Altschul, S. F., Gish, W., Miller, W., Myers, E. W., & Lipman, D. J. (1990). Basic local alignment search tool. *Journal of Molecular Biology, 215*(3), 403–410.

Alvarez, M., Huygens, D., Fernandez, C., Gacitúa, Y., Olivares, E., Saavedra, I., & Valenzuela, E. (2009). Effect of ectomycorrhizal colonization and drought on reactive oxygen species metabolism of *Nothofagus dombeyi* roots. *Tree Physiology, 29*(8), 1047–1057.

Ameye, M., Van Meulebroek, L., Meuninck, B., Vanhaecke, L., Smagghe, G., Haesaert, G., & Audenaert, K. (2020). Metabolomics reveal induction of ROS production and glycosylation events in wheat upon exposure to the green leaf volatile Z-3-hexenyl acetate. *Frontiers in Plant Science, 11*, 596271.

Arthikala, M., Sànchez-López, R., Nava, N., Santana, O., Cárdenas, L., & Quinto, C. (2014). RbohB, a *Phaseolus vulgaris* NADPH oxidase gene, enhances symbiosome number, bacteroid size, and nitrogen fixation in nodules and impairs mycorrhizal colonization. *New Phytologist, 202*(3), 886–900.

Bago, B., & Bécard, G. (2002). Bases of the obligate biotrophy of arbuscular mycorrhizal fungi. In S. Gianinazzi, H. Schüepp, J. M. Barea, & K. Haselwandter (Eds.). *Mycorrhizal technology in agriculture: From genes to bioproducts* (pp. 33–48). Basel: Birkhäuser.

Bai, X.-N., Hao, H., Hu, Z.-H., & Leng, P.-S. (2021). Ectomycorrhizal inoculation enhances the salt tolerance of *Quercus mongolica* seedlings. *Plants, 10*(9), 1790.

Baptista, P., Martins, A., Pais, M. S., Tavares, R. M., & Lino-Neto, T. (2007). Involvement of reactive oxygen species during early stages of ectomycorrhiza establishment between *Castanea sativa* and *Pisolithus tinctorius*. *Mycorrhiza, 17*(3), 185–193.

Baxter, A., Mittler, R., & Suzuki, N. (2014). ROS as key players in plant stress signalling. *Journal of Experimental Botany, 65*(5), 1229–1240.

Becquer, A., Guerrero-Galan, C., Eibensteiner, J., Houdinet, G., Bücking, H., Zimmermann, S. D., & Garcia, K. (2019). The ectomycorrhizal contribution to tree nutrition. *Advances in Botanical Research, 89*, 77–126.

Bedard, K., Lardy, B., & Krause, K. (2007). NOX family NADPH oxidases: Not just in mammals. *Biochimie, 89*(9), 1107–1112.

Bedini, A., Mercy, L., Schneider, C., Franken, P., & Lucic-Mercy, E. (2018). Unraveling the initial plant hormone signaling, metabolic mechanisms and plant defense triggering the endomycorrhizal symbiosis behavior. *Frontiers in Plant Science, 9*, 1800.

Bellion, M., Courbot, M., Jacob, C., Blaudez, D., & Chalot, M. (2006). Extracellular and cellular mechanisms sustaining metal tolerance in ectomycorrhizal fungi. *FEMS Microbiology Letters, 254*(2), 173–181.

Belmondo, S., Calcagno, C., Genre, A., Puppo, A., Pauly, N., & Lanfranco, L. (2016). The *Medicago truncatula MtRbohE* gene is activated in arbusculated cells and is involved in root cortex colonization. *Planta, 243*, 251–262.

Belousov, V., Fradkov, A., Lukyanov, K., Staroverov, D. B., Shakhbazov, K. S., Terskikh, A. V., & Lukyanov, S. (2006). Genetically encoded fluorescent indicator for intracellular hydrogen peroxide. *Nature Methods, 3*, 281–286.

Benjamin, G., Pandharikar, G., & Frendo, P. (2022). Salicylic acid in plant symbioses: Beyond plant pathogen interactions. *Biology, 11*(6), 861.

Berger, A., Brouquisse, R., Pathak, P. K., Hichri, I., Inderjit, I., Bhatia, S., Boscari, A., Igamberdiev, A. U., & Gupta, K. J. (2018). Pathways of nitric oxide metabolism and operation of phytoglobins in legume nodules: Missing links and future directions. *Plant Cell and Environment, 41*, 2057–2068.

Bertoni, G. (2019). Perception of ectomycorrhizal signals by poplar induces root colonization. *Plant Cell, 31*(10), 2283–2284.

Biteau, B., Labarre, J., & Toledano, M. B. (2003). ATP-dependent reduction of cysteine–sulphinic acid by *S. cerevisiae* sulphiredoxin. *Nature, 425*(6961), 980–984.

Bödeker, I. T. M., Nygren, C. M. R., Taylor, A. F. S., Olson, Å., & Lindahl, B. D. (2009). Class II peroxidase-encoding genes are present in a phylogenetically wide range of ectomycorrhizal fungi. *ISME Journal, 3*(12), 1387–1395.

Bolchi, A., Ruotolo, R., Marchini, G., Vurro, E., di Toppi, L. S., Kohler, A., & Ottonello, S. (2011). Genome-wide inventory of metal homeostasis-related gene products including a functional phytochelatin synthase in the hypogeous mycorrhizal fungus *Tuber melanosporum*. *Fungal Genetics and Biology, 48*(6), 573–584.

Bonfante, P., & Venice, F. (2020). Mucoromycota: Going to the roots of plant-interacting fungi. *Fungal Biology Reviews, 34*(2), 100–113.

Bouffaud, M.-L., Herrmann, S., Tarkka, M. T., Bönn, M., Feldhahn, L., & Buscot, F. (2020). Oak displays common local but specific distant gene regulation responses to different mycorrhizal fungi. *BMC Genomics, 21*(1), 399.

Bunyard, B. (2022). *The lives of fungi: A natural history of our planet's decomposers*. Princeton, NJ: Princeton University Press,.

Burke, R. M., & Cairney, J. W. G. (1998). Carbohydrate oxidases in ericoid and ectomycorrhizal fungi: A possible source of Fenton radicals during the degradation of lignocellulose. *New Phytologist, 139*(4), 637–645.

Cameron, D. D., Neal, A. L., van Wees, S. C. M., & Ton, J. (2013). Mycorrhiza-induced resistance: More than the sum of its parts? *Trends in Plant Science, 18*(10), 539–545.

Cárdenas, L., & Quinto, C. (2008). Reactive oxygen species (ROS) as early signals in root hair cells responding to rhizobial nodulation factors. *Plant Signaling & Behavior, 3*(12), 1101–1102.

Cárdenas, L., Martínez, A., Sánchez, F., & Quinto, C. (2008). Fast, transient and specific intracellular ROS changes in living root hair cells responding to Nod factors (NFs). *Plant Journal, 56*(5), 802–813.

Castaño, J. D., Zhang, J., Anderson, C. E., & Schilling, J. S. (2018). Oxidative damage control during decay of wood by brown rot fungus using oxygen radicals. *Applied and Environmental Microbiology, 84*(22) e01937-18.

Chen, D. M., Taylor, A. F. S., Burke, R. M., & Cairney, J. W. G. (2001). Identification of genes for lignin peroxidases and manganese peroxidases in ectomycorrhizal fungi. *New Phytologist, 152*(1), 151–158.

Chen, S., Hawighorst, P., Sun, J., & Polle, A. (2014). Salt tolerance in Populus: Significance of stress signaling networks, mycorrhization, and soil amendments for cellular and whole-plant nutrition. *Environmental and Experimental Botany, 107*, 113–124.

Chiapello, M., Martino, E., & Perotto, S. (2015). Common and metal-specific proteomic responses to cadmium and zinc in the metal tolerant ericoid mycorrhizal fungus *Oidiodendron maius* Zn. *Metallomics*, 7(5), 805–815.

Chot, E., & Reddy, M. S. (2022). Role of ectomycorrhizal symbiosis behind the host plants ameliorated tolerance against heavy metal stress. *Frontiers in Microbiology, 13*, 855473.

Cope, K. R., Bascaules, A., Irving, T. B., Venkateshwaran, M., Maeda, J., Garcia, K., & Ané, J.-M. (2019). The ectomycorrhizal fungus *Laccaria bicolor* produces lipochitooligosaccharides and uses the common symbiosis pathway to colonize *Populus* roots. *Plant Cell, 31*(10), 2386–2410.

Correa-Aragunde, N., Foresi, N., & Lamattina, L. (2015). Nitric oxide is a ubiquitous signal for maintaining redox balance in plant cells: Regulation of ascorbate peroxidase as a case study. *Journal of Experimental Botany, 66*(10), 2913–2921.

Courtade, G., & Aachmann, F. L. (2019). Chitin-active lytic polysaccharide monooxygenases. In Q. Yang, & T. Fukamizo (Eds.). *Targeting chitin-containing organisms* (pp. 115–129). Singapore: Springer.

Courty, P.-E., Buee, M., Diedhiou, A. G., Frey-Klett, P., Le Tacon, F., Rineau, F., & Garbaye, J. (2010). The role of ectomycorrhizal communities in forest ecosystem processes: New perspectives and emerging concepts. *Soil Biology & Biochemistry, 42*(5), 679–698.

Dachuan, Y., & Jinyu, Q. (2021). The physiological response of ectomycorrhizal fungus *Lepista sordida* to Cd and Cu stress. *PeerJ, 9*, e11115.

Dat, J. F., Pellinen, R., Tom Beeckman, Van De Cotte, B., Langebartels, C., Kangasjärvi, J., & van Breusegem, F. (2003). Changes in hydrogen peroxide homeostasis trigger an active cell death process in tobacco. *The Plant Journal, 33*(4), 621–632.

Davletova, S., Schlauch, K., Coutu, J., & Mittler, R. (2005). The zinc-finger protein Zat12 plays a central role in reactive oxygen and abiotic stress signaling in Arabidopsis. *Plant Physiology, 139*(2), 847–856.

Deng, C., Zhu, Z., Liu, J., Zhang, Y., Zhang, Y., Yu, D., & Chen, S. (2021). Ectomycorrhizal fungal strains facilitate Cd^{2+} enrichment in a woody hyperaccumulator under co-existing stress of cadmium and salt. *International Journal of Molecular Sciences, 22*(21), 11651.

Ditengou, F. A., Müller, A., Rosenkranz, M., Felten, J., Lasok, H., van Doorn, M. M., & Polle, A. (2015). Volatile signalling by sesquiterpenes from ectomycorrhizal fungi reprogrammes root architecture. *Nature Communications, 6*(1), 6279.

Dreischhoff, S., Das, I. S., Jakobi, M., Kasper, K., & Polle, A. (2020). Local responses and systemic induced resistance mediated by ectomycorrhizal fungi. *Frontiers in Plant Science, 11*, 590063.

Drula, E., Garron, M.-L., Dogan, S., Lombard, V., Henrissat, B., & Terrapon, N. (2022). The carbohydrate-active enzyme database: Functions and literature. *Nucleic Acids Research, 50*(D1), D571–D577.

El-Shetehy, M., Wang, C., Shine, M. B., Yu, K., Kachroo, A., & Kachroo, P. (2015). Nitric oxide and reactive oxygen species are required for systemic acquired resistance in plants. *Plant Signaling & Behavior, 10*(9), e998544.

Erb, M. (2018). Volatiles as inducers and suppressors of plant defense and immunity—Origins, specificity, perception and signaling. *Current Opinion in Plant Biology, 44*, 117–121.

Fernandez, J., & Wilson, R. A. (2014). Characterizing roles for the glutathione reductase, thioredoxin reductase and thioredoxin peroxidase-encoding genes of *Magnaporthe oryzae* during rice blast disease. *PLoS One, 9*(1), e87300.

Fichman, Y., & Mittler, R. (2020). Rapid systemic signaling during abiotic and biotic stresses: Is the ROS wave master of all trades? *The Plant Journal, 102*(5), 887–896.

Fichman, Y., Miller, G., & Mittler, R. (2019). Whole-plant live imaging of reactive oxygen species. *Molecular Plant, 12*, 1203–1210.

Floudas, D., Binder, M., Riley, R., Barry, K., Blanchette, R. A., Henrissat, B., & Hibbett, D. S. (2012). The Paleozoic origin of enzymatic lignin decomposition reconstructed from 31 fungal genomes. *Science, 336*(6089), 1715–1719.

Frank, L., Wenig, M., Ghirardo, A., Krol, A., Vlot, A. C., Schnitzler, J., & Rosenkranz, M. (2021). Isoprene and β-caryophyllene confer plant resistance via different plant internal signalling pathways. *Plant, Cell & Environment, 44*(4), 1151–1164.

Gafur, A., Schützendübel, A., Langenfeld-Heyser, R., Fritz, E., & Polle, A. (2004). Compatible and incompetent *Paxillus involutus* isolates for ectomycorrhiza formation in vitro with poplar (*Populus × canescens*) differ in H_2O_2 production. *Plant Biology, 6*(1), 91–99.

Gange, A. C., Gane, D. R. J., Chen, Y., & Gong, M. (2005). Dual colonization of *Eucalyptus urophylla* S.T. Blake by arbuscular and ectomycorrhizal fungi affects levels of insect herbivore attack. *Agricultural and Forest Entomology,* 7(3), 253–263.

Garcia, K., Delaux, P., Cope, K. R., & Ané, J. (2015). Molecular signals required for the establishment and maintenance of ectomycorrhizal symbioses. *New Phytologist, 208*(1), 79–87.

Garcia-Iriepa, C., Marazzi, M., & Navizet, I. (2020). The role of CO_2 detachment in fungal bioluminescence: Thermally vs. excited state induced pathways. *Physical Chemistry Chemical Physics, 22*(46), 26787–26795.

Garnica-Vergara, A., Barrera-Ortiz, S., Muñoz-Parra, E., Raya-González, J., Méndez-Bravo, A., Macías-Rodríguez, L., Ruiz-Herrera, L. F., & López-Bucio, J. (2016). The volatile 6-pentyl-2H-pyran-2-one from Trichoderma atroviride regulates Arabidopsis thaliana root morphogenesis via auxin signaling and ETHYLENE INSENSITIVE 2 functioning. *New Phytologist, 209*, 1496–1512.

Gaspar, M. L., & Pawlowska, T. E. (2022). Innate immunity in fungi: Is regulated cell death involved? *PLoS Pathogens, 18*(5), e1010460.

Gilroy, S., Suzuki, N., Miller, G., Choi, W.-G., Toyota, M., Devireddy, A. R., & Mittler, R. (2014). A tidal wave of signals: Calcium and ROS at the forefront of rapid systemic signaling. *Trends in Plant Science, 19*(10), 623–630.

Gramss, G., Kirsche, B., Voigt, K.-D., Günther, Th, & Fritsche, W. (1999). Conversion rates of five polycyclic aromatic hydrocarbons in liquid cultures of fifty-eight fungi and the concomitant production of oxidative enzymes. *Mycological Research, 103*(8), 1009–1018.

Hage, H., Rosso, M.-N., & Tarrago, L. (2021). Distribution of methionine sulfoxide reductases in fungi and conservation of the free-methionine-R-sulfoxide reductase in multicellular eukaryotes. *Free Radical Biology and Medicine, 169*, 187–215.

Huang, H., Ullah, F., Zhou, D.-X., Yi, M., & Zhao, Y. (2019). Mechanisms of ROS regulation of plant development and stress responses. *Frontiers in Plant Science, 10*, 800.

Huang, Y.-M., Zou, Y.-N., & Wu, Q.-S. (2017). Alleviation of drought stress by mycorrhizas is related to increased root H_2O_2 efflux in trifoliate orange. *Scientific Reports*, 7(1), 42335.

Jacob, C., Courbot, M., Brun, A., Steinman, H. M., Jacquot, J.-P., Botton, B., & Chalot, M. (2001). Molecular cloning, characterization and regulation by cadmium of a superoxide dismutase from the ectomycorrhizal fungus *Paxillus involutus*. *European Journal of Biochemistry, 268*(11), 3223–3232.

Kaling, M., Schmidt, A., Moritz, F., Rosenkranz, M., Witting, M., Kasper, K., & Polle, A. (2018). Mycorrhiza-triggered transcriptomic and metabolomic networks impinge on herbivore fitness. *Plant Physiology, 176*(4), 2639–2656.

Kasper, K., Abreu, I. N., Feussner, K., Zienkiewicz, K., Herrfurth, C., Ischebeck, T., & Polle, A. (2022). Multi-omics analysis of xylem sap uncovers dynamic modulation of poplar defenses by ammonium and nitrate. *The Plant Journal, 111*(1), 282–303.

Ke, H.-M., Lee, H.-H., Lin, C.-Y. I., Liu, Y.-C., Lu, M. R., Hsieh, J.-W. A., & Tsai, I. J. (2020). *Mycena* genomes resolve the evolution of fungal bioluminescence. *Proceedings of the National Academy of Sciences, 117*(49), 31267–31277.

Kemppainen, M., Duplessis, S., Martin, F., & Pardo, A. G. (2009). RNA silencing in the model mycorrhizal fungus *Laccaria bicolor*: Gene knock-down of nitrate reductase results in inhibition of symbiosis with *Populus*. *Environmental Microbiology, 11*(7), 1878–1896.

Kohler, A., Kuo, A., Nagy, L. G., Morin, E., Barry, K. W., Buscot, F., & Martin, F. (2015). Convergent losses of decay mechanisms and rapid turnover of symbiosis genes in mycorrhizal mutualists. *Nature Genetics, 47*(4), 410–415.

Kothamasi, D., Wannijn, J., Van Hees, M., Nauts, R., Van Gompel, A., Vanhoudt, N., & Vandenhove, H. (2019). Exposure to ionizing radiation affects the growth of ectomycorrhizal fungi and induces increased melanin production and increased capacities of reactive oxygen species scavenging enzymes. *Journal of Environmental Radioactivity, 197*, 16–22.

Kumari, A., Pathak, P. K., Loake, G. J., & Gupta, K. J. (2019). The PHYTOGLOBIN-NO cycle regulates plant mycorrhizal symbiosis. *Trends in Plant Science, 24*, 981–983.

Kuruthukulangarakoola, G. T., Zhang, J., Albert, A., Winkler, B., Lang, H., Buegger, F., Gaupels, F., Heller, W., Michalke, B., Sarioglu, H., Schnitzler, J. P., Hebelstrup, K. H., Durner, J., & Lindermayr, C. (2017). Nitric oxide-fixation by non-symbiotic haemoglobin proteins in *Arabidopsis thaliana* under N-limited conditions. *Plant Cell and Environment, 40*, 36–50.

Labourel, A., Frandsen, K. E. H., Zhang, F., Brouilly, N., Grisel, S., Haon, M., & Berrin, J.-G. (2020). A fungal family of lytic polysaccharide monooxygenase-like copper proteins. *Nature Chemical Biology, 16*(3), 345–350.

Langenfeld-Heyser, R., Gao, J., Ducic, T., Tachd, P., Lu, C., Fritz, E., & Polle, A. (2007). *Paxillus involutus* mycorrhiza attenuate NaCl-stress responses in the salt-sensitive hybrid poplar *Populus × canescens*. *Mycorrhiza, 17*(2), 121–131.

Lebreton, A., Zeng, Q., Miyauchi, S., Kohler, A., Dai, Y.-C., & Martin, F. M. (2021). Evolution of the mode of nutrition in symbiotic and saprotrophic fungi in forest ecosystems. *Annual Review of Ecology, Evolution, and Systematics, 52*(1), 385–404.

Li, Jiali, Li, C., Tsuruta, M., Matsushita, N., Goto, S., Shen, Z., & Lian, C. (2022). Physiological and transcriptional responses of the ectomycorrhizal fungus Cenococcum geophilum to salt stress. *Mycorrhiza, 32*(3–4), 327–340.

Li, Jing, Bao, S., Zhang, Y., Ma, X., Mishra-Knyrim, M., Sun, J., & Chen, S. (2012). *Paxillus involutus* strains MAJ and NAU mediate K+/Na+ homeostasis in ectomycorrhizal *Populus × canescens* under sodium chloride stress. *Plant Physiology, 159*(4), 1771–1786.

Lindermayr, C., & Durner, J. (2015). Interplay of reactive oxygen species and nitric oxide: Nitric oxide coordinates reactive oxygen species homeostasis. *Plant Physiology, 167*(4), 1209–1210.

Lindermayr, C., Saalbach, G., & Durner, J. (2005). Proteomic identification of S-nitrosylated proteins in Arabidopsis. *Plant Physiology, 137*(3), 921–930.

Liu, B., Dong, P., Zhang, X., Feng, Z., Wen, Z., Shi, L., & Chen, Y. (2022). Identification and characterization of eight metallothionein genes involved in heavy metal tolerance from the ectomycorrhizal fungus *Laccaria bicolor*. *Environmental Science and Pollution Research, 29*(10), 14430–14442.

Liu, H., Chen, H., Ding, G., Li, K., & Ren, Q. (2020). Identification of candidate genes conferring tolerance to aluminum stress in *Pinus massoniana* inoculated with ectomycorrhizal fungus. *BMC Plant Biology, 20*(1), 521.

Lofgren, L. A., Nguyen, N. H., Vilgalys, R., Ruytinx, J., Liao, H., Branco, S., & Kennedy, P. G. (2021). Comparative genomics reveals dynamic genome evolution in host specialist ectomycorrhizal fungi. *New Phytologist, 230*(2), 774–792.

Luo, Z.-B., Janz, D., Jiang, X., Göbel, C., Wildhagen, H., Tan, Y., & Polle, A. (2009). Upgrading root physiology for stress tolerance by ectomycorrhizas: Insights from metabolite and transcriptional profiling into reprogramming for stress anticipation. *Plant Physiology, 151*(4), 1902–1917.

Luo, Z.-B., Li, K., Gai, Y., Göbel, C., Wildhagen, H., Jiang, X., & Polle, A. (2011). The ectomycorrhizal fungus (*Paxillus involutus*) modulates leaf physiology of poplar towards improved salt tolerance. *Environmental and Experimental Botany, 72*(2), 304–311.

Luo, Z.-B., Wu, C., Zhang, C., Li, H., Lipka, U., & Polle, A. (2014). The role of ectomycorrhizas in heavy metal stress tolerance of host plants. *Environmental and Experimental Botany, 108*, 47–62.

Lyall, R., Nikoloski, Z., & Gechev, T. (2020). Comparative analysis of ROS network genes in extremophile eukaryotes. *International Journal of Molecular Sciences, 21*(23), 9131.

Martin, F. M., Harrison, M., Lennon, S., Lindahl, B., Öpik, M., Polle, A., & Selosse, M.-A. (2018). Cross-scale integration of mycorrhizal function. *New Phytologist, 220*, 941–946.

Martin, F., Aerts, A., Ahrén, D., Brun, A., Danchin, E., Duchaussoy, F., & Grigoriev, I. (2008). The genome of *Laccaria bicolor* provides insights into mycorrhizal symbiosis. *Nature, 452*(7183), 88.

Martínez-Medina, A., Pescador, L., Fernández, I., Rodríguez-Serrano, M., García, J. M., Romero-Puertas, M. C., & Pozo, M. J. (2019). Nitric oxide and phytoglobin PHYTOGB1 are regulatory elements in the *Solanum lycopersicum–Rhizophagus irregularis* mycorrhizal symbiosis. *New Phytologist, 223*, 1560–1574.

Maruta, T., Sawa, Y., Shigeoka, S., & Ishikawa, T. (2016). Diversity and evolution of ascorbate peroxidase functions in chloroplasts: More than just a classical antioxidant enzyme? *Plant & Cell Physiology, 57*(7), 1377–1386.

Matsui, K. (2016). A portion of plant airborne communication is endorsed by uptake and metabolism of volatile organic compounds. *Current Opinion in Plant Biology, 32*, 24–30.

Mattila, H., Österman-Udd, J., Mali, T., & Lundell, T. (2022). Basidiomycota fungi and ROS: Genomic perspective on key enzymes involved in generation and mitigation of reactive oxygen species. *Frontiers in Fungal Biology, 3*, 837605.

McCord, J., & Fridovich, I. (1969). Superoxide dismutase. *Journal of Biological Chemistry, 244*, 6049–6065.

Miloradovic van Doorn, M., Merl-Pham, J., Ghirardo, A., Fink, S., Polle, A., Schnitzler, J., & Rosenkranz, M. (2020). Root isoprene formation alters lateral root development. *Plant, Cell & Environment, 43*(9), 2207–2223.

Mittler, R. (2017). ROS are good. *Trends in Plant Science, 22*(1), 11–19.

Mittler, R., & Blumwald, E. (2015). The roles of ROS and ABA in systemic acquired acclimation. *The Plant Cell, 27*(1), 64–70.

Miyauchi, S., Kiss, E., Kuo, A., Drula, E., Kohler, A., Sánchez-García, M., & Martin, F. M. (2020). Large-scale genome sequencing of mycorrhizal fungi provides insights into the early evolution of symbiotic traits. *Nature Communications, 11*(1), 5125.

Moisan, K., Raaijmakers, J. M., Dicke, M., Lucas-Barbosa, D., & Cordovez, V. (2021). Volatiles from soil-borne fungi affect directional growth of roots. *Plant Cell & Environment, 44*, 339–345.

Moreau, M., Lindermayr, C., Durner, J., & Klessig, D. F. (2010). NO synthesis and signaling in plants—Where do we stand? *Physiologia Plantarum, 138*(4), 372–383.

Morel, M., Kohler, A., Martin, F., Gelhaye, E., & Rouhier, N. (2008). Comparison of the thiol-dependent antioxidant systems in the ectomycorrhizal *Laccaria bicolor* and the saprotrophic *Phanerochaete chrysosporium*. *New Phytologist, 180*(2), 391–407.

Mucha, J., Napierała-Filipiak, A., Gabała, E., Pawłowski, T. A., & Zadworny, M. (2019). Redistribution of iron and hydrogen peroxide in *Pinus sylvestris* roots in response to trophically diverse fungi. *European Journal of Plant Pathology, 153*(4), 1275–1286.

Münzenberger, B., Otter, T., Polle, A., & Wüstrich, D. (1997). Peroxidase and laccase activities in mycorrhizal and non-mycorrhizal fine roots of Norway spruce (*Picea abies*) and larch (*Larix decidua*). *Canadian Journal of Botany, 75*(6), 932–938.

Nath, M., Bhatt, D., Prasad, R., Gill, S. S., Anjum, N. A., & Tuteja, N. (2016). Reactive oxygen species generation, scavenging and signaling during plant-arbuscular mycorrhizal and *Piriformospora indica* interaction under stress condition. *Frontiers in Plant Science, 7*, 1574.

Naznin, H. A., Kiyohara, D., Kimura, M., Miyazawa, M., Shimizu, M., & Hyakumachi, M. (2014). Systemic resistance induced by volatile organic compounds emitted by plant growth-promoting fungi in *Arabidopsis thaliana*. *PLoS One, 9*(1), e86882.

Oddsdottir, E. S., Eilenberg, J., Sen, R., Harding, S., & Halldorsson, G. (2010). Early reduction of *Otiorhynchus* spp. larval root herbivory on *Betula pubescens* by beneficial soil fungi. *Applied Soil Ecology, 45*(3), 168–174.

Ott, T., Fritz, E., Polle, A., & Schützendübel, A. (2002). Characterisation of antioxidative systems in the ectomycorrhiza-building basidiomycete *Paxillus involutus* (Bartsch) Fr. and its reaction to cadmium. *FEMS Microbiology Ecology, 42*(3), 359–366.

Pellegrin, C., Morin, E., Martin, F. M., & Veneault-Fourrey, C. (2015). Comparative analysis of secretomes from ectomycorrhizal fungi with an emphasis on small-secreted proteins. *Frontiers in Microbiology, 6*, 1278.

Pfabel, C., Eckhardt, K.-U., Baum, C., Struck, C., Frey, P., & Weih, M. (2012). Impact of ectomycorrhizal colonization and rust infection on the secondary metabolism of poplar (*Populus trichocarpa* x *deltoides*). *Tree Physiology, 32*(11), 1357–1364.

Plett, J. M., Daguerre, Y., Wittulsky, S., Vayssières, A., Deveau, A., Melton, S. J., & Martin, F. (2014). Effector MiSSP7 of the mutualistic fungus Laccaria bicolor stabilizes the *Populus* JAZ6 protein and represses jasmonic acid (JA) responsive genes. *Proceedings of the National Academy of Sciences, 111*(22), 8299–8304.

Plett, J. M., Kemppainen, M., Kale, S. D., Kohler, A., Legué, V., Brun, A., & Martin, F. (2011). A secreted effector protein of *Laccaria bicolor* is required for symbiosis development. *Current Biology, 21*(14), 1197–1203.

Plett, J. M., Khachane, A., Ouassou, M., Sundberg, B., Kohler, A., & Martin, F. (2014). Ethylene and jasmonic acid act as negative modulators during mutualistic symbiosis between *Laccaria bicolor* and *Populus* roots. *New Phytologist, 202*(1), 270–286.

Pritsch, K., & Garbaye, J. (2011). Enzyme secretion by ECM fungi and exploitation of mineral nutrients from soil organic matter. *Annals of Forest Science, 68*(1), 25–32.

Punja, Z. K., & Zhang, Y. Y. (1993). Plant chitinases and their roles in resistance to fungal diseases. *Journal of Nematology, 25*(4), 526–540.

Qi, Y., Zhao, N., Liu, J., & Huang, J. (2016). Biochemical responses of ten ectomycorrhizal fungal isolates to manganese. *Water, Air, & Soil Pollution, 227*(12), 1–10.

Reis, F. S., Sandrina, A., Heleno, S. A., Barros, L., Sousa, M. J., Martins, A., Santos-Buelga, C., & Ferreira, I. C. F. R. (2011). Toward the antioxidant and chemical characterization of mycorrhizal mushrooms from Northeast Portugal. *Journal of Food Science, 76*, C824–C830. https://doi.org/10.1111/j.1750-3841.2011.02251.xF

Rekhter, D., Lüdke, D., Ding, Y., Feussner, K., Zienkiewicz, K., Lipka, V., & Feussner, I. (2019). Isochorismate-derived biosynthesis of the plant stress hormone salicylic acid. *Science, 365*(6452), 498–502.

Riedlmeier, M., Ghirardo, A., Wenig, M., Knappe, C., Koch, K., Georgii, E., & Vlot, A. C. (2017). Monoterpenes support systemic acquired resistance within and between plants. *The Plant Cell, 29*(6), 1440–1459.

Rieske, L. K., Rhoades, C. C., & Miller, S. P. (2003). Foliar chemistry and gypsy moth, *Lymantria dispar* (L.), herbivory on pure American Chestnut, *Castanea dentata* (Fam: Fagaceae), and a disease-resistant hybrid. *Environmental Entomology, 32*(2), 359–365.

Rivera Pérez, C. A., Janz, D., Schneider, D., Daniel, R., & Polle, A. (2022). Transcriptional landscape of ectomycorrhizal fungi and their host provides insight into N uptake from forest soil. *mSystems,* 7(1), e0095721.

Salzer, P., Hubner, B., Sirrenberg, A., & Hager, A. (1997). Differential effect of purified spruce chitinases and β-1,3-glucanases on the activity of elicitors from ectomycorrhizal fungi. *Plant Physiology, 114*(3), 957–968.

Salzer, P., Hebe, G., & Hager, A. (1997). Cleavage of chitinous elicitors from the ectomycorrhizal fungus *Hebeloma crustuliniforme* by host chitinases prevents induction of K^+ and Cl^- release, extracellular alkalinization and H_2O_2 synthesis of *Picea abies* cells. *Planta, 203*(4), 470–479.

Salzer, P., Hebe, G., Reith, A., Zitterell-Haid, B., Stransky, H., Gaschler, K., & Hager, A. (1996). Rapid reactions of spruce cells to elicitors released from the ectomycorrhizal fungus *Hebeloma crustuliniforme,* and inactivation of these elicitors by extracellular spruce cell enzymes. *Planta, 198*(1).

Sauter, M., & Hager, A. (1989). The mycorrhizal fungus *Amanita muscaria* induces chitinase activity in roots and in suspension-cultured cells of its host *Picea abies*. *Planta, 179*(1), 61–66.

Schickler, H., & Chet, I. (1997). Heterologous chitinase gene expression to improve plant defense against phytopathogenic fungi. *Journal of Industrial Microbiology and Biotechnology, 19*(3), 196–201.

Schützendübel, A., & Polle, A. (2002). Plant responses to abiotic stresses: Heavy metal-induced oxidative stress and protection by mycorrhization. *Journal of Experimental Botany, 53*, 1351–1365.

Schwacke, R., & Hager, A. (1992). Fungal elicitors induce a transient release of active oxygen species from cultured spruce cells that is dependent on Ca^{2+} and protein-kinase activity. *Planta, 187*(1).

Sebastiana, M., da Silva, A. B., Matos, A. R., Alcântara, A., Silvestre, S., & Malhó, R. (2018). Ectomycorrhizal inoculation with *Pisolithus tinctorius* reduces stress induced by drought in cork oak. *Mycorrhiza, 28*(3), 247–258.

Shah, F., Schwenk, D., Nicolás, C., Persson, P., Hoffmeister, D., & Tunlid, A. (2015). Involutin is an Fe^{3+} reductant secreted by the ectomycorrhizal fungus *Paxillus involutus*

during Fenton-based decomposition of organic matter. *Applied and Environmental Microbiology, 81*(24), 8427–8433.
Shi, L., Zhao, X., Zhong, K., Jia, Q., Shen, Z., Zou, J., & Chen, Y. (2022). Physiological mechanism of the response to Cr(VI) in the aerobic denitrifying ectomycorrhizal fungus *Pisolithus* sp.1. *Journal of Hazardous Materials, 429*, 128318.
Shi, L., Dong, P., Song, W., Li, C., Lu, H., Wen, Z., & Chen, Y. (2020). Comparative transcriptomic analysis reveals novel insights into the response to Cr(VI) exposure in Cr (VI) tolerant ectomycorrhizal fungi *Pisolithus* sp. 1 LS-2017. *Ecotoxicology and Environmental Safety, 188*, 109935.
Sivaprakasam Padmanaban, P. B., Rosenkranz, M., Zhu, P., Kaling, M., Schmidt, A., Schmitt-Kopplin, P., Polle, A., & Schnitzler, J.-P. (2022). Mycorrhiza-tree-herbivore interactions: Alterations in poplar metabolome and volatilome. *Metabolites, 12*(2), 93.
Smirnoff, N. (2018). Ascorbic acid metabolism and functions: A comparison of plants and mammals. *Free Radical Biology and Medicine, 122*, 116–129.
Smith, S. E., & Read, D. J. (2010). *Mycorrhizal symbiosis* (3rd ed.). London: Academic Press.
Strullu-Derrien, C., Selosse, M., Kenrick, P., & Martin, F. M. (2018). The origin and evolution of mycorrhizal symbioses: From palaeomycology to phylogenomics. *New Phytologist, 220*(4), 1012–1030.
Sun, W., Yang, B., Zhu, Y., Wang, H., Qin, G., & Yang, H. (2022). Ectomycorrhizal fungi enhance the tolerance of phytotoxicity and cadmium accumulation in oak (*Quercus acutissima* Carruth.) seedlings: Modulation of growth properties and the antioxidant defense responses. *Environmental Science and Pollution Research, 29*(5), 6526–6537.
Sun, X., Bonfante, P., & Tang, M. (2015). Effect of volatiles versus exudates released by germinating spores of Gigaspora margarita on lateral root formation. *Plant Physiology and Biochemistry, 97*, 1–10.
Takemoto, D., Tanaka, A., & Scott, B. (2007). NADPH oxidases in fungi: Diverse roles of reactive oxygen species in fungal cellular differentiation. *Fungal Genetics and Biology, 44*(11), 1065–1076.
Tanaka, T., Izawa, S., & Inoue, Y. (2005). GPX2, encoding a phospholipid hydroperoxide glutathione peroxidase homologue, codes for an atypical 2-Cys peroxiredoxin in *Saccharomyces cerevisiae*. *Journal of Biological Chemistry, 280*(51), 42078–42087.
Tian, W., Wang, C., Gao, Q., Li, L., & Luan, S. (2020). Calcium spikes, waves and oscillations in plant development and biotic interactions. *Nature Plants, 6*(7), 750–759.
Trappe, J. M. (2005). A.B. Frank and mycorrhizae: The challenge to evolutionary and ecologic theory. *Mycorrhiza, 15*(4), 277–281.
Tsang, C. K., Liu, Y., Thomas, J., Zhang, Y., & Zheng, X. F. S. (2014). Superoxide dismutase 1 acts as a nuclear transcription factor to regulate oxidative stress resistance. *Nature Communications, 5*(1), 3446.
Vandenabeele, S., Vanderauwera, S., Vuylsteke, M., Rombauts, S., Langebartels, C., Seidlitz, H. K., & van Breusegem, F. (2004). Catalase deficiency drastically affects gene expression induced by high light in *Arabidopsis thaliana*. *The Plant Journal, 39*(1), 45–58.
Vandhana, T. M., Reyre, J., Sushmaa, D., Berrin, J., Bissaro, B., & Madhuprakash, J. (2022). On the expansion of biological functions of lytic polysaccharide mono-oxygenases. *New Phytologist, 233*(6), 2380–2396.
Vanzo, E., Merl-Pham, J., Velikova, V., Ghirardo, A., Lindermayr, C., Hauck, S. M., & Schnitzler, J.-P. (2016). Modulation of protein S-nitrosylation by isoprene emission in poplar. *Plant Physiology, 170*(4), 1945–1961.
Veneault-Fourrey, C., Commun, C., Kohler, A., Morin, E., Balestrini, R., Plett, J., & Martin, F. (2014). Genomic and transcriptomic analysis of *Laccaria bicolor* CAZome reveals insights into polysaccharides remodelling during symbiosis establishment. *Fungal Genetics and Biology, 72*, 168–181.

Villalobos-Escobedo, J. M., Esparza-Reynoso, S., Pelagio-Flores, R., López-Ramírez, F., Ruiz-Herrera, L. F., López-Bucio, J., & Herrera-Estrella, A. (2020). The fungal NADPH oxidase is an essential element for the molecular dialog between *Trichoderma* and *Arabidopsis*. *The Plant Journal, 103*(6), 2178–2192.

Vishwanathan, K., Zienkiewicz, K., Liu, Y., Janz, D., Feussner, I., Polle, A., & Haney, C. H. (2020). Ectomycorrhizal fungi induce systemic resistance against insects on a non-mycorrhizal plant in a CERK1-dependent manner. *New Phytologist, 228*(2), 728–740.

Vlot, A. C., Sales, J. H., Lenk, M., Bauer, K., Brambilla, A., Sommer, A., & Nayem, S. (2021). Systemic propagation of immunity in plants. *New Phytologist, 229*(3), 1234–1250.

Wenig, M., Ghirardo, A., Sales, J. H., Pabst, E. S., Breitenbach, H. H., Antritter, F., & Vlot, A. C. (2019). Systemic acquired resistance networks amplify airborne defense cues. *Nature Communications, 10*(1), 3813.

Werner, S., Polle, A., & Brinkmann, N. (2016). Belowground communication: Impacts of volatile organic compounds (VOCs) from soil fungi on other soil-inhabiting organisms. *Applied Microbiology and Biotechnology, 100*, 8651–8665.

Wrzaczek, M., Brosché, M., & Kangasjärvi, J. (2013). ROS signaling loops—Production, perception, regulation. *Current Opinion in Plant Biology, 16*(5), 575–582.

Yu, Y., Gui, Y., Li, Z., Jiang, C., Guo, J., & Niu, D. (2022). Induced systemic resistance for improving plant immunity by beneficial microbes. *Plants, 11*(3), 386.

Zámocký, M., Gasselhuber, B., Furtmüller, P. G., & Obinger, C. (2014). Turning points in the evolution of peroxidase–catalase superfamily: Molecular phylogeny of hybrid heme peroxidases. *Cellular and Molecular Life Sciences, 71*(23), 4681–4696.

Zhang, F., Anasontzis, G. E., Labourel, A., Champion, C., Haon, M., Kemppainen, M., & Martin, F. (2018). The ectomycorrhizal basidiomycete *Laccaria bicolor* releases a secreted β-1,4 endoglucanase that plays a key role in symbiosis development. *New Phytologist, 220*(4), 1309–1321.

Zhang, X., Li, X., Wu, C., Ye, L., Kang, Z., & Zhang, X. (2019). Exogenous nitric oxide and phosphorus stress affect the mycorrhization, plant growth, and associated microbes of *Carya illinoinensis* seedlings colonized by *Tuber indicum*. *Frontiers in Microbiology, 10*, 2634.

Zhang, Y., Sa, G., Zhang, Y., Zhu, Z., Deng, S., Sun, J., & Chen, S. (2017). *Paxillus involutus*-facilitated Cd^{2+} influx through plasma membrane Ca^{2+}-permeable channels is stimulated by H_2O_2 and H^+-ATPase in ectomycorrhizal *Populus* × *canescens* under cadmium stress. *Frontiers in Plant Science, 7*, 1975.

Zou, Y.-N., Wu, Q.-S., & Kuča, K. (2021). Unravelling the role of arbuscular mycorrhizal fungi in mitigating the oxidative burst of plants under drought stress. *Plant Biology, 23*(S1), 50–57.

CHAPTER NINE

Systemic acquired acclimation, network acquired acclimation and cellular light memory in plants – Molecular, biochemical, and physiological mechanisms

Stanisław Karpiński[a,*], and Magdalena Szechyńska-Hebda[a,b,*]
[a]Warsaw University of Life Sciences – SGGW, Poland
[b]Plant Breeding and Acclimatization Institute – National Research Institute, Radzików, Błonie, Poland
*Corresponding authors. e-mail address: stanislaw_karpinski@sggw.edu.pl; m.szechynska@ihar.edu.pl

Contents

Abstract

Global warming and climate change lead to increased evapotranspiration and desertification, thus reducing the availability of suitable soils for agricultural cultivation. Furthermore, high light, drought, heat shock, UV stress, and disease have become the most relevant limiting factors to crop productivity. Plant responses to these adverse conditions promote energy dissipation as heat, fluorescence, and stomatal closure, that lead to inhibition of photosynthesis, induction of photorespiratory, burst of reactive oxygen species (ROS), changes in hormonal and electrical signalling (ES). In turn, these processes are altogether a part of conditionally regulated cellular light memory (CLM), acclimation, and foliar cell death (CD), thus leading to altered growth and yield. Recently, we showed that ES and ROS induce a new acclimation phenomenon called 'network acquired acclimation' (NAA), which is necessary for the induction of systemic acquired acclimation (SAA) and systemic acquired resistance (SAR) within a plant community. In this book chapter, we describe the molecular, physiological, and biochemical bases of these mechanisms.

Advances in Botanical Research, Volume 105
ISSN 0065-2296, https://doi.org/10.1016/bs.abr.2022.11.005

1. Introduction

Global warming is associated with the coexistence of a number of abiotic and biotic stress factors and therefore can greatly affect agricultural productivity. The occurrence of abiotic stresses can alter crop plant susceptibility to pathogens, insects, and reduce competitive ability with weeds. On the other hand, biotic stress can modify the tolerance of plants to temperature and light extremes (cold – heat, low light – high light), unfavourable water conditions (drought – flooding), suboptimal access to minerals (low content, salinity, heavy metals in excess), elevated CO_2 (eCO_2), O_2 (low-high, normoxia, hypoxia), etc. Climate change is making it worse because rapid environmental fluctuations coincide with the trends in modern breeding i.e., favouring individual traits related to quality and yield, not common crop competitiveness in fluctuating field conditions. It is known that old varieties are characterized by greater stability under changing conditions during the growing season and, despite generally lower yield, they win the competition with modern high-yielding varieties during stress. Among others, this results from the fact that the acclimation mechanisms and adaptive responses to single biotic or abiotic stress are extensively characterized, while relatively little information is available on the combined biotic and abiotic stress response. Since single stress occurs rarely in natural environment, thus key processes and signalling pathways can only be understood correctly, provided the integrated response of plants to multiple stresses will be studied under natural field conditions. Integrated physiological and genetic factors, with particular roles for ROS homeostasis and the antioxidative enzyme systems are involved in both, the functioning of plants under varying environmental conditions and achieving successive stages of plant ontogeny. Therefore, genetic improvement of crop yield will need to be integrative, and multiple stress and signalling pathways in the plant should be considered, at least by: (1) defining mechanisms of excess excitation energy (EEE), elements of the sensory systems, and signalling pathways, at a level of one plant, which lead to local and systemic short- and long-term acclimation to abiotic stress (SAA, systemic acquired acclimation) and resistance to pathogens (SAR, systemic acquired resistance), or at a level of the whole plant community (NAA, network acquired acclimation); (2) complex recognition of the events and scenarios leading to SAA, SAR, NAA; and (3) formulation of new hypotheses concerning the integration of retrograde signalling involved in such processes (Fryer et al., 2003; Karpiński et al., 1999;

Mittler, 2017; Mühlenbock et al., 2008; Szechyńska-Hebda et al., 2010, 2017, 2022a).

2. On the most important matters in the life of plants in a 4-dimensional view

The availability of light can be considered as a factor representing an uncompromising "to be or not to be" for autotrophic photosynthetic organisms, including plants. Energy captured from sunlight during photosynthesis is the main source of energy for higher plants, and therefore, plants by definition need light energy for metabolism and growth. The core developmental steps during the lifecycle of the plant (germination, photomorphogenesis, flowering, senescence) are determined by light conditions (and a derivative of light, namely temperature). Growth without light, at least temporary, is possible, e.g., of etiolated shoots or cell culture in vitro, and even of complete ontogeny as in the case of plant albino *Monotropa uniflora* (Wallace, 2012), but these plants need to use alternative energy sources.

Natural light conditions in the field change very dynamically. Quality of light in a canopy, seasonal day length, or temporal variations in the amount of radiation suitable for photosynthesis; all affect the physiology and morphology of the plant. Other abiotic and biotic factors also alter the final light impact and light-dependent plant responses. During the vegetation period, plants are exposed to a combination of stressful conditions, e.g., high light and cold or drought (winter and early spring, respectively), high light, heat, and drought (summer), low light and flooding (autumn), or abiotic stresses combined with pathogen infection (all seasons). Such highly variable conditions have two functional consequences for plants. First, the demand for light energy depends on the current physiological state of the plant. As an example, EEE, i.e. an amount of the absorbed light energy that cannot be used for CO_2 fixation, results from high light (800–2000 μmol m^2 s^{-1}) in moderate temperature but also from lower light intensity (approx. 150 μmol m^2 s^{-1}) during any environmental stress that impacts the rate of photosynthesis, such as cold, drought, mechanical damage or pathogen infection (Bailey-Serres & Voesenek, 2008; Chaves & Oliveira, 2004; Hirel et al., 2007; Nakashima & Yamaguchi-Shinozaki, 2006; Szechyńska-Hebda et al., 2016; Wang et al., 2003). Furthermore, EEE is induced in plants acclimated to low light after exposure to 500 μmol photons m^2 s^{-1}, while the plants acclimated to high light maintain efficient PSII functioning and photosynthetic electron transport even over 500 μmol photons m^2 s^{-1} (Tian et al., 2017). Plants with dysfunctional or deregulated PsbS protein could tolerate a light intensity of only 350 mmol μmol photons m^2 s^{-1}, while plants with overexpressed levels of PsbS tolerate even light of 600 μmol photons m^2 s^{-1} (Ruban & Belgio, 2014). Second, the unbalance between the available light energy and its current use in chloroplasts is a much more common state than the optimum of the energy during photochemical reactions. In the context of plant metabolism, optimum occurs within a very narrow range of values for each environmental factor. Thus, the stress induced by light in the background of other factors is a natural physiological state that determines plant metabolism at the level of molecules, cells, and whole organism.

Plants have evolved specific mechanisms allowing for flexible functioning in both, the low supply of light energy and in conditions of excess energy. On the other hand, the plant (cell/chloroplast) has a defined and limited pool of responses. Therefore, it is not surprising that all: light-triggered signalling pathways, acclimation to abiotic stresses, and defence reactions, can be induced in a similar way (e.g., by programmed cell death responses, PCD) leading to cross-tolerance. General priming signals (e.g., light changes) may prepare the plant to respond in a more selective way to subsequent, stimulus-specific signals (e.g., pathogen infection). Furthermore, plants co-evolved with other biological systems (e.g., other plants, animals, pathogens) in changing light conditions; thus, common signals can be effective and universal ways to transmit information about changing light and light-dependent responses within organism communities (at least within plant communities via NAA). On the other hand, specificity of signals can be encoded in the spatial and temporal dynamics of the signal or its strength (regulated by gene expression or synthesis level) not only its type. In all cases, a number of specific mechanisms and signalling systems have been proposed to operate in combination. Among others, overlap mechanisms were found between EEE-induced SAA to abiotic stress and SAR to pathogens/biotic stress, and further integrated responses at the level of NAA. All of above, i.e., SAA, SAR, and NAA are regulated by the genetic mechanisms that comprise a number of the same pathways: (1) overproduction of ROS and antioxidative systems (Baker, 2008; Fichman & Mittler, 2020, 2021; Fichman et al., 2021; Górecka et al., 2020; Mühlenbock et al., 2008; Szechyńska-Hebda et al., 2010, 2017, 2022a) resulted from changes in the PSII and the photosynthetic electron transport chain (Arias et al., 2003; Frenkel et al.,

2009; Karpiński et al., 1999); (2) induction of non-photochemical quenching (NPQ), fluorescence, and thermal dissipation with feedback in ROS homeostasis (Ciszak et al., 2015; Górecka et al., 2020; Szechyńska-Hebda et al., 2010, 2017); (3) the activation of signalling pathways dependent on salicylic acid, jasmonic acid, ethylene, glutathione (Ball et al., 2004; Clarke et al., 2004; Kunkel & Brooks, 2002; Sharma et al., 1996), protein kinase cascades, calcium gradients (Atkinson & Urwin, 2012; Witoń et al., 2016, 2021), and the photoelectrochemical signalling (Szechyńska-Hebda et al., 2010, 2017, 2022a); (4) regulation of the same Quantitative Trait Loci, which are associated with the efficiency of photosynthesis during acclimation, resistance to pathogens, and biomass accumulation (Szechyńska-Hebda et al., 2011, 2015), light- and photosynthesis-dependent genes (Chang et al., 2004; Szechyńska-Hebda et al., 2015), and transcription factors functioning as activators or repressors (Barah et al., 2016); (5) induction of PCD and hypersensitivity reactions (Luna et al., 2012; Mateo et al., 2004; Mühlenbock et al., 2008; Szechyńska-Hebda et al., 2010).

In this view, the light acclimation processes are cascades of reactions regulated on a time-dependent scale. The fastest process occurring in nanoseconds is the sensing of stress factors by plant chloroplasts and, more specifically, by the oxygen-evolving reaction centres of photosystem II (PSII) (Styring et al., 2012). The absorption of light energy and then the quantum events that depend on the status and size of the antenna in PSII can trigger secondary redox reactions in photosystems, e.g., electrical charge separation, chloroplast lumen acidification, and activation of the xanthophyll cycle by means of NPQ, redox reactions between the photosynthetic electron carriers (electron transport), and formation of ROS in nano- to milliseconds period. Subsequently, this influences the redox status of glutathione, ascorbate, hormones, sugars, and ion signal transduction pathways on the microsecond to second time scale. Within seconds – minutes, some of these signals trigger stomatal closure under MITOGEN-ACTIVATED PROTEIN KINASE 4 (MPK4) regulation, photorespiration, ROS, and ethylene. In the next seconds-minutes, LESION SIMULATING DISEASE 1 (LSD1), by controlling ENHANCED DISEASE SUSCEPTIBILITY 1 (EDS1)/PHYTOALEXIN DEFICIENT 4 (PAD4)/ETHYLENE INSENSITIVE2 (EIN2), and antioxidative enzymes (SODs, CATs, APXs), limits ROS and ethylene production, and thus regulates genetically dependent CLM and PCD within minutes – hours – days (Karpinski et al., 2013; Szechyńska-Hebda et al., 2022a, 2022b; Witoń et al., 2021; Wituszyńska et al., 2013). Time-related cascades (minutes – hours – days) optimize plant photosynthesis, transpiration, trigger CLM signals to develop SAA, SAR, and NAA in the three-dimensional space of different cells, tissues, organs, and even different plants in the community. Recently published data suggests that plants, with the help of above described mechanisms, perform biological processing (computation) in order to achieve successful competitiveness and fitness.

The above data demonstrate the evident need to revise the traditional approaches at least in the case of: (1) EEE is not only a negative factor damaging cell structures and causing the death of tissues and whole plant; (2) considering a time scale, EEE should be considered as a holistic factor inducing acclimation and resistance mechanisms to survive under adverse environmental conditions, and under following stresses likely to occur in the future; (3) bearing in mind, that plant is a multicellular organism with a unique 3D structure, commonly living in complex ecosystems, EEE-dependent events should be explained as a hub interconnecting different signalling networks to compute and control growth and development processes as well as acclimation and defence responses. (4) EEE-dependent events are not only related to the chemical (ROS, hormons) and electrical retrograde signalling but also to the physical retrograde signalling of heat generated via NPQ and Eq and osmotic stress.

Together, this indicates that plants function as three-dimensional systems in time; perception of external factors and response to the environment by modulation of genetically encoded traits are four-dimensional (4D, Fig. 1). Playing with the 4th dimension (3D space and time) has generally not been considered in plant biology and plant molecular experiments. Therefore, it is important to rethink, how to obtain the correct conclusions in the 4D system of cells and tissue behaviour on intra- and inter-organismal levels. Understanding the kinetics of 3D and 4D 'Omics' can clarify many of the results in signalling sciences. The basis is to take into account that the plant consists of three interlinked spatial networks. A continuum of protoplast membranes connecting with plasmodesmata (first network) is set in an apoplast network built by tightly connecting cell walls, within which water solutions can diffuse freely (second network) (Kwiatkowska & Burian, 2022) and surface of a plant connecting various organs of the same plant or neighbouring plants with use of surface electrical and gaseous signals (Szechyńska-Hebda et al., 2022a). Plants integrate a communication of organs with different physiological functions (shoot system consisting of stem, leaves, flowers, fruits, as well as root

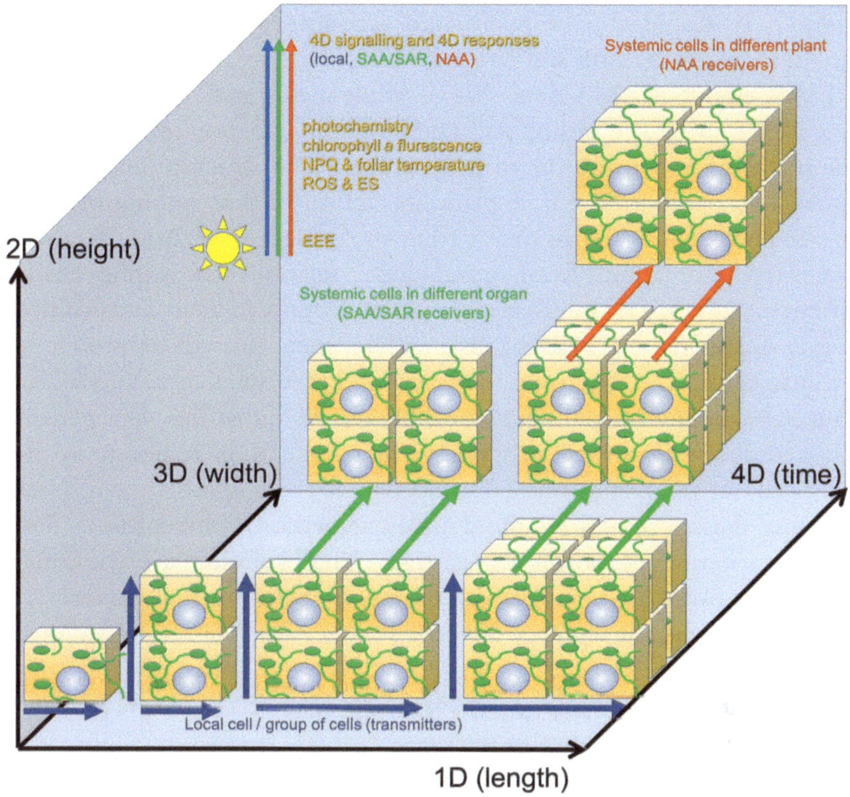

Fig. 1 Simplified model of 4D signalling and 4D responses. The energy of light is the most important factor for the functioning of plants. Light energy is used for photochemistry, but the excess excitation energy (EEE) is redirected to chlorophyll *a* fluorescence, non-photochemical quenching (NPQ), heat dissipation, reactive oxygen species (ROS), and electrical signalling (ES). They are the signals, which determine short- and long-term plant responses, i.e., local response (blue arrows) in a cell or group of cells; systemic acquired acclimation (SAA) and systemic acquired resistance (SAR) within a plant (green arrows); and network acquired acclimation (NAA) between different plants (red arrows). The responses are recognized in a four-dimensional context ('4D signalling and 4D responses'): 3D space dimensions, by any combination of three chosen from width/breadth, height/depth, length; and fourth dimension as the time. Local signal transduction is enabled by membrane continuity, i.e., between different chloroplasts through stromules (green circles and lines inside cells) and between different cells through plasma membrane extension in plasmodesmata (Bobik & Burch-Smith, 2015). SAA and SAR signal transduction between different organs is enabled (among others) by membrane connection (plasmodesmata) of bundle sheath cells (Szechyńska-Hebda et al., 2010); while NAA signal transduction requires plant membrane connections through a closing circuit (Szechyńska-Hebda et al., 2022a).

systems). It should also be taken into account, that cells have encoded evolutionary 3D division and differentiation (they expand) (Mina & Bissell, 2017; Whitewoods & Coen, 2017), while being stiff enough to resist turgor exerted by the protoplast. These opposing factors result in specific 3D geometry and shape developed in time. The structure and shape of tissues, organs, and the whole plant are critical in determining biological function across all scales of order; from molecule to cell, from tissues and organs to the organism, from an individual organism that is influenced by external factors to an inter-organismal interaction in field and landscape levels. Achieving an optimal balance between growth, reproduction, acclimatization, and defence under stress and then suitable modification of plant 3D structure (shape, size) is possible provided that the plant perceives and transduces the signals properly in time, and then responds to them holistically in space and time (4D, Fig. 1).

All of these factors drive the dynamic evolution of mechanisms determining various plant morphological traits. The evolution will continue to progress, as plants have to adapt to changing environmental conditions in the field or they are under a directed selection to follow innovations in breeding.

From the above statements, an extremely elegant and simple scheme of 'the most important matters in the life of plants' has emerged: light energy as the most important factor of plant functioning is simultaneously a central determinant of short- and long-term reactions; these in turn are the most important triggers of a wide range of adaptive mechanisms, ensuring the achievement of the primary purpose of plant life, that is, Darwinian competitiveness and evolutionary success in the 4D biological world.

3. On the principles of photoinhibition and photoprotection

The initial photophysical and photochemical processes during photosynthesis are provided by photosystem II (PSII), a supramolecular complex that contains numerous protein subunits and various cofactors, embedded within the thylakoid membranes. The supercomplex forms a homodimer with two-fold symmetry. Each monomer contains a core complex composed of the α- and β-subunits of cytochrome b559 and four large intrinsic subunits (PSII reaction centre core: D1, D2, CP43, and CP47). Twelve intrinsic subunits of low molecular mass (PsbE, PsbF,

PsbH, PsbI, PsbJ, PsbK, PsbL, PsbM, PsbTc, PsbW, PsbX, and PsbZ) surround the core of PSII. The additional four extrinsic subunits are attached to the luminal surface (PsbO, PsbP, PsbQ, and PsbTn) (Umena et al., 2011; Wei et al., 2016). In the outer region of the core, Light-Harvesting Complex II (LHCII) is located. LHCII complexes consist of the major LHCII trimers (Lhcb1, Lhcb2, Lhcb3) and minor LHCII monomers (CP29, CP26 and CP24); and are further associated with a large number of cofactors, including chlorophylls, carotene, xanthophylls, haem, Mn_4CaO_5 cluster, plastoquinone, and lipid molecules (Wei et al., 2016). A functional feature of such structure is the ability to absorb light energy by a photosynthetic antenna (i.e., chlorophylls, carotenoids), transfer it to PSII reaction centres, and induce charge separation in the chlorophyll pair P680 (Sacharz et al., 2017). It induces the production of $P680^+$ and the donation of the released electron to the pool of plastoquinone (PQ) (Lima-Melo et al., 2021; Miyake, 2010). $P680^+$ then oxidizes water and evolves O_2. The energy of four photons of light is used by the oxygen-evolving complex (OEC) to drive the splitting of two water molecules and, in turn, provide four electrons to the electron transfer chain.

The light-harvesting systems exhibit great efficiency: large part of the absorbed photons are utilized during charge separation (Saccon et al., 2020). On the other hand, with increasing light intensity, photosynthetic reaction centres are rapidly saturated, reducing the fraction of energy used in photosynthesis (Ruban & Belgio, 2014). The excess energy is not used for photochemical reactions, but it has to be dissipated as heat and fluorescence and leads to the excessive electric charge separation and in consequence to the formation of reactive oxygen compounds. Singlet oxygen ($^1O_2$) is formed by excitation energy transfer from triplet chlorophylls to O_2. On the PSII electron donor side, two-electron oxidation of H_2O leads to the generation of hydrogen peroxide (H_2O_2), further oxidized to the superoxide anion radical ($O_2^{\bullet-}$), or reduced to the hydroxyl radical ($HO^{\bullet}$). On the PSII electron acceptor side, a one-electron reduction of O_2 generates $O_2^{\bullet-}$, which initiates the formation of H_2O_2 and $HO^{\bullet}$ (Li & Kim, 2022). EEE-induced ROS formation causes photooxidative damage to the PSII reaction centre within seconds or minutes of excess light exposure. The primary target is the D1 protein; the specific amino acid residues of the D1 (and D2) proteins are oxidized by $HO^{\bullet}$ and $O_2^{\bullet-}$ during EEE (Li & Kim, 2022). Since D1 is essential for Mn-cluster binding and the splitting of water molecules by the oxygen-evolving complex (OEC) (Ware et al., 2015), then its damage reduces the PSII yield. However such damage of the PSII core proteins is evolutionary advantage since it is protecting the chloroplasts and plants cells from electrons overloud, generation of ROS in excess and rapid inactivation of photosystem I (PSI). Inactivation of PSI and its reaction centres (P700) can be induced to a similar extent to that observed in the case of PSII and is dangerous due to the much slower recovery rate of PSI (Huang et al., 2018). The proposed mechanism of PSI inactivation consists of the reduction of oxygen molecules on the acceptor side of PSI by electrons supplied from PSII and then the destruction of iron-sulphur centres by ROS (Roach et al., 2020). Thus, the electron flow from PSII to PSI seems to be more essential than the redox state of the PSI. A wide range of molecular responses to high light, manifested by inefficiency in PSII and PSI functioning, reduced capacity of the photosynthetic linear electron flow, decreased quantum yield of CO_2 assimilation, is known as photoinhibition, which finally, under extreme conditions, leads to death of cells, tissues, or whole organism.

Plants have developed diverse short-term (regulation) and long-term (acclimation) photoprotection mechanisms that control the input of light energy into reaction centres to prevent photoinhibition. The short-term mechanisms initiated within the thylakoid membrane can be divided into photochemical (acting as an electron sink) or non-photochemical (occur prior to photolysis in PSII) reactions (Murchie & Ruban, 2020). The photochemical mechanisms consist of the main electron sink reactions, i.e., photosynthetic carbon reduction (the Calvin-Benson cycle) and photorespiratory carbon oxidation cycles (Chen et al., 2022; Lima-Melo et al., 2021; Miyake, 2010), and alternative electron flows. Two cyclic electron flow (CEF) systems around PSI increase the electron transfer from PSI to plastoquinone i.e., gradient regulation 5 (PGR5)/PRG5-like photosynthetic phenotype 1 (PGRL1)-mediated pathway and the NADH dehydrogenase-like complex-dependent pathway (Ma, 2021). They generate a ΔpH across the thylakoid membrane and produce ATP without the accumulation of NADPH in chloroplasts. CEFs optimize photosynthesis and plant growth under repeated light fluctuations (Lima-Melo et al., 2021; Miyake, 2010). CEF linked to qE can promote the repair of photodamaged PSII, whereas CEF independent on qE can suppress photodamage to PSII (Takahashi & Badger, 2011). An alternative electron flow also consists of the water–water cycle (the Mehler–ascorbate peroxidase pathway), an electron flow from water through PSII and PSI to molecular oxygen. The superoxide anion is disproportionated by superoxide dismutase (Cu/ZnSOD) to H_2O_2, and H_2O_2 is converted by ascorbate peroxidase (thylakoid-APX) back into the water. It shunts electrons through the photosynthetic apparatus and maintains the pH gradient essential for the function of the zeaxanthin cycle even when CO_2 fixation is inhibited (Chen et al., 2022; Ma, 2021).

Non-photochemical quenching of chlorophyll a fluorescence (NPQ) is the process that arises in the photosynthetic membrane and is not photochemical in origin. NPQ promptly transforms the harmful excess energy into heat to protect PSII from the absorption of excess light (Ruban & Belgio, 2014). The dominant kinetic component of NPQ is energy-dependent quenching (qE), which has been proposed to originate in the PSII core complex within 20–60 s (Dall'Osto et al., 2017; Walter & Kromdijk, 2022). qE depends on the acidification of the thylakoid lumen. Δ pH across the thylakoid membrane can be sensed by the photosystem II subunit S protein (PsbS) and a lumen-localized violaxanthin de-epoxidase (VDE) enzyme, which converts violaxanthin to zeaxanthin. The second component of NPQ, zeaxanthin-dependent quenching (qZ) (Walter & Kromdijk, 2022), turns on and relaxes in tens of minutes, exclusively in PSII (Tian et al., 2017). qZ requires the formation of zeaxanthin, but is independent of lumen pH once zeaxanthin is present. It was proposed that the formation of Chl-xanthophyll excited states serve as traps for Chl^*, and then charge transfer events occur in a Chl a–Zea heterodimer, followed by charge recombination to the ground state. The interaction between Chl *a* and Zea could be promoted by a conformational change from the interaction of protonated PsbS with CP29 or between PsbS and LHCII, forming a Zea–PsbS/LHCII complex at the interface. $^1Chl^*$ quenching can also occur by the formation of a transient Chl^--Lut^+ state (Dall'Osto et al., 2017; Walter & Kromdijk, 2022). The third component, state transitions (qT), displays kinetics similar to those of qZ, but functions to balance the energy absorption between PSI and PSII. The mechanism involves reversible phosphorylation and uncoupling of LHCII from PSII, which migrate to the PSI. The balance excitation energy of PSII and PSI requires the state transition kinase i.e., serine/threonine-protein kinase (STN7). qT is inhibited in high light via stromal redox regulation. The fourth NPQ component relaxes more slowly than those described above, and is referred to as photoinhibition quenching (qI). Although qI is the result of oxidative photoinactivation of PSII reaction centres (Ware et al., 2015; Yu et al., 2021), the amount of NPQ due to qI very often cannot be explained by the amount of reaction centre damage. Therefore, it was indicated that other slowly relaxing photoprotective mechanisms contribute to qI. Its induction is light intensity dependent but does not require a trans-thylakoid pH gradient, zeaxanthin formation, or LHC phosphorylation, and it appears to be unrelated to the PSII damage-and-repair cycle. The formation of this NPQ component is prevented by the Suppressor of Quenching 1 (SOQ1), a large protein with multiple functional domains found in the thylakoid membranes. Plants without SOQ1 have a higher NPQ that is very slow to relax. This additional NPQ in *soq1* is the result of a decrease in the chlorophyll excited-state lifetime and not pigment bleaching. Under full sunlight, all NPQ components can convert up to 80% of absorbed photons into heat (Yu et al., 2021). Although plants induce the quenching mechanism rapidly under excess light stress, recovery of PSII antennae from the quenched to the unquenched state can take up to half an hour in low light conditions (e.g., when clouds shade the sunlight). Therefore, the photoprotection mechanisms themselves do not ensure optimal plant function under fluctuating light conditions. More important is how quickly plants can switch between photoinhibition and photoprotection mechanisms. It was shown that the recessive *Arabidopsis* mutant *npq4–1* has lost the ability to optimize the fluorescence decay time up to 2 h after excess light episode. Therefore, functional PsbS is also required for optimization of absorbed photon fate and optimal light acclimation responses after excess light stress. Similarly, slow relaxation of NPQ is safe for plants; however, following a rapid drop in the intensity of light, there is a delay in the decline of NPQ. An overprotective period, in which NPQ is higher than optimal, transiently depressed light harvesting, photosynthetic quantum yield of CO_2 fixation, and the overall productivity of the plant (Chen et al., 2022; Ruban & Belgio, 2014; Walter & Kromdijk, 2022). Using genetic engineering techniques, it was shown that enhancing the transcription of the genes that encode the enzyme violaxanthin de-epoxidase (VDE), the enzyme zeaxanthin epoxidase (ZEP), and the PsbS protein is needed to induce NPQ faster in high light, and relax NPQ faster, when the plant is in shade (Walter & Kromdijk, 2022).

More recent studies showed that photochemistry, chlorophyll *a* fluorescence, and heat losses (leaf temperature increase) are in direct competition with each other; consequently, increase in one process results reductions of the other ones under constant light condition. CLM, SAA, and NAA are regulated by the rate of heat dissipation and fluorescence. However, it is not known, how much calories of heat are radiated back from leaves (chloroplasts) to the environment during photosynthesis and during its inhibition due to EEE stress. Higher EEE dissipation and heat emission from plants results in lower photosynthesis efficiency (CO_2 assimilation).

Under excess light, the overproduction of ROS in chloroplasts must be balanced by their scavenging. Therefore, effective photoprotection also depends on the rapid activation of multiple antioxidant enzymes (i.e., superoxide dismutase, ascorbate peroxidase and peroxiredoxin, see below) and non-enzymatic antioxidants (water-soluble ascorbate and membrane-bound α-tocopherol, carotenoids, such as zeaxanthin, neoxanthin, and lutein). Antioxidants may play an integral role not only in controlling ROS but also in transmitting oxidative signals. EEE generates elevated ROS that allows induction of PCD/HR and thus SAA and SAR, therefore balance between ROS production and scavenging must be strictly controlled (Shigeoka & Maruta, 2014). The type of molecule, its concentration, and cellular localization, or a combination of all of these, need to be regulated to direct specific signalling pathways. *APX1* and *APX2* are key components of systemic signalling and robust marker of SAA and SAR, however, many other enzymes also regulate the integration of systemic responses, PCD, retrograde signalling. A specific expression pattern of the genes encoding antioxidative enzymes in *Arabidopsis thaliana* was shown during EEE induced locally. The genes encoding chloroplast enzymes *FSD2, FSD3, sAPX, GPX6, GPX7,* and enzymes localized in various cellular compartments i.e., *APX1, APX2, APX3, CAT1, CAT1, GR1, DHAR1, MDAR1, MDAR4* were specifically upregulated during and after EEE. Surprisingly, a large number of genes were down-regulated in leaves undergoing SAA, mainly *GPX7* and *DHAR1,* and to a lesser extent *PrxB, PRXIIF,* and *CAT1* were induced. Similarly, FSD2, FSD3, but not FSD1, appeared to be essential for photoprotection during EEE (Szechyńska-Hebda & Karpiński, 2013; Szechyńska-Hebda et al., 2013). Indeed, *Arabidopsis fsd2* and *fsd3*, but not the *fsd1* mutants, showed slow growth, pale green colour, abnormal chloroplasts, and in the case of double mutant *fsd2 fsd3,* even an albino phenotype (Kuo et al., 2013; Szechyńska-Hebda & Karpiński, 2013). H_2O_2 reduction during EEE can proceed via two pathways. The ascorbate-dependent water–water cycle implicates thylakoid-bound and stromal ascorbate peroxidase (tAPX, sAPX) coupled to monodehydroascorbate reductase (MDHAR) or dehydroascorbate reductase (DHAR) and glutathione reductase (GR) (Asada, 1999; Szechyńska-Hebda & Karpiński, 2013). Outside chloroplasts, MDAR1 and MDAR4 are involved in the regeneration of reduced ascorbate. The genes encoding some of the enzyme isoforms were upregulated after EEE, with the exception of *tAPX* (thylakoid APX bound in the vicinity of PSI, together with PSI-associated SOD) (Szechyńska-Hebda et al., 2013). Consistent with these results, chloroplast bleaching was shown under oxidative stress for the *sapx* mutant, while *tapx* was only partially impaired (Kangasjärvi et al., 2008; Szechyńska-Hebda & Karpiński, 2013). It seems that sAPX is important during long-term responses; *sAPX* transcripts are unresponsive during the first 30 min following transfer to high light and then rapidly increased. At the same time, the *tAPX* transcript does not change and after 24 h a significant drop in the amount of tAPX protein was observed (Oelze et al., 2012; Szechyńska-Hebda & Karpiński, 2013). Furthermore, the ascorbate independent pathway uses peroxiredoxins (PRX) and thiol-based peroxide detoxification mechanisms; however, the PRX-provided catalytic mechanism is activated primarily in leaves that undergo SAA, not in leaves directly exposed to EEE (Szechyńska-Hebda et al., 2013). Additionally, GPXs, which use both, GSH and TRX, as reducing substrates (Herbette et al., 2002; Szechyńska-Hebda & Karpiński, 2013), provide an alternative pathway to the water–water cycle during EEE and SAA. In the GPX system, the regeneration of reduced TRX is linked to the PET chain through ferredoxin-TRX reductases or NADPH-dependent TRX reductases. Constitutively reduced expression of GPX7 led to an increase in leaf levels of ascorbic acid, glutathione, and salicylic acid (Chang et al., 2009).

Although short-term photoprotective mechanisms can help reduce the excitation pressure on photosynthetic systems, their damages are unavoidable under oxidative stress. Therefore, plants have evolved long-term responses to maintain photosynthetic performance. Among others, this involves multi-step repair of photodamaged PSII proteins by replacement of them with newly synthesized molecules. The major steps in higher plants include: synthesis and reassembly of new D1, re-incorporation of CP43, re-attachment of OEC, migration of the PSII core monomer back to grana stacks, and dimerization into PSII core dimers and reformation of PSII-LHCII supercomplexes (Lu, 2016). Repair results from a light-dependent regulation of gene expression occurring at the transcriptional, translational, and post-translational levels. Moderate light intensities induce this process, but under conditions of excess light, the step of protein translation is depressed. The EEE-dependent changes in the ATP:ADP ratio and stromal redox potential, reduction of the Calvin cycle activity, and ROS overproduction, were shown to cause inhibition of PSII repair, and thus accelerate photoinhibition (Takahashi & Badger, 2011). Together with the repair, the rearrangement of proteins in thylakoid membranes occurs during and after the excess light exposure of the plant. Increasing light intensity leads to a decrease in LHCII abundance, an increase in the amount of cytochrome b6f complexes. As a consequence, acclimation to excess light lowers the light-harvesting capacity and increases the electron-transport capacity. Apart from molecular photoprotection pathways, some mechanistic processes can minimize the absorption of light in excess, e.g., avoidance associated with the movement of leaves and chloroplasts, accumulation of UV-absorbing compounds in the leaf cuticle and epidermis (phenolics).

4. Survive and become stronger: EEE induces acclimation (SAA) and resistance (SAR)

There is a view that photoinhibition is a process that plants should be protected against. However, photoinhibition may be a means by which plants sustain photoprotection (Adams et al., 2008) or develop acclimation and resistance to stresses that can occur in the future (Roden & Ingle, 2009; Szechyńska-Hebda et al., 2010, 2015). Only after the loss of a great number of PSII complexes photosynthetic capacity is limited (Ruban & Belgio, 2014). In most cases, the effective photoprotection mechanisms and PSII repair prevent a substantial loss of functional PSII complexes, and the extent of PSII inactivation usually does not exceed 50% (moderate photoinhibition) under natural field conditions (Kornyeyev et al., 2006). In some cases, those plants experiencing the higher level of photoinhibition were found to exhibit improved growth during the subsequent recovery period (Adams et al., 2008). Recent reports have shown direct effects of excess light-induced photoinhibition on both, the defence response in the host plant and the virulence of the pathogen (Roden & Ingle, 2009). Even if EEE is followed by PCD (hypersensitive responses), it can be a part of the acclimation response that occurs in the leaves exposed to EEE directly and in unexposed leaves (Szechyńska-Hebda et al., 2010). This pre-acclimation of the shaded leaves during or after local EEE, known as SAA to EL, also induces SAR against pathogens. EEE induced by white and 650–680 nm wavelength light (red light), but not 450 nm (blue light) and 700 nm (far-red light) wavelength light can regulate many of the genes involved simultaneously in acclimation and defence responses (Mühlenbock et al., 2008; Szechyńska-Hebda et al., 2010) (Fig. 2).

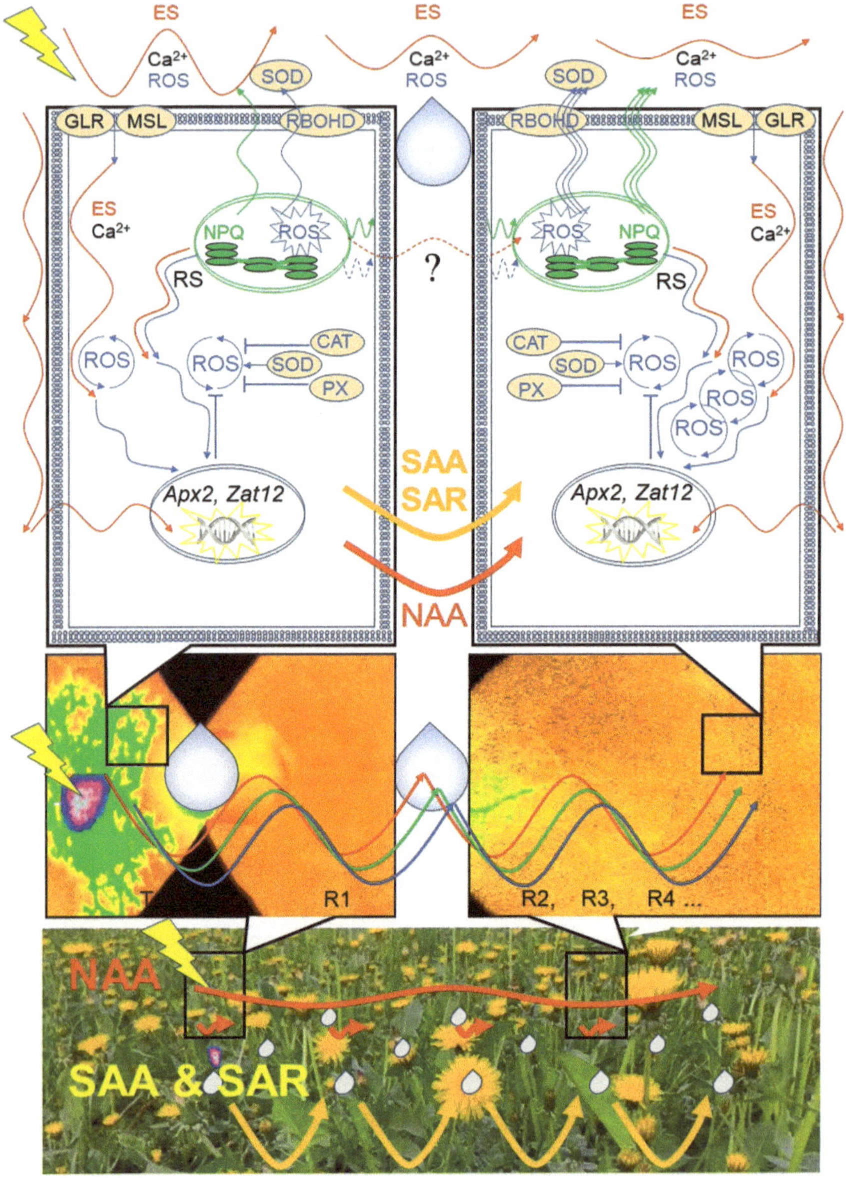

(caption on next page)

Since, the amount of absorbed energy, physiological status of the cells, and the spatial and temporal spread of damage during photoinhibition are interdependent, the scenario of plant local and systemic responses can be as follows: (1) *de novo* synthesis of cell molecules that are damaged by ROS during local EEE stress leads to local cell defence and acclimation, prevents PCD and does not trigger SAA and SAR; (2) local PCD of a small group of cells (microlesions) induced by EEE triggers specific retrograde signalling and leads to holistic plant responses, SAA, and SAR; (3) failure of the acclimation and defence responses during EEE leads to CD (necrosis) spread, enhancing host plant susceptibility to abiotic and biotic stress, and whole plant death. The scenario does not result from the mechanism of cell death itself, but rather the regulation of cell death exploits the differences between the availabilities of cell death pathways. The numerous signals are constantly integrated by the cell to produce a local fate decision: to activate the cell death program (scenario 2) or not (scenario 1); and once activation is completed, cells have to face the problem of how to adjust the sensitivity of distinct cells to prevent the uncontrolled spread of a local lesion (scenario 3). The model of signalling pathways that regulate EEE acclimation, cell death, and immune defence (scenario no 2) initially consists of (described above) local changes in PSII, redox status of the components of the photosynthetic electron transport chain (e.g., plastoquinone and plastid terminal oxidase activity), acidification of the chloroplast lumen, modifications in NPQ and cyclic electron transport, ROS, redox status of

◀ **Fig. 2** A model of SAA, SAR, NAA (Szechyńska-Hebda et al., 2022a). The panels represent two different cells (scheme), leaves (NPQ photos made with FluorCam PSI, Czech Republic), and different dandelion plants living in a community (field photos). Aboveground electrical signalling (ES) communicates danger (excess light, wound) and induces systemic acquired acclimation (SAA) and systemic acquired resistance (SAR) in individual plants and network acquired acclimation (NAA) between injured plant and receiver plant (R, uninjured, receiving electrical signalling, ES) in wet conditions. ES drives spatio-temporal changes in energy quenching (NPQ), reactive oxygen species (ROS), Ca^{2+} fluxes, and retrograde signalling (RS), which in turn, depends on the regulation of RESPIRATORY BURST OXIDASE HOMOLOGUE D (RBOHD), superoxide dismutase (SOD), and catalase (CAT) activities, GLUTAMATE RECEPTOR-LIKE (GLR), and MECHANOSENSITIVE CHANNEL OF SMALL CONDUCTANCE-LIKE 10 (MSL10). The autopropagation of ROS can occur in the receiver cells. It regulates gene responses, e.g., the *ZING FINGER PROTEIN 12* (*ZAT12*) and *ASCORBATE PEROXIDASE 2* (*APX2*). Red lines, ES-dependent signalling; blue lines, ROS-dependent signalling; green lines, NPQ-dependent signalling; yellow arrow, wounding; red dashed line, hypothetical ES between cells of different leaves.

glutathione, ascorbate, hormones (SA, ABA, ET, IAA), sugars, and ion (Ca^{2+}) signal transduction pathways (for review see e.g., Karpinski et al., 2013; Szechyńska-Hebda & Karpiński, 2013). LSD1, EDS1, PAD4, EIN2, and MPK4 constitute a regulatory hub for all pathways (Karpinski et al., 2013; Mühlenbock et al., 2008; Szechyńska-Hebda et al., 2010; Witoń et al., 2016, 2021; Wituszyńska et al., 2013).

According to the concept of local and systemic response scenarios, the control of the local cell death borders is crucial to the outcome. Analysis of *lsd1* mutants showed that PCD could be activated autonomously and survival signals are required to restrict lesion size (Czarnocka et al., 2017; Wituszyńska et al., 2013). The light and redox status of the PQ pool induces the hypersensitive response during effector-triggered immunity. *lsd1* mutants under ambient light conditions developed localized chlorosis after DCMU treatment (oxidation of the PQ pool) and runaway cell death after DBMIB treatment (reduction of the PQ pool). This was associated with the induction of gene markers for SAR (*PR1*) and a marker for PCD (*PRXcb*), which preceded the development of cell death. Plants grown in the dark have impaired lesion formation in response to incompatible bacterial and viral pathogens (Mühlenbock et al., 2008). In the same way, plants inoculated with avirulent strains (e.g., PsmE4326, avrRpm1) display improved resistance to subsequent infection with virulent strains (e.g., PsmE4326). However, when the primary infection was performed in the dark, acquired resistance was abolished (Roden & Ingle, 2009). The induction of ethylene and H_2O_2 in response to redox changes of the PQ pool is under negative control of LSD1 and under positive control of EDS1 and PAD4. Single and double mutant analysis confirmed that LSD1 suppressed EDS1- and PAD4-dependent ROS and ethylene signalling and that EIN2 acts downstream of EDS1 and PAD4 (Muhlenbock et al., 2008) but only in low light ambient laboratory conditions (Wituszyńska et al., 2013). Signalling gradients that establish cell death control borders at sites of pathogen recognition have been demonstrated (Coll et al., 2014).

Disruptions of the key physiological pathways could predetermine the uncontrolled spread of PCD at the level of the cell wall and or cell wall modification by ROS. LSD1 was shown to interact directly with proteins involved in cell cycle and cell wall formation, as well as control the expression of some genes involved in these processes (Czarnocka et al., 2017). Cell number (cell cycle), leaf mesophyll cell size (differentiation), and cell wall properties (lignification), all of them were controlled by LSD1 (Czarnocka et al., 2017; Szechyńska-Hebda et al., 2016). The leaves of the mutant *lsd1* contained a significantly higher number of palisade and spongy mesophyll cells, and these cells were generally smaller and more compact compared to those of wild-type plants. In contrast, the *LSD1-OE* line (*35Spro:LSD1-GFP*) had significantly larger cells, and the number of palisade and spongy mesophyll cells per unit of leaf area decreased. These differences in cell number and area influenced the size of the rosettes and their dry weight, both of which were significantly reduced in the *lsd1* mutant and increased in the *LSD1-OE* plants (Czarnocka et al., 2017). Similarly, the compaction of the tissue wood structure consisting of a large number of small cells was also observed for *PAD4-RNAi* and *LSD1-RNAi* transgenic aspen lines (Szechynska-Hebda et al., 2016). Similarly, as in *Arabidopsis*, *PAD4* suppression caused deregulation of ROS metabolism, the accumulation of H_2O_2, increased activity of the manganese superoxide dismutase isoform (MnSOD) and catalase (CAT), photosynthetic efficiency. It results in increased growth rate by approximately 5–8% per season under field conditions (Ślesak et al., 2015). The changes in leaves and shoots were also positively correlated with wood formation and its physical and chemical properties. A higher rate of cell division (wider xylem ring resulted from a larger number of individual cell layers), increased thickness of the secondary cell wall (mainly due to the deposition of successive layers of cellulose, but not lignification), decreased average tracheid size and their number were found. These characteristics determined different physical and mechanical parameters of the wood of transgenic plants compared to wild-type, i.e., higher density, reduced swelling during tissue hydration, and greater flexibility of the wood. The results confirmed that *PAD4*, which is involved in defence processes and pathogenesis, is also a universal regulator of acclimation (Szechyńska-Hebda et al., 2016).

The cell wall structure is an outcome of photosynthesis-dependent and sugars-related metabolism, furthermore, the cell wall is modified during and after abiotic and biotic (oxidative) stresses, and finally, the cell wall plays a crucial role in the defence and acclimation responses. Therefore, direct physiological and molecular links between EEE/PCD and cell wall structure should exist, and the specific cell wall structure can play a crucial role during EEE-dependent induction of PCD. Hypersensitive responses require specific cell wall properties that allow cells to isolate pathogens from other tissues. On the other hand, cell wall specific pattern influences the basic plant resistance. The cell wall properties that determine defence against pathogen include: chemical cellulose interaction with other components that reinforce the cell wall and enhance resistance to pathogen's digesting enzymes; physical permeability influencing pathogen enzymes diffusion; and mechanical strength and resistance to pathogen pressure at the molecular and cellular level. Taken together, these results suggest that plants perform integrated biological processing in response to external biotic and abiotic stimuli, and that some parts of the regulation system are the pathways providing the chemical and physical state of the cell wall. In stress conditions, plants adjust their metabolisms and the molecular mechanisms that can balance energy distribution between processes: repair of cell wall damage made by pathogens, and strategies that help to avoid stresses in the future i.e., basal defence by strengthened cell wall. Information about locally induced EEE, cell unbalanced redox status and PCD (microlesions) is communicated to distant cells and organs via photoelectrophysiological signalling (PEPS), which consists of electrical waves, followed by changes in NPQ, ROS, expression of *APX 2* (Szechyńska-Hebda et al., 2017) and the calcium (Ca^{2+}) wave (Storti et al., 2018). Systemically, PEPS affects multiple physiological processes e.g., photorespiration, transpiration, changes in ATP content, photosynthesis, transcription, and translation of specific proteins, the synthesis of hormones such as ethylene, salicylic acid, jasmonic acid, and ROS autopropagation. Finally, the adjustment of these signalling components leads to the induction of SAA to abiotic stresses and SAR to pathogens in distant organs and whole plants (Karpiński & Szechyńska-Hebda, 2010; Karpinski et al., 2013; Storti et al., 2018).

According to the mechanisms described above, there should be a photosynthesis-dependent physiological and genetic crosstalk between light acclimation/light-induced resistance triggered by high light under moderate temperature and light acclimation/light-induced resistance triggered by lower light under low temperature. Indeed, in evergreen species that down-regulate photosynthesis during winter, there is a strong correlation between the capacity for photosynthesis and the pre-dawn PSII efficiency during winter and during the transition period from winter through spring. Furthermore, the reactions specific for EEE and cell death can be induced in winter triticale by both high-intensity light at a temperature close to the optimal of plant growth and light of lower intensity under conditions of lower temperature (cold). In both cases, EEE cannot be used for CO_2 assimilation, while it can be used to activate the integrated control mechanisms of acclimation, resistance, and development (Szechyńska-Hebda et al., 2007, 2010, 2013; Szechyńska-Hebda & Karpinski, 2013).

The integration of resistance mechanisms against the pathogen with the acclimation to low temperatures (in the light intensity optimal for growth), was also confirmed in the experimental system: winter triticale – fungal pathogen *Microdochium nivale*. Analysis revealed that treatment of young seedlings 28 days at low temperatures under light conditions induced resistance to infection, but this resistance was dependent on the genotype of the plant. Induction of the immune response at the physiological level (Szechyńska-Hebda et al., 2015) and genetic level (Szechyńska-Hebda et al., 2011, 2015), was observed only for certain cultivars or lines of the mapping population, despite earlier cold hardening of all genotypes. One of the factors determining effective resistance was fast germination and leaf development, and as a consequence, early achievement of physiological maturity of chloroplasts, effective photosynthesis, and a supply of energy for the production of the metabolites involved in plant defence. The mechanism of the processes also involved the adjustment of the reactions in the PSII. The genotypes capable of developing the cold-induced resistance showed a higher maximum quantum yield of PSII and a more efficient integration of the primary photochemistry of light reactions with the dark reactions (performance index). The same genotypes during hardening induced the active mechanism of protection against oxidative stress, involving at least peroxidases that eliminate excess H_2O_2. Adjustment of photosynthesis and stress acclimation has enabled fast plant growth and avoidance of developmental stages sensitive to fungal infection. The same mechanisms allowed quick regrowing of plants during the post-disease period. Traits related to photosynthesis-dependent cold acclimation and cold-induced resistance; biomass accumulation and growth; as well as antioxidative system, were integrated not only at the physiological but also at the genetic level. Certain groups of genes were regulated by photosynthesis-induced signals. Chromosomes 5B and 7R were found to contain QTLs that shared the same positions within traits related to photosynthesis effectiveness during cold hardening, cold-dependent leaves development, fresh and dry weight accumulation, cold-regulated peroxidase activity, and cold-induced resistance against the fungal pathogen *M. nivale*. The dependence of the responses between cold acclimation and cold-induced resistance as well as the role of photosynthesis and the antioxidative action of peroxidases was also confirmed by a meta-analysis of microarray experiments. 83 transcripts related to PSII had a very similar expression profile when a plant treated with cold was compared to a plant infected with fungal pathogens. Similarly, 46 transcripts related to peroxidases showed complementary modulation during cold acclimation and defence responses, with the exception that infection had influenced transcript level more effectively. It seems that both ascorbate peroxidases and glutathione peroxidases play a crucial role in these processes.

Signals under EEE induced by cold in moderate light (Szechyńska-Hebda et al., 2007, 2010, 2013; Szechyńska-Hebda & Karpinski, 2013) may be also responsible for the integration of immune and acclimation mechanisms, at the level of basal defence. Strengthening cell walls through the appropriate modification of their chemical and physical properties protects the plant holistically: under the attack of the pathogen (provide a defence barrier) and during rapid lowering of the temperature (prevent water loss and protect against tissue disruption by crystals forming at temperatures below zero degrees). The resistance to the fungal pathogen *M. nivale* depended on the potential of winter triticale to fortify the tissues of young leaves (Szechyńska-Hebda et al., 2013). The linear increase in resistance was found along with the prolongation of cold treatment from 0 to 98 days. Microscopic analysis allowed to observe a specific fungal hyphae growth depending on the degree of resistance. The strategy of the fungus enabling access to the nutrients of the resistant host was to penetrate the leaf tissue through stomata and subsequent biotrophic growth of single hyphae in intercellular spaces and the formation of single haustoria within the mesophyll cell walls. In contrast, destructive necrotrophic infection was observed in the leaves of susceptible plants, where hyphae grew directly through the epidermis, mesophyll, and even vascular tissue. Different ways of infection were determined by changes in the physical and chemical properties. The surface of resistant leaves prevented both, adhesion of *M. nivale* mycelium and direct penetration of hyphae. The leaves of resistant plants growing at low temperatures were characterized by very high hydrophobicity. This was a consequence of the formation of numerous small structures on the surface of the leaf, as well as the presence of substances, such as lipids, waxes, phenols, etc., which prevented the direct wetting of the surface. These features of the leaf adversely affected the growth of fungal hyphae, strongly depend on the presence of moisture. Furthermore, the cells walls below the epidermis (mesophyll, vascular tissue) had reduced permeability, which is important to limit the diffusion of fungal enzymes that digest cell walls of the host. The specific stability of the cell wall was associated with a specific orientation of the long cellulosic microfibrils in tight dense crystalline complexes.

As described above, there is a complex regulation that leads to SAR. However, it is important to underline, that SAR is activated after a plant is exposed to elicitors from pathogens or artificial chemical stimuli. The active host plant-pathogen interaction is required to propagate disease resistance because molecular reactions that can be activated by pathogens are repressed in non-pathogenic conditions. Pathogen makes the plant more susceptible via the production of virulence factors (some effector proteins, pathogen-produced phytohormones, some toxins, and some siderophores) (Gao et al., 2015). The first layer of host defence involves the recognition of pathogen-/microbe-associated molecular patterns (PAMPs/MAMPs) by the plant's innate immune systems pattern recognition receptors (PRRs). It triggers downstream signalling leading to the activation of basal resistance termed PAMP-triggered immunity. The *avr* factors are in turn recognized by the plant-encoded resistance (R) proteins, which confer more robust resistance named R gene- or effector-triggered immunity, associated with hypersensitive response (PCD). Among the signals contributing are e.g., SA and methylated derivative of SA, azelaic acid, amino acid derivative pipecolic acid, auxin, the phosphorylated sugars, nitric oxide, and ROS, galactolipids, factors contributing to cuticle formation, and the lipid transfer proteins (Newman et al., 2013). However vast majority of experiments for R genes and proteins functions definition were made in low light ambient laboratory conditions and functions of these genes and proteins was never tested in excess light conditions or in UV stress, which naturally occurs in field but not in the laboratory conditions.

5. Integration of cellular light memory, photoelectrochemical signalling, and redox retrograde signalling

Photosynthetic organisms have evolved signalling mechanisms from the chloroplasts to the nucleus to induce the expression of the appropriate gene set in changing light conditions, light wavelength-specific photoelectrophysiological signalling, and memory of excess light episodes (Szechyńska-Hebda et al., 2010). Potential mechanisms for systemic spread of information to other cells, tissues, plant organs, and integration of these signals at the level of the whole organism can involve an electric signal generated in the chloroplasts, transmitted by the integrated membrane system of the individual chloroplasts, nucleus, endoplasmic reticulum, and cell membrane. For the model plant *Arabidopsis thaliana,* excess light induces specific signal from the chloroplasts named PhotoElectroPhysiological Signalling (PEPS), both locally (cells treated with an excess of light) and systemically (shaded cells of the same rosette). The intercellular transduction of PEPS involves wave-like changes in: ROS, expression of ascorbate peroxidase *APX2,* and NPQ. PEPS propagates in leaves and a flowering stem as waves shifted in phase. Leaf patches consisting of cells with reduced NPQ values, high level of H_2O_2, and induction of *APX:LUC* expression interweave with regions consisting of cells with increased NPQ, low level of H_2O_2, and downregulation of *APX:LUC* genes. Furthermore, in the *apx2–1* mutant, the deregulation of the electrical signal transduction mechanisms and systemic responses was found. These processes were associated with redox alterations in PSII, photosynthetic electron transport chain, and glutathione metabolism during chloroplast treatment with excess light. The application of DCMU inhibitor (3-(3,4-dichlorophenyl)−1,1-dimethylurea) in a single leaf exposed to excess light blocked the local and systemic induction of *APX1:LUC, APX2:LUC*, NPQ, and the electrical signal. Similarly, deregulation of PEPS and changes in the PQ pool and its redox state were observed in recessive mutants *cad2* and *rax1* associated with glutathione metabolism. The potential ways of inter-organelle and intracellular PEPS transduction were identified (Fig. 1). Local and systemic transduction of PEPS is related to the status of chloroplasts (EEE, photosynthesis changes, ROS, APX2, and NPQ), which are interconnected by stromules and driven by the actin cytoskeleton and ethylene signalling (Bobik & Burch-Smith, 2015; Gray et al., 2012; Szechyńska-Hebda et al., 2006). Furthermore, PEPS is transduced by the plasma membrane, and then membrane continuity through plasmodesmata between different cells, including bundle sheath cells in leaf veins. Mechanical damage to these cells (but not to the mesophyll cells) or the use of $LaCl_3$ ion channel inhibitor blocked the transduction of the electrical signal related to changes in light. Furthermore, it was indicated that the EEE induces metabolic changes not only in real-time, but that the information about the excess energy can be ‘used’ for a longer period of time after exposure to light. This long-term effect enables the optimization of the photosynthesis process during acclimation to excess light, but also the simultaneous induction of defence mechanisms against pathogens. In the leaves infected with the bacterial pathogen *Pseudomonas syringae* pv. *tomato* DC3000, disease development was limited only in the case of infection after exposure to excess light, while exposure to light applied one hour after infection did not induce defence mechanisms. Furthermore, a longer period of time between exposure to excess light and infection (in the range of 8–24 h) resulted in more effective inhibition of pathogen growth. In a similar way, plants show enhanced tolerance to UV-C-induced photoinhibition and CD upon induction of CLM by EEE episodes (Górecka et al., 2020). Wild-type plants exposed to 1 h of blue or red light in excess, after low light recovery were significantly more tolerant to UV-C, unlike the *npq4–1* mutant and *oePsbS* plants (Ciszak et al., 2015). At the transcriptomic level, CLM was differentially regulated in wild-type, *npq-4*, and *oePsbS* (Górecka et al., 2020); and robust molecular markers of CD and EEE stress responses (e.g., APX2, ERFs, TIFYs, MYBs, DREBs, and HSPs), and uncharacterized genes (T22K18_16) were involved in molecular induction of CLM, SAA and NAA, and cross-tolerance to UV-C stress.

The integration of acclimation and resistance mechanisms is closely dependent on the qualitative light composition. The pattern of transduced PEPS and its individual components (electrical signal, APX, H_2O_2, NPQ), the rate of systemic propagation of PEPS as well as the effect of cellular 'memory' is specific for blue (450 nm), red (650 nm), and white light, despite the same amount of energy was provided. The most effective in the induction of acclimation to EEE and resistance to bacterial pathogens was red light, followed by white light. Therefore, (1) PEPS transduction to cells that have not experienced stress, induces changes that are retained up to 24 h by the plant in order to 'prepare' the whole organism for potential future stress; (2) physiological changes are sufficiently versatile to induce both acclimation responses to abiotic stresses and defence responses to biotic stress.

CLM bases on dynamics of physiological processes in plant leaves. The spatial pattern of stomatal conductance and quantum redox changes in PSII (EEE dissipation as heat and fluorescence), reflected by a wavy, dynamic, and discrete in time and in space character of changes of NPQ, ROS, ES, could be mathematically described by functions similar to sin and 1-sin (Peak et al., 2004; Szechyńska-Hebda et al., 2010, 2022a). This complex pattern of physiological or molecular changes indicates that EEE incidents and spectral composition of light are differentially 'memorized' in each cell. The pattern of changes and 'memorized' information can therefore determine general leaf response and improve plant survival chances.

The CLM mechanism (Szechyńska-Hebda et al., 2010; Szechyńska-Hebda & Karpiński, 2013) enables survival of plants under multifactor stress. The sum of CLM responses is associated with the specific transcription of a number of nuclear genes, for example, *LHC* (*LIGHT-HARVESTING COMPLEX*), *RbcS* (*RIBULOSE-1,5-BISPHOSPHATE CARBOXYLASE/OXYGENASE SMALL SUBUNIT*), *ELIP* (*EARLY LIGHT-INDUCIBLE PROTEIN*), *APX1 and APX2*, (*ASCORBIC PEROXIDASE 1* and *2*), *PR1* (*PATHOGENESIS-RELATED GENE 1*). The presence of chloroplasts and their physiological state determine these processes, as CLM and PCD occurs only in cells containing fully developed chloroplasts. LSD1-dependent signalling pathways involve regulation of hypersensitivity reactions and then CLM via: (1) proteins encoded by the resistance genes (R-genes) NBS-LRR (NUCLEOTIDE-BINDING SITE LEUCINE-RICH REPEAT proteins); these proteins recognize pathogen effectors and activate defence responses; (2) integrated $O_2^{\bullet-}$ production processes by membrane *At*rbohD (NADPH-dependent oxidoreductases) and salicylic acid; (3) control of transcription factors; LSD1 itself can also play a role of such a transcription factor and regulate many genes involved in CLM, SAA, and SAR, as well as other processes dependent on PCD and/or acclimation, e.g. aerenchyma formation and tissue aging; (4) changes of NPQ, in the level of ethylene, salicylic acid, H_2O_2, antioxidative system activity and stomatal conductance; (5) integration of LSD1 function with two proteins involved in basal plant resistance to pathogens, i.e., EDS1 and PAD4; their cooperation influence on the level of ROS, salicylic acid and ethylene as well as is responsible for tissue-specific cell death, e.g. in the process of aerenchyma formation in roots exposed to hypoxia stress.

6. NAA – The most elegant communication system between plants

Recently, a mechanism triggered by an SAA, called Network Acquired Acclimation (NAA), was discovered. The mechanism enables plant communication by transmitting electrical signals on the surface between the plant leaves in contact. It has been proven that if two plants grow side by side and their leaves touch each other, if one leaf is injured it will send electrical information to the leaf of other plant. The condition is that the electrical circuit is closed, for example, high humidity or dewdrops connecting these leaves are present on the leaf surface. Although transmitted differently, that is, on the leaf surface, electric signalling leading to NAA is most likely similar to a slow wave potential (SWP) or system potential (SP) (Barres et al., 1976; Farmer et al., 2020; Górecka et al., 2020;

Kurenda et al., 2019; Sukhova & Sukhov, 2021; Zimmermann et al., 2009). SAA electric signalling is recorded extracellularly or intracellularly not on leaf surface as NAA (the latter with reversed amplitude; Szechyńska-Hebda et al., 2010). However, in both cases (SAA, NAA), the electric signals: (1) began with a small (several microvolts) and transient membrane hyperpolarization, and some pulses of depolarization with a period of ~50 s occurred within the slow repolarization phase; (2) moved within different areas of the same leaf, from organ to organ, and from plant to plant; (3) its amplitude is modulated (interdependent ion fluxes), from which the plant can gain information about the nature and intensity of the injury; (4) needs to be to some extent a self-propagating signal, as ES did not disappear completely with distance, but only weakened in amplitude, which was dependent on changes in H^+ pump activity; (5) $LaCl_3$, a Ca^{2+} antagonist that can be used as an inhibitor of Ca^{2+}-dependent signalling, prevents ES propagation, systemic changes in PSII, systemic stomatal responses, and induction of acclimation responses in directly exposed leaves, as well as in systemic leaves undergoing SAA and NAA; (6) carries information that can regulate photosynthesis and light absorption by photosystems (Białasek et al., 2017; Devireddy et al., 2018; Szechyńska-Hebda et al., 2010, 2022a). The transmission of electrical signal to induce SAA, can be accompanied by changes in hydraulic, calcium, and ROS waves, transcripts, metabolites, and phytohormones, stomata closure/opening and other physiological responses between local and systemic tissues of the same plant (Fichman & Mittler, 2020, 2021; Gilroy et al., 2016; Kollist et al., 2018; Mittler, 2017; Sukhova & Sukhov, 2021; Szechyńska-Hebda et al., 2010; Tian et al., 2020). These signals can be transmitted also between two different plants as part of NAA signalling. However, intra-organismal ES detected on the surface of leaves has been reported to mediate also changes in the status of ions, pH, sugars, ROS, and turgor; changes in extracellular electric potential, and electric field; or even the induction of micro-vibrations that can be sensed by mechano-sensor channels that alter Ca^{2+} levels (Pogson et al., 2008; Szechyńska-Hebda et al., 2022a; Tian et al., 2020).

ES during SAA is transmitted internally through the vascular system, among other, bundle sheath cells and adjacent parenchyma cells, with a velocity of ~0.5–5 mm s^{-1} (Gilroy et al., 2016; Nguyen et al., 2018; Salvador-Recatalà et al., 2014; Szechyńska-Hebda et al., 2010, 2022a). Similarly, ES spread on the leaf surface with a velocity of ~5 mm s^{-1} between different plants, or within a network of plants to induce NAA. Most importantly, both types of electric signals can induce a range of systemic molecular, biochemical, and physiological responses: specific physiological changes in Chl *a* fluorescence (NPQ, ΦPSII, qP, F_v/F_m, F_v/F_o, and F_v'/F_m'), biochemical changes in phytohormones (SA, JA, SA, and ABA) and ROS levels, and molecular alterations in gene expression (*APX2* and *ZAT12* induction), between transmitter and receiver plants.

Photosynthesis-dependent ES is transferred between individual chloroplasts, cells, organs, and plants through a complex pattern of physiological or molecular changes, that are differentially 'memorized' in each cell (as described above, Peak et al., 2004; Szechyńska-Hebda et al., 2010, 2022a). The current flow regulates the pattern of NPQ and ROS in the contacting leaf (in the signal receiver) and in other non-touching leaves of this plant, then the second plant, third plant, etc. Plants that touch each other form a network of connections and they can communicate electrically. The network of electrical connections can theoretically be considered to modulate the distribution of absorbed energy at different scales of complexity. In one chloroplast there are several thousand photosystems and one cell has several or several dozen chloroplasts. One leaf is made up of tens of thousands of cells, and a plant has thousands of cells. Therefore, the network of electrical connections in contacting plants on a meadow or in a forest reaches trillions of possible virtual connections. Experiments on such scale have not been performed, and there are probably other signals that can interfere and prevent the very long-range NAA, however, the mechanism seems to be universal for plants (inter-species).

7. Future prospects

Understanding of gene function, involved in the molecular and physiological mechanisms such as heat production (EEE dissipation), electrical signalling, ROS chloroplast retrograde signalling, CLM, and subsequent acclimation and defence responses (SAA, SAR, and NAA) is currently the most important goal for scientists. In particular, new knowledge is required to find answers to the following important questions: (1) How and to what extent does the architecture of LHCB determine PsbS- and PQ-pool-dependent electrical, hormonal, and ROS signalling? (2) How do plants molecularly regulate the energy balance between photochemistry and energy dissipation as heat and fluorescence? (3) How do plants regulate CLM, SAA, NAA, CD, photosynthesis, transpiration, and stomatal conductance? (4) How do NPQ, ROS, and ES wave propagation in SAA and NAA regulate nuclear- and chloroplast-encoded gene expression and ROS scavenger enzyme activity? (5) How and to what extent do heat shock, UV, and EEE episodes disturb photo-synthesis/photoinhibition stoichiometry, thus dissipation of EEE as heat and fluorescence? (6) How much heat is radiated to the environment by leaves of different plant species during photosynthesis/ photoinhibition? (7) Although NAA could be considered as a new communication mechanism within plant ecosystems, however, whether NAA is a 'side effect' of SAA or conveys an evolutionary advantage to plants living within the community, and to what extent can ES carry specific information and determine specific responses, are also open questions.

References

Adams, W. W., III, Zarter, C. R., Mueh, K. E., Amiard, V., & Demmig-Adams, B. (2008). Energy dissipation and photoinhibition: A continuum of photoprotection. In B. Demmig-Adams, W. W. Adams, & A. K. Mattoo (Vol. Eds.), *Photoprotection, photoinhibition, gene regulation, and environment. Advances in photosynthesis and respiration: 21*. Dordrecht: Springer.

Arias, M. C., Lenardon, S., & Taleisnik, E. (2003). Carbon metabolism alterations in sunflower plants infected with the sunflower chlorotic mottle virus. *Phytopathologische Zeitschrift, 151*, 267–273.

Asada, K. (1999). The water–water cycle in chloroplasts: Scavenging of active ogy and dissipation of excess photons. *Annual Review of Plant Physiology and Plant Molecular Biology, 50*, 601–639.

Atkinson, N. J., & Urwin, P. E. (2012). The interaction of plant biotic and abiotic stresses: From genes to field. *Journal of Experimental Botany, 63*, 3523–3544.

Bailey-Serres, J., & Voesenek, L. A. (2008). Flooding stress: Acclimations and genetic diversity. *Annual Review of Plant Biology, 59*, 313–339.

Baker, N. R. (2008). Chlorophyll fluorescence: A probe of photosynthesis in vivo. *Annual Review of Plant Biology, 59*, 89–113.

Ball, L., Accotto, G. P., Bechtold, U., Creissen, G., Funck, D., Jimenez, A., et al. (2004). Evidence for a direct link between glutathione biosynthesis and stress defense gene expression in *Arabidopsis*. *The Plant Cell, 16*, 2448–2462.

Barah, P., Mahantesha, N. B. N., Jayavelu, N. D., Sowdhamini, R., Shameer, K., & Bones, A. M. (2016). Transcriptional regulatory networks in Arabidopsis thaliana during single and combined stresses. *Nucleic Acids Research, 44*, 3147–3164.

Barres, S. J., Perry, T. J. S., & Pickard, B. G. (1976). Mediation of rapid electrical, metabolic, transpirational, and photosynthetic changes by factors released from wounds. I. variation potentials and putative action potentials in intact plants. Miscellaneous Publication *Canadian Journal of Botany, 54*, 2642–2671.

Białasek, M., Górecka, M., Mittler, R., & Karpiński, S. (2017). Evidence for the involvement of electrical, calcium and ROS signalling in the systemic regulation of non-photochemical quenching and photosynthesis. *Plant and Cell Physiology, 58*, 207–215.

Bobik, K., & Burch-Smith, T. M. (2015). Chloroplast signaling within, between and beyond cells. *Frontiers in Plant Science, 6*, 781.

Chang, C. C., Ślesak, I., Jordá, L., Sotnikov, A., Melzer, M., Miszalski, Z., et al. (2009). *Arabidopsis* chloroplastic glutathione peroxidases play a role in cross talk between photooxidative stress and immune responses. *Plant Physiology, 150*, 670–683.

Chang, C. C. C., Ball, L., Fryer, M., Baker, N. R., Karpiński, S., & Mullineaux, P. (2004). Induction of *ASCORBATE PEROXIDASE 2* expression in wounded *Arabidopsis* leaves does not involve known wound-signalling pathways but is associated with changes in photo- synthesis. *Plant Journal, 38*, 499–511.

Chaves, M. M., & Oliveira, M. M. (2004). Mechanisms underlying plant resilience to water deficits: Prospects for water-saving agriculture. *Journal of Experimental Botany, 55*, 2365–2384.

Chen, T., Fang, Y., Jiang, Q., Dykes, G. F., Lin, Y., Price, G. D., et al. (2022). Incorporation of functional Rubisco activases into engineered carboxysomes to enhance carbon fixation. *ACS Synthetic Biology, 11*, 154–161.

Ciszak, K., Kulasek, M., Barczak, A., Grzelak, J., Maćkowski, S., & Karpiński, S. (2015). PsbS is required for systemic acquired acclimation and post-excess-light-stress optimization of chlorophyll fluorescence decay times in *Arabidopsis*. *Plant Signalling and Behaviour, 10*, e982018.

Clarke, S. M., Mur, L. A. J., Wood, J. E., & Scott, I. M. (2004). Salicylic acid dependent signalling promotes basal thermotolerance but is not essential for acquired thermotolerance in *Arabidopsis thaliana*. *Plant Journal, 38*, 432–447.

Coll, N., Smidler, A., Puigvert, M., et al. (2014). The plant metacaspase AtMC1 in pathogen-triggered programmed cell death and aging: functional linkage with autophagy. *Cell Death Differ, 21*, 1399–1408. https://doi.org/10.1038/cdd.2014.50

Czarnocka, W., Van Der Kelen, K., Willems, P., Szechyńska-Hebda, M., Shahnejat-Bushehri, S., Balazadeh, S., et al. (2017). The dual role of *LESION SIMULATING DISEASE 1* as a condition-dependent scaffold protein and transcription regulator. *Plant Cell and Environment, 40*, 2644–2662.

Dall'Osto, L., Cazzaniga, S., Bressan, M., Paleček, D., Židek, K., Niyogi, K. K., et al. (2017). Two mechanisms for dissipation of excess light in monomeric and trimeric light-harvesting complexes. *Nature Plants, 3*, 17033.

Devireddy, A. R., Zandalinas, S. I., Gómez-Cadenas, A., Blumwald, E., & Mittler, R. (2018). Coordinating the overall stomatal response of plants: Rapid leaf-to-leaf communication during light stress. *Science Signalling, 11*, eaam9514.

Farmer, E. E., Gao, Y. Q., Lenzoni, G., Wolfender, J. L., & Wu, Q. (2020). Wound- and mechanostimulated electrical signals control hormone responses. *New Phytologist, 227*, 1037–1050.

Fichman, Y., & Mittler, R. (2020). Rapid systemic signalling during abiotic and biotic stresses: Is the ROS wave master of all trades? *Plant Journal, 102*, 887–896.

Fichman, Y., & Mittler, R. (2021). Integration of electric, calcium, reactive oxygen species and hydraulic signals during rapid systemic signalling in plants. *Plant Journal, 107*, 7–20.

Fichman, Y., Myers, R. J., Grant, D. G., & Mittler, R. (2021). Plasmodesmata-localized proteins and ROS orchestrate light-induced rapid systemic signalling in *Arabidopsis*. *Science Signalling, 23*, eabf0322.

Frenkel, M., Külheim, C., Jänkänpää, H. J., Skogström, O., Dall'Osto, L., Ågren, J., et al. (2009). Improper excess light energy dissipation in *Arabidopsis* results in a metabolic reprogramming. *BMC Plant Biology, 9*, 12.

Fryer, M. J., Ball, L., Oxborough, K., Karpinski, S., Mullineaux, P. M., & Baker, N. R. (2003). Control of *Ascorbate Peroxidase 2* expression by hydrogen peroxide and leaf water status during excess light stress reveals a functional organisation of *Arabidopsis* leaves. *Plant Journal, 33*, 691–705.

Gao, Q.-M., Zhu, S., Kachroo, P., & Kachroo, A. (2015). Signal regulators of systemic acquired resistance. *Frontiers in Plant Sciences, 6*, 228.

Gilroy, S., Białasek, M., Suzuki, N., Górecka, M., Devireddy, A. R., Karpiński, S., et al. (2016). ROS, calcium, and electric signals: Key mediators of rapid systemic signalling in plants. *Plant Physiology, 171*, 1606–1615.

Górecka, M., Lewandowska, M., Dąbrowska-Bronk, J., Białasek, M., Barczak-Brzyżek, A., Kulasek, M., et al. (2020). Photosystem II 22kDa protein level a prerequisite for excess light-inducible memory, cross-tolerance to UV-C, and regulation of electrical signalling. *Plant Cell and Environment, 43*, 649–661.

Gray, J. C., Hansen, M. R., Shaw, D. J., Graham, K., Dale, R., Smallman, P., et al. (2012). Plastid stromules are induced by stress treatments acting through abscisic acid. *Plant Journal, 69*, 387–398.

Herbette, S., Lenne, C., Leblanc, N., Julien, J. L., Drevet, J. R., & Roeckel-Drevet, P. (2002). Two GPX-like proteins from *Lycopersicon esculentum* and *Helianthus annuus* are antioxidant enzymes with phospholipid hydroperoxide glutathione peroxidase and thioredoxin peroxidase activities. *European Journal of Biochemistry, 269*, 2414–2420.

Hirel, B., Le Gouis, J., Ney, B., & Gallais, A. (2007). The challenge of improving nitrogen use efficiency in crop plants: towards a more central role for genetic variability and quantitative genetics within integrated approaches. *Journal of Experimental Botany, 58*, 2369–2387.

Huang, W., Yang, Y. J., Zhang, S. B., & Liu, T. (2018). Cyclic electron flow around photosystem I promotes ATP synthesis possibly helping the rapid repair of photodamaged photosystem II at low light. *Frontiers in Plant Science, 9*, 239.

Kangasjärvi, S., Lepistö, A., Hännikäinen, K., Piippo, M., Luomala, E. M., Aro, E. M., et al. (2008). Diverse roles for chloroplast stromal and thylakoid-bound ascorbate peroxidases in plant stress responses. *Biochemical Journal, 412*, 275–285.

Karpiński, S., & Szechyńska-Hebda, M. (2010). Secret life of plants. *Plant Signalling and Behaviour, 5*, 1391–1394.

Karpiński, S., Szechyńska-Hebda, M., Wituszyńska, W., & Burdiak, P. (2013). Light acclimation, retrograde signalling, cell death and immune defences in plants. *Plant, Cell and Environment, 36*(4), 736–744.

Karpiński, S., Reynolds, H., Karpinska, B., Wingsle, G., Creissen, G., & Mullineaux, P. (1999). Systemic signalling and acclimation in response to excess excitation energy in *Arabidopsis*. *Science, 284*, 654–657.

Kollist, H., Zandalinas, S. I., Sengupta, S., Nuhkat, M., Kangasjärvi, J., & Mittler, R. (2018). Rapid responses to abiotic stress: Priming the landscape for the signal transduction network. *Trends in Plant Science, 24*, 25–37.

Kornyeyev D., Logan B. A., Tissue D. T., Allen R. D., Holaday A. S. (2006). Compensation for PSII photoinactivation by regulated non-photochemical dissipation influences the impact of photoinactivation on electron transport and CO2 assimilation. 47 437–446. 10.1093/pcp/pcj010.

Kunkel, B. N., & Brooks, D. M. (2002). Cross talk between signalling pathways in pathogen defense. *Current Opinion in Plant Biology, 5*, 325–331.

Kuo, W. Y., Huang, C. H., Liu, A. C., Cheng, C. P., Li, S. H., Chang, W. C., et al. (2013). CHAPERONIN 20 mediates iron superoxide dismutase (FeSOD) activity independent of its co-chaperonin role in *Arabidopsis* chloroplasts. *New Phytologist, 197*, 99–110.

Kurenda, A., Nguyen, C. T., Chételat, A., Stolz, S., & Farmer, E. E. (2019). Insect-damaged *Arabidopsis* moves like wounded *Mimosa pudica*. *Proceedings of the National Academy of Science of the USA, 116*, 26066–26071.

Kwiatkowska, D., & Burian, A. (2022). Plant biomechanics—A natural transition from molecular to organ scale. *International Journal of Molecular Sciences, 23*, 7575.

Li, M., & Kim, C. (2022). Chloroplast ROS and stress signalling. *Plant Communications, 3*, 100264.

Lima-Melo, Y., Kılıç, M., Aro, E. M., & Gollan, P. J. (2021). Photosystem I inhibition, protection and signalling: Knowns and unknowns. *Frontiers in Plant Science, 12*, 791124.

Lu, Y. (2016). Identification and roles of photosystem II assembly, stability, and repair factors in *Arabidopsis*. *Frontiers in Plant Science, 7*, 168.

Luna, E., Bruce, T. J., Roberts, M. R., Flors, V., & Ton, J. (2012). Next-generation systemic acquired resistance. *Plant Physiology, 158*, 844–853.

Mateo, A., Muhlenbock, P., Rusterucci, C., Chi-Chen, C., Miszalski, Z., Karpinska, B., et al. (2004). The *LESION SIMULATING DISEASE (LSD1)* gene is required for acclimation to conditions that promote excess excitation energy. *Plant Physiology, 136*, 2818–2830.

Mina, J., & Bissell, J. (2017). Goodbye flat biology – Time for the 3rd and the 4th dimensions. *Cell Science, 130*, 3–5.

Mingzhu Ma, Yifei Liu, Chunming Bai, Yunhong Yang, Zhiyu Sun, Xinyue Liu, Siwei Zhang, Xiaori Han and Jean Wan Hong Yong, The Physiological Functionality of PGR5/PGRL1-Dependent Cyclic Electron Transport in Sustaining Photosynthesis.

Mittler, R. (2017). ROS are good. *Trends in Plant Science, 22*, 11–19.

Miyake, C. (2010). Alternative electron flows (water–water cycle and cyclic electron flow around PSI) in photosynthesis: Molecular mechanisms and physiological functions. *Plant and Cell Physiology, 51*, 1951–1963.

Mühlenbock, P., Szechyńska-Hebda, M., Płaszczyca, M., Baudo, M., Mateo, A., Mullineaux, P. M., et al. (2008). Chloroplast signalling and LESION SIMULATING DISEASE1 regulate crosstalk between light acclimation and immunity in *Arabidopsis*. *The Plant Cell, 20*, 2339–2356.

Murchie, E. H., & Ruban, A. V. (2020). Dynamic non-photochemical quenching in plants: From molecular mechanism to productivity. *The Plant Journal, 101*, 885–896.

Nakashima, K., & Yamaguchi-Shinozaki, K. (2006). Regulons involved in osmotic stress-responsive and cold stress-responsive gene expression in plants. *Physiologia Plantarum, 126*, 62–71.

Newman, M. A., Sundelin, T., Nielsen, J. T., & Erbs, G. (2013). MAMP (microbe-associated molecular pattern) triggered immunity in plants. *Frontiers in Plant Sciences, 4*, 139.

Nguyen, C. T., Kurenda, A., Stolz, S., Chételat, A., & Farmer, E. E. (2018). Identification of cell populations necessary for leaf-to-leaf electrical signalling in a wounded plant. *Proceedings of the National Academy of Science of the United States, 115*, 101780–101832.

Oelze, M. L., Vogel, M. O., Alsharafa, K., Kahmann, U., Viehhauser, A., Maurino, et al. (2012). Efficient acclimation of the chloroplast antioxidant defence of *Arabidopsis thaliana* leaves in response to a 10- or 100-fold light increment and the possible involvement of retrograde signals. *Journal of Experimental Botany, 63*, 1297–1313.

Peak, D., West, J. D., Messinger, S. M., & Mott, K. A. (2004). Evidence for complex, collective dynamics and emergent, distributed computation in plants. *Proceedings of the National Academy of Science of the United States, 101*, 918–922.

Pogson, B. J., Woo, N. S., & Förster, B. (2008). Small, plastid signalling to the nucleus and beyond. *Trends in Plant Science, 13*, 602–609.

Roach, T., Na, C. S., Stöggl, W., & Krieger-Liszkay, A. (2020). The non-photochemical quenching protein LHCSR3 prevents oxygen-dependent photoinhibition in *Chlamydomonas reinhardtii*. *Journal of Experimental Botany, 71*, 2650–2660.

Roden, L. C., & Ingle, R. A. (2009). Lights, rhythms, infection: The role of light and the circadian clock in determining the outcome of plant–pathogen interactions. *The Plant Cell, 21*, 2546–2552.

Ruban, A., & Belgio, E. (2014). The relationship between maximum tolerated light intensity and photoprotective energy dissipation in the photosynthetic antenna: Chloroplast gains and losses. *Philosophical Transactions of the Royal Society B Biological Sciences, 369*, 20130222.

Saccon, F., Giovagnetti, V., Shukla, M. K., & Ruban, A. V. (2020). Rapid regulation of photosynthetic light harvesting in the absence of minor antenna and reaction centre complexes. *Journal of Experimental Botany, 71*, 3626–3637.

Sacharz, J., Giovagnetti, V., Ungerer, P., Mastroianni, G., & Ruban, A. V. (2017). The xanthophyll cycle affects reversible interactions between PsbS and light-harvesting complex II to control non-photochemical quenching. *Nature Plants, 3*, 1–9.

Salvador-Recatalà, V., Tjallingii, W. F., & Farmer, E. E. (2014). Real-time, in vivo intracellular recordings of caterpillar-induced depolarization waves in sieve elements using aphid electrodes. *New Phytologist, 203*, 674–684.

Sharma, Y. K., Leon, J., Raskin, I., & Davis, K. R. (1996). Ozone-induced responses in *Arabidopsis* thaliana: The role of salicylic acid in the accumulation of defense-related transcripts and induced resistance. *Proceedings of the National Academy of Science of the USA, 93*, 5099–5104.

Shigeoka, S., & Maruta, T. (2014). Cellular redox regulation, signalling, and stress response in plants. *Bioscience Biotechnology and Biochemistry, 78*, 1457–1470.

Ślesak, I., Szechyńska-Hebda, M., Fedak, H., Sidoruk, N., Dąbrowska-Bronk, J., Witoń, D., et al. (2015). PHYTOALEXIN DEFICIENT 4 affects reactive oxygen species metabolism, cell wall and wood properties in hybrid aspen (*Populus tremula* L.× *tremuloides*). *Plant Cell and Environment, 38*, 1275–1284.

Storti, M., Costa, A., Golin, S., Zottini, M., Morosinotto, T., & Alboresi, A. (2018). Systemic calcium wave propagation in Physcomitrella *patens*. *Plant Cell and Physiology, 59*, 1377–1384.

Styring, S., Sjöholm, J., & Mamedov, F. (2012). Two tyrosines that changed the world: Interfacing the oxidizing power of photochemistry to water splitting in photosystem II. *Biochimica et Biophysica Acta – Bioenergetics, 1817*, 76–87.

Sukhova, E., & Sukhov, V. (2021). Electrical signals, plant tolerance to actions of stressors, and programmed cell death: is interaction possible? *Plants (Basel), 10*, 1704.

Szechyńska-Hebda, M., & Karpiński, S. (2013). Light intensity-dependent retrograde signalling in higher plants. *Journal of Plant Physiology, 170*, 1501–1516.

Szechyńska-Hebda, M., Lewandowska, M., & Karpiński, S. (2017). Electrical signalling, photosynthesis and systemic acquired acclimation. *Frontiers in Physiology, 8*, 684.

Szechyńska-Hebda, M., Hebda, M., Mirek, M., & Miernik, K. (2016). Cold-induced changes in cell wall stability determine the resistance of winter triticale to fungal pathogen *Microdochium nivale*. *Journal of Thermal Analysis and Calorimetry, 126*, 77–90.

Szechyńska-Hebda, M., Wędzony, M., Dubas, E., Kieft, H., & Van Lammeren, A. (2006). Visualisation of microtubules and actin filaments in fixed BY-2 suspension cells using an optimised whole mount immunolabelling protocol. *Plant Cell Reports, 25*, 758–766.

Szechyńska-Hebda, M., Kruk, J., Górecka, M., Karpińska, B., & Karpiński, S. (2010). Evidence for light wavelength-specific photoelectrophysiological signalling and memory of excess light episodes in *Arabidopsis*. *The Plant Cell, 22*, 2201–2218.

Szechyńska-Hebda, M., Skrzypek, E., Dąbrowska, G., Biesaga-Kościelniak, J., Filek, M., & Wędzony, M. (2007). The role of oxidative stress induced by growth regulators in the regeneration process of wheat. *Acta Physiologiae Plantarum, 29*, 327–337.

Szechyńska-Hebda, M., Wędzony, M., Tyrka, M., Gołębiowska, G., Chrupek, M., Czyczyło-Mysza, I., et al. (2011). Identifying QTLs for cold-induced resistance to *Microdochium nivale* in winter triticale. *Plant Genetic Resources, 9*, 296–299.

Szechyńska-Hebda, M., Hebda, M., Mierzwinski, D., Kuczynska, P., Mirek, M., Wedzony, W., et al. (2013). Effect of cold-induced changes in physical and chemical leafproperties on the resistance of winter triticale (*xTriticosecale*) to the fungal pathogen *Microdochium nivale*. *Plant Pathology, 62*, 867–878.

Szechyńska-Hebda, M., Wąsek, I., Gołębiowska-Pikania, G., Dubas, E., Żur, I., & Wędzony, M. (2015). Photosynthesis-dependent physiological and genetic crosstalk between cold acclimation and cold-induced resistance to fungal pathogens in triticale (*Triticosecale* Wittm.). *Journal of Plant Physiology, 177*, 30–43.

Szechyńska-Hebda, M., Lewandowska, M., Witoń, D., Fichman, Y., Mittler, R., & Karpiński, S. M. (2022). Aboveground plant-to-plant electrical signalling mediates network acquired acclimation. *The Plant Cell, 34*, 3047–3065.

Takahashi, S., & Badger, M. (2011). Photoprotection in plants: A new light on photosystem II damage. *Trends in Plant Science, 16*, 53–60.

Tian, W., Wang, C., Gao, O., Li, L., & Luan, S. (2020). Calcium spikes, waves and oscillations in plant development and biotic interactions. *Nature Plants, 6*, 750–759.

Tian, Y., Sacharz, J., Ware, M. A., Zhang, H., & Ruban, A. V. (2017). Effects of periodic photoinhibitory light exposure on physiology and productivity of *Arabidopsis* plants grown under low light. *Journal of Experimental Botany, 68*, 4249–4262.

Umena, Y., Kawakami, K., Shen, J. R., & Kamiya, N. (2011). Crystal structure of oxygen-evolving photosystem II at a resolution of 1.9 Å. *Nature, 473*, 55–60.

Wallace, G. D. (2012). *Monotropa uniflora*. *Jepson Flora Project*. Jepson eFlora,2022. ⟨https://ucjeps.berkeley.edu/eflora/eflora_display.php?tid=33968⟩ accessed on October 24.

Walter, J., & Kromdijk, J. (2022). Here comes the sun: How optimization of photosynthetic light reactions can boost crop yields. *Journal of Integrative Plant Biology, 64*, 564–591.

Wang, W., Vinocur, B., & Altman, A. (2003). Plant responses to drought, salinity and extreme temperatures: Towards genetic engineering for stress tolerance. *Planta, 218*, 1–14.

Ware, M. A., Belgio, E., & Ruban, A. V. (2015). Comparison of the protective effectiveness of NPQ in *Arabidopsis* plants deficient in PsbS protein and zeaxanthin. *Journal of Experimental Botany, 66*, 1259–1270.

Wąsek, I., Dyda, M., Gołębiowska, G., et al. (2022b). Quantitative trait loci and candidate genes associated with freezing tolerance of winter triticale (× Triticosecale Wittmack). *J Appl Genetics, 63*, 15–33. https://doi.org/10.1007/s13353-021-00660-1

Wei, X., Su, X., Cao, P., Liu, X., Chang, W., Li, M., et al. (2016). Structure of spinach photosystem II-LHCII supercomplex at 3.2 Å resolution. *Nature, 534*, 69–74.

Whitewoods, C., & Coen, E. (2017). Growth and development of three-dimensional plant form. *Current Biology, 27*, R910–R918.

Witoń, D., Gawroński, P., Czarnocka, W., Ślesak, I., Rusaczonek, A., Sujkowska-Rybkowska, M., et al. (2016). Mitogen activated protein kinase 4 (MPK4) influences growth in *Populus tremula* L.× *tremuloides*. *Environmental and Experimental Botany, 130*, 189–205.

Witoń, D., Sujkowska-Rybkowska, M., Dąbrowska-Bronk, J., Czarnocka, W., Bernacki, M., Szechyńska-Hebda, et al. (2021). MITOGEN-ACTIVATED PROTEIN KINASE 4 impacts leaf development, temperature, and stomatal movement in hybrid aspen. *Plant Physiology, 186*, 2190–2204.

Wituszyńska, W., Ślesak, I., Vanderauwera, S., Szechyńska-Hebda, M., Kornaś, A., Van Der Kelen, K., et al. (2013). Lesion simulating disease1, enhanced disease susceptibility1, and phytoalexin deficient4 conditionally regulate cellular signalling homeostasis, photosynthesis, water use efficiency, and seed yield in *Arabidopsis*. *Plant Physiology, 161*, 1795–1805.

Yu, G., Pan, X., Hao, J., Shi, L., Zhang, Y., Wang, J., et al. (2021). Structure of SOQ1 lumenal domains identifies potential disulfide exchange for negative regulation of photoprotection, qH. *bioRxiv*. https://doi.org/10.1101/2021.03.16.435614

Zimmermann, M. R., Maischak, H., Mithöfer, A., Boland, W., & Felle, H. H. (2009). System potentials, a novel electrical long-distance apoplastic signal in plants, induced by wounding. *Plant Physiology, 149*, 1593–1600.

CHAPTER TEN

Plant aquaporins: crossroads of hydrogen peroxide signaling

Jahed Ahmed, and François Chaumont*
Louvain Institute of Biomolecular Science and Technology, Université catholique de Louvain, Louvain-la-Neuve, Belgium
*Corresponding author. e-mail address: francois.chaumont@uclouvain.be

Contents

Abstract

Reactive oxygen species (ROS), including hydrogen peroxide (H_2O_2), are important inter- and intracellular signaling molecules involved in various plant physiological processes under biotic and abiotic stress conditions. They are produced in different subcellular compartments such as apoplast, plasma membrane, chloroplasts, peroxisomes, and mitochondria. Under stress, H_2O_2 is mainly produced in the apoplast by plasma membrane localized respiratory oxygen burst oxidase homologs (RBOHs), dismutases, and cell wall peroxidases. The function of H_2O_2 as a signal requires its diffusion through the plasma membrane or internal membranes, a process that is facilitated by the presence of channels named aquaporins (AQPs). In this chapter, we will discuss the mechanisms by which AQPs facilitate and regulate H_2O_2 membrane diffusion, and the relevance of their contribution in organellar, cellular, and cell-to-cell H_2O_2 signaling processes in the context of abiotic and biotic stresses.

Advances in Botanical Research, Volume 105
ISSN 0065-2296, https://doi.org/10.1016/bs.abr.2022.11.003

1. Introduction

Reactive oxygen species (ROS) play a central role in plant abiotic and biotic stress responses. Among the different ROS species, H_2O_2 is most stable and is a key signal to trigger downstream responses acting in plant defence and tolerance mechanisms. H_2O_2 is produced in most cell compartments. Apoplastic H_2O_2 production is induced in response to abiotic and biotic stimuli i.e., temperature, UV irradiation, ozone, phytohormones (ABA, ethylene, etc.), salt stress, flg22, viruses, and chitin (Castro et al., 2021; Guo et al., 2019; Hossain et al., 2015). H_2O_2 is produced mainly by cell wall peroxidases and plasma membrane localized respiratory burst oxidase homologs (RBOHs) (Daudi et al., 2012). But H_2O_2 is also produced in other organelles such as chloroplasts, mitochondria, and peroxisomes (Fig. 1). In chloroplasts, under high light intensity, O_2 is photoreduced to the superoxide anion ($O_2^{\cdot -}$) (Melher reaction) at the acceptor side of photosystem I during the Water-Water cycle. Reduction of $O_2^{\cdot -}$ then leads to the formation of H_2O_2 through the action of superoxide dismutases (Asada, 1999; Mullineaux et al., 2018). This process

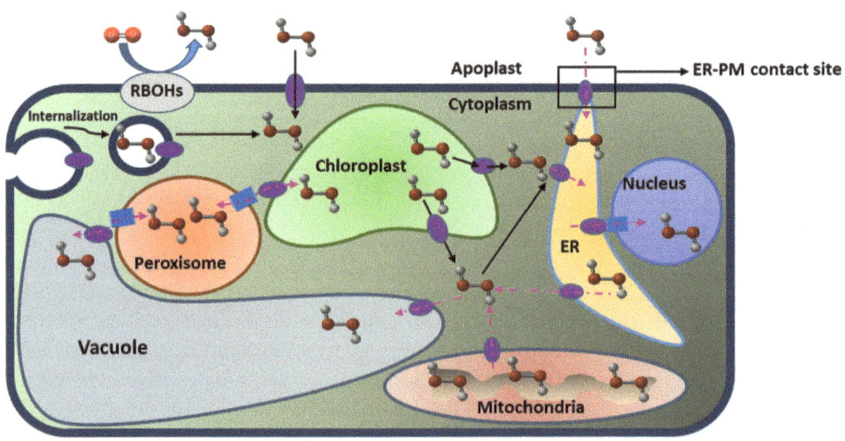

Fig. 1 Control of H_2O_2 (red and white sphere) cell homoeostasis, including production and transport. Extracellular H_2O_2 production is induced by the cell wall bound peroxidases and respiratory burst oxidase homologs (RBOHs). Its diffusion to the cell uses AQPs (shown in purple ball). Major intracellular H_2O_2 production sites such as mitochondria, peroxisomes, and chloroplasts were shown. Intraorganelle H_2O_2 movement is proposed to be mediated by AQPs (shown in purple oval) and putative channels in peroxisomal and nuclear membranes (shown in blue cylinder). Black solid arrow indicates transport; magenta dashed arrow indicates putative transport.

is important in triggering chloroplast-nucleus retrograde signaling and modification of gene expression and the associated stress response (Castro et al., 2021; Slesak et al., 2007). During CO_2 fixation in chloroplasts, 2-phosphoglycolate is produced by the oxygenase activity of RuBisCo, and is converted into glycolate that enters peroxisomes, where it is transformed into glyoxylate, generating H_2O_2. The electron transport chain complexes located in the mitochondrial inner membrane are the main sources of singlet O_2, which is later dismutated to H_2O_2 by the mitochondrial superoxide dismutase. Besides, a flavin adenine dinucleotide oxidase present in the endoplasmic reticulum (ER) could also generate H_2O_2 by reducing oxygen (Aller & Meyer, 2013; Tu & Weissman, 2002). While there are antioxidant enzymes such as catalases and peroxidases that scavenge excess H_2O_2 and regulate its cellular homoeostasis, there are environmental circumstances that perturb its balance, such as drought, high temperature, and/or pathogen attack (Smirnoff & Arnaud, 2019; Waszczak et al., 2018). Excessive H_2O_2 triggers oxidative damage to DNA, lipids, and proteins (Castro et al., 2021). Therefore, ROS production, distribution, and scavenging in plant cells need to be tightly regulated.

H_2O_2 transport between the cell compartments is an inherent component of H_2O_2 homoeostasis and ROS signaling. H_2O_2 can permeate cellular membranes by simple diffusion, but this process is slow. The involvement of water channels, named aquaporins (AQPs), in H_2O_2 membrane transport was first demonstrated by heterologous expression in yeast (Bienert et al., 2007; Bienert & Chaumont, 2014), and then *in planta* (Tian et al., 2016). The water channel activity of the first mammalian AQP was reported in 1992 (Preston & Agre, 1992), and this groundbreaking discovery was recognized by the Nobel Prize in Chemistry to Peter Agre in 2003. AQPs are found in all life kingdoms including plants. While 13 AQP isoforms have been identified in mammals, vascular plant genomes encode many more isoforms (> 30 to more than 100) that differ in terms of tissue and cell distribution, subcellular localization, transport selectivity, and regulation properties (Laloux et al., 2018). The water channel activity of the first plant AQP (AtTIP1;1) was demonstrated in 1993 (Maurel et al., 1993). AQPs have attracted considerable interest in plants due to the diverse substrate specificity that is essential for plant growth, development, and physiology. In this chapter, we will provide a comprehensive overview of plant AQPs and their roles in cellular H_2O_2 distribution and signaling upon stress.

2. Plant AQP subfamilies

Plant AQPs play various important physiological roles in long and short distance movement of water, small solutes such as gases (CO_2, NH_3...), and metalloids (Bienert & Chaumont, 2011; Chaumont & Tyerman, 2014). Recent research has revealed that AQPs also govern the transmembrane diffusion of H_2O_2 and even cations (Bienert et al., 2007, 2014; Byrt et al., 2016; Dynowski et al., 2008; Tian et al., 2016). Thus AQP research has spread to diverse plant science fields, from membrane biophysics to cell biology including stress responses to signaling.

Vascular plant AQPs are highly diverse and divided into five different subfamilies: (1) plasma membrane intrinsic proteins (PIPs); (2) tonoplast intrinsic proteins (TIPs); (3) Nodulin-26-like intrinsic proteins (NIPs); (4) small basic intrinsic proteins (SIPs), and (5) X intrinsic proteins (XIPs) (Chaumont et al., 2001; Danielson & Johanson, 2008; Johanson et al., 2001). To date, the latter subfamily has not been found in Brassicaceae and in monocots (Bienert et al., 2011; Danielson & Johanson, 2008). Interestingly, two additional subfamilies were found in the moss *Physcomitrella patens* called hybrid intrinsic proteins (HIPs) and GlpF-like intrinsic proteins (GIPs), which may have evolved from bacterial origin by a horizontal gene transfer (Danielson & Johanson, 2008; Gustavsson et al., 2005). However, *Selaginella moellendorffii* (spike mosses) have only HIPs along with PIPs, TIPs, NIPs, SIPs, and XIPs (Anderberg et al., 2012). A new subfamily called large intrinsic protein (LIP) was identified in algae that is lacking in seed plants (Khabudaev et al., 2014).

3. AQP structure and solute movement

Three dimensional structures of human and some bacterial AQPs provide a valuable resource to investigate the molecular mechanisms that govern solute movement through the pore and their regulation. In plants, the spinach SoPIP2;1 structure was obtained at high-resolution and, interestingly, revealed open and closed states (Nyblom et al., 2009; Tornroth-Horsefield et al., 2006), meaning that the channel activity can be regulated by a gating mechanism. Similarly, the AtTIP2;1 crystal structure also suggested a possible gating mechanism for ammonia transport (Kirscht et al., 2016).

The general structure of AQPs is highly conserved from bacteria to human. AQPs comprise six transmembrane (TM) α-helices (TM1-TM6),

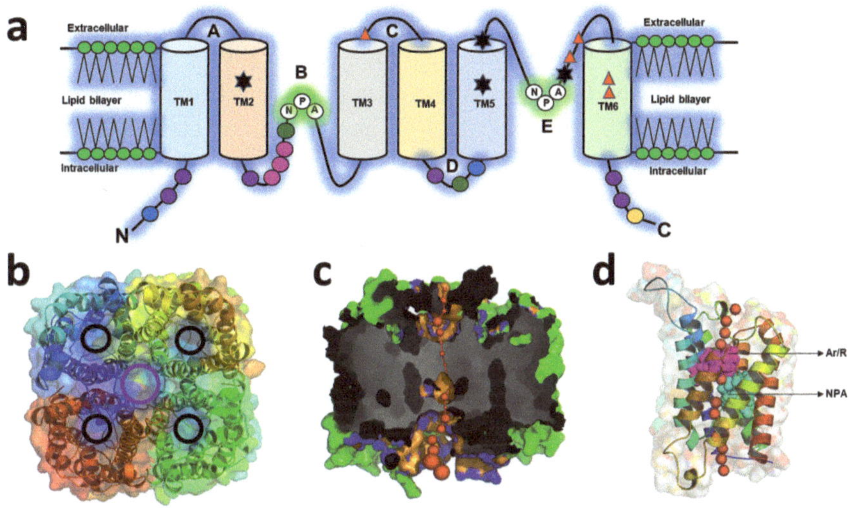

Fig. 2 Plant AQP general structure. (A) The AQP topology consists of six TM α-helices (TM1-TM6), which are linked by five loops (loops A–E), with both the N and C-termini located on the membrane cytoplasmic side. Several posttranslational modification sites are indicated including potential phosphorylation sites (purple), pH sensing residues (green), Ca^{2+} responsive residues (blue), mechano-sensing residues (magenta), and cGMP sensing residue (yellow). The aromatic arginine (ar/R) residues are shown in black stars and Froger's positions are shown in red triangle. (B) Tetrameric assembly of maize plasma membrane AQP (PIP2;5); individual and proposed central pore are indicated with a black and purple circle, respectively. (C) Cross section of a representative maize AQP (PIP2;5) tetramer and the path of the central pore is shown in a red ball. The figure was generated with PoreWalker software. (D) Monomeric maize PIP2;5 3D structure and the pore are indicated with red spheres. Ar/R filter residues are shown in magenta spheres and NPA motifs are shown in cyan spheres.

which are connected by five loops (loops A–E), and both the N and C-termini are located in the cytoplasm (Fig. 2A). Loops B and E form short α-helices each containing the very conserved Asn Pro Ala (NPA) motif and dip halfway into the membrane. AQPs form tetramers, but each monomer functions as a channel (Fig. 2B and D). However, the AQP tetrameric assembly forms a central pore that is predicted to be a possible channel for CO_2 and cations in some AQPs (Henderson et al., 2022; Tyerman et al., 2021; Yool & Weinstein, 2002) (Fig. 2B and C).

The monomer substrate selectivity is associated with two pore regions: a central region and the aromatic/arginine (ar/R) constriction filter, even if other channel parts, including TM domains can also participate in AQP channel activity (Berny et al., 2016; Chaumont et al., 2005; Maurel et al., 2008). The first constriction occurs at the pore center by close apposition of the two asparagine residues of the NPA motifs of the loops B and E that preclude proton (H^+) permeation (Murata et al., 2000). The ar/R constriction filter is formed at the extracellular pore aperture by four residues present in TM2, TM5, and loop E (LE1 and LE2), respectively (Fu et al., 2000; Sui et al., 2001) (Fig. 2D).

The wide range of substrate conductance observed in plant AQPs is assumed to be the result of changes at the ar/R constriction filter (Azad et al., 2012; Bansal & Sankararamakrishnan, 2007; Gupta & Sankararamakrishnan, 2009; Wallace & Roberts, 2005). However, a group of five moderately conserved amino acid residues at the so-called Froger's positions are also important in substrate selectivity (Froger et al., 1998; Heymann & Engel, 2000). More recently, some substrate specificity determining positions have been recommended by combining the plant AQP sequences with the experimentally identified substrates (Ahmed et al., 2020; Azad et al., 2016; Hove & Bhave, 2011).

The specificity and movement of solutes through AQPs is determined by the size and volume of the substrate and by the physicochemical properties of the amino acid residues placed along the pore. Water molecules move as a file through the narrow pore and this process is assisted by a series of backbone carbonyl groups and hydrophilic side chains lining the pore. At the pore center, water is reoriented by the dipoles created by the two half helices of loop B and E. This reorientation breaks the water file connectivity and prevents proton transport via a Grotthus mechanism (Kosinska Eriksson et al., 2013). As H_2O_2 has a slightly larger dipole moment/polarity and molecular diameter compared with water and is able to form hydrogen bonds as water, it was hypothesized to diffuse through AQPs and to mimic the movement of water molecules. Molecular dynamics simulation using SoPIP2;1 indeed indicated that the channel may facilitate transport of H_2O_2 and even protonated superoxide radicals (HO_2) (Cordeiro, 2015).

4. Determination of H_2O_2 channel activity of AQPs

The first prediction suggesting that H_2O_2 and water diffuse in parallel through plant cell membranes came from experiments and mathematical modeling performed in internodal cells of the freshwater algae *Chara corallina* (Henzler & Steudle, 2000). The authors determined the H_2O_2 permeability coefficient, which is a measure of the speed with which a compound crosses a membrane, and found that it was very close to that of water. Also, treatment with the often-used AQP blocker $HgCl_2$, significantly reduced the influx of both molecules into the cells (Henzler & Steudle, 2000).

Experimental evidence demonstrating that specific AQPs facilitate H_2O_2 diffusion came from heterologous expression of AQPs in yeast and testing the strain growth sensitivity to increasing H_2O_2 amounts in the external medium (Bienert et al., 2007). The presence of H_2O_2 in the medium leads to a decrease in growth and cell survival, depending on the ability of the strain to detoxify ROS. Expression of human hAQP8 and Arabidopsis AtTIP1;1 and AtTIP1;2 reduced yeast growth and survival on medium containing H_2O_2 (Bienert et al., 2007). This sensitivity was not due to a decreased ROS detoxification ability of the strain, demonstrating that the growth and survival impairments were due to the facilitated diffusion of H_2O_2 into the cells by the expressed AQPs (Bienert et al., 2006). Using a similar yeast expression system, an increasing number of plant AQPs from four subfamilies (PIPs, TIPs, NIPs, and XIPs) were reported as H_2O_2 channels and are listed in Table 1. Actually, it was proposed that all water-permeable AQPs are H_2O_2 channels, but with variable permeabilities according to the isoform and pore diameter (Almasalmeh et al., 2014). However, this may not be true for all AQPs. For instance, we demonstrated that the maize water channel ZmPIP2;5 was able to facilitate H_2O_2 diffusion in yeast, in contrast to ZmPIP1;2, a channel that, when coexpressed with PIP2;5, assembles in heterotetramers and acts as an active water channel (Berny et al., 2016; Bienert et al., 2014; Fetter et al., 2004). Surprisingly, an inactive water channel can also facilitate H_2O_2 membrane diffusion, as demonstrated for tobacco NtXIP1;1. This isoform has no water channel activity when expressed in the plasma membrane of *Xenopus* oocytes, but it increases the yeast cell sensitivity to H_2O_2 (Bienert et al., 2011). Other AQPs with water channel activity seem not to facilitate H_2O_2 membrane diffusion when expressed in yeast, such as Arabidopsis AtPIP2;3, AtPIP2;6 and AtPIP2;8 or rice and barley PIPs except HvPIP2;5 (Hooijmaijers et al., 2012; Rhee et al., 2017). However, negative results have to be interpreted with caution, and verification of the aquaporin expression, abundance and localization is required. For instance, most PIPs belonging to the PIP1 subfamily need to interact with PIP2 to reach the plasma membrane (see Section 5.2). Furthermore, to express maize PIPs in yeast cells, codon optimization of the *PIP* cDNAs was necessary to detect the protein (Bienert et al., 2014). Nevertheless, these data might indicate

Table 1 AQPs facilitating H_2O_2 diffusion.

AQP name	Subcellular localization	Model used to test the channel activity	Physiological roles	References
AtPIP1;4	PM	Yeast	Plant immunity	Tian et al. (2016)
AtPIP2;1	PM	Yeast, Stomata	Stomatal closure (H_2O_2 Signaling)	Dynowski et al. (2008), Hooijmaijers et al. (2012), Rodrigues et al. (2017)
AtPIP2;2	PM	Yeast	?	Hooijmaijers et al. (2012)
AtPIP2;3	PM	Yeast	?	Dynowski et al. (2008), Israel et al. (2022)
AtPIP2;4	PM	Yeast Reconstitution in liposomes	?	Hooijmaijers et al. (2012), Wang et al. (2020a)
AtPIP2;5	PM	Yeast	?	Hooijmaijers et al. (2012)
AtPIP2;7	PM	Yeast	?	Hooijmaijers et al. (2012)
GmPIP2;13	PM	*N. benthamiana*	Plant immunity	Ai et al. (2021)
HvPIP2;5	PM	Yeast	?	Rhee et al. (2017)
NbPIP2;2	PM	*N. benthamiana*	Plant immunity	Ai et al. (2021)

(continued on next page)

Table 1 *(continued)*

AQP name	Subcellular localization	Model used to test the channel activity	Physiological roles	References
OsPIP2;2	PM	Yeast	Plant immunity	Zhang et al. (2022)
OsPIP2;3	PM	Yeast	?	Zhang et al. (2022)
OsPIP2;6	PM	Yeast	?	Zhang et al. (2022)
SoPIP2;1	PM	Reconstitution in liposomes	?	Wang et al. (2020a)
TaPIP2;10	PM	Yeast	Plant immunity	Wang et al. (2021)
ZmPIP2;5	PM	Yeast Stomata	Stomatal movement	Bienert et al. (2014), Ding et al. (2022)
AtTIP2;3	Tonoplast	Yeast	?	Dynowski et al. (2008)
TgTIP2;1	Tonoplast	Yeast	?	Azad et al. (2012)
TgTIP4;1	Tonoplast	Yeast	?	Azad et al. (2012)
AtNIP1;2	ER	Yeast	?	Dynowski et al. (2008)
NtXIP1;1	PM	Yeast	?	Bienert et al. (2011)
SlXIP1;1	PM	Yeast	?	Bienert et al. (2011)

that still unknown structural determinants can influence the channel pore properties and make it specific for water and/or H_2O_2. The terms "peroxoporin/peroxiporin" were also proposed for AQPs mediating H_2O_2 movement (Henzler & Steudle, 2000; Sies, 2014). However, as these AQPs are also channels for water and/or other solutes, this name can be misleading and should be used carefully depending on the context.

While knowledge about the H_2O_2 permeability of plant AQPs was mostly obtained after their expression in yeast, a direct H_2O_2 transport assay in plant cells would be an added value tool. Recently, we took advantage of the H_2O_2 fluorescent probe HyPer to develop an inducible genetic system to evaluate AQP H_2O_2 transport in tobacco BY-2 suspension plant cells. This plant cell system allowed us to confirm the H_2O_2 transport ability of ZmPIP2;5 and AtPIP2;1 (unpublished data). Furthermore, it has recently been shown that reconstitution of purified AQP proteoliposomes provides an accurate quantitative measurement of AQP driven H_2O_2 transport (Wang et al., 2020a, 2020b). However, this assay required the expression and purification of AQPs, a process that is not straightforward for membrane proteins.

While we focus in this chapter on plant AQPs and their roles in cellular H_2O_2 membrane diffusion for which recent interesting data have been obtained, this specific role has also been documented in different organisms, including mammals and bacteria (Bienert & Chaumont, 2014). This is not surprising as AQPs share important structural determinants, even if differences in specific amino acid residues within the pore are also observed (Montiel et al., 2020; Wang et al., 2020a). The involvement of mammalian AQPs (AQP0, AQP1, AQP3, AQP5, AQP8, AQP9 and AQP11) as peroxiporins and thus activators of intracellular downstream signaling events has been extended to many diverse physiological processes such as cell survival, cell migration, cancer growth, sterol production, ER redox homeostasis, myocardial remodeling, mitochondrial function, sperm motility, lens transparency, etc. (Bestetti et al., 2020; Bienert & Chaumont, 2014; Chauvigné et al., 2021; Danielli et al., 2019; Hara-Chikuma et al., 2015; Krüger et al., 2021; Montiel et al., 2020; Rodrigues et al., 2019; Varadaraj and Kumari, 2020). Several bacterial or fungal AQPs act also as peroxiporins, such as LpGlpF isoforms in *Lactobacillus plantarum*, the H_2O_2-inducible So-AqpA from *Streptococcus oligofermentans*, which facilitates cellular H_2O_2 efflux and reduces oxidative stress, or LbAQP1 from the ectomycorrhizal fungi *Laccaria bicolor* (Bienert et al., 2014; Navarro-RóDenas et al., 2015; Tong et al., 2019). These results obtained in different organisms can certainly provide interesting clues to better understand the function and regulation of plant AQPs as peroxiporins.

5. Plant AQP regulation

An important concept to keep in mind is that the flow can be bidirectional. This because solute movement through AQPs is passive and only dependent on the solute concentration gradient on both membrane sides. This can have implications when considering H_2O_2 movement according to the location of its production and its gradient. It is probably why these channels are highly regulated at different levels to control the membrane permeability to solutes (Chaumont & Tyerman, 2014). In addition to *AQP* gene expression regulation, there are different post-translational mechanisms that contribute to membrane permeability to water and H_2O_2.

5.1 Gating

The channel activity of plant AQPs is regulated by gating, and this mechanism is dependent on pH, divalent cations, and phosphorylation events (Hedfalk et al., 2006; Tornroth-Horsefield et al., 2006; Verdoucq et al., 2008). A unified model of pore opening and closing was proposed based on SoPIP2;1 molecular structure (Nyblom et al., 2009; Tornroth-Horsefield et al., 2006). PIPs are highly water permeable at neutral or basic pH but largely impermeable at acidic pH, which is sensed by the protonation state of a highly conserved His residue located in loop D (His193 in SoPIP2;1) (Nyblom et al., 2009; Tournaire-Roux et al., 2003). This protonation induced a conformational change of loop D and the occlusion of the pore by a Leu residue (Leu 197 in SoPIP2;1). The channel opening and closing are also controlled by phosphorylation and dephosphorylation of two serine residues located in loop B and the C-terminus (Ser115 and Ser274), respectively, and by interactions between hydrophobic residues from loop D, the N-terminus Asp28 and Glu31 residues, and loop B (Ser115 and Arg118) (Tornroth-Horsefield et al., 2006). Cd^{2+} and/or Ca^{2+} can also bind to the N-terminus and influence the gating mechanism through a network involved in ionic interactions and hydrogen bonding (Tornroth-Horsefield et al., 2006).

Oxidative gating of plant AQPs was also proposed. A reduction of the plasma membrane osmotic water permeability by HO radicals produced by a Fenton reaction was reported in *C. corallina,* maize leaf parenchyma and root cortex cells, and *Trandenscantia fluminensis* epidermal cells, an inhibition which was reversible after removing the radicals (Henzler et al., 2004; Kim & Steudle, 2009; Ye & Steudle, 2006; Ye et al., 2008). The exact mechanism was unknown, but a possible direct action of the radicals within the pore was proposed, even if an indirect effect on the lipid bilayer oxidation affecting AQPs on the outside was not excluded (Henzler et al., 2004). The oxidation of disulfide bonds shown to link the extracellular loops A of two PIP monomers within a tetramer seems not to be responsible for the observed osmotic permeability inhibition, as tested in swelling assays using *Xenopus* oocytes or by molecular dynamics simulations (Bienert et al., 2012; Cordeiro, 2015). Actually, the absence of the effect of H_2O_2 preincubation on the osmotic water permeability of oocytes expressing AtPIP2;1, AtPIP1;2 and AtTIP1;1, suggested that the water permeability inhibition observed in Arabidopsis root cortical cells upon incubation with H_2O_2 is due to another cellular process, the internalization of AQPs present in the plasma membrane (see below) (Boursiac et al., 2008).

5.2 AQP membrane abundance and trafficking

AQPs are co-translationally inserted into the ER membrane. From there, they follow the secretory pathway towards the Golgi apparatus and the trans-Golgi network (TGN), to reach their final destinations, mostly the vacuole or the plasma membrane (Chevalier & Chaumont, 2015). However, AQPs appeared to have a large variety of subcellular localization patterns, including chloroplasts and mitochondria (Fig. 1) (Møller et al., 2020; Soto et al., 2010; Uehlein et al., 2008). While the trafficking of AQPs into the latter organelles has not been studied, the transport of PIPs into the secretory pathway to the plasma membrane was thoroughly investigated and shown to depend on several protein-protein interactions and post-translational modifications. Also, while in the plasma membrane, AQPs can be internalized through an endocytic route (Hachez et al., 2013).

After synthesis in the rough ER, PIPs assemble as tetramers that might be stabilized by a disulfide bond between the loop A of two adjacent PIP monomers (Bienert et al., 2012; Zelazny et al., 2007). While most PIP1s are retained in the ER when expressed alone, PIP2s are moving forward to the secretory pathway to reach the plasma membrane (Zelazny et al., 2007). PIP2 ER exit is mostly dependent on a diacidic motif present in their N-terminus, a motif known to interact with the Sec24 coat protein of COPII vesicles

(Sorieul et al., 2011; Zelazny et al., 2007), and also on a motif (LxxxA) found in TM3 (Chevalier et al., 2014). When PIP1s are co-expressed with PIP2s, they physically interact in the ER within heterotetramers, and the complexes continue their journey to the plasma membrane, where both PIP1s and PIP2s exhibit channel activities (Berny et al., 2016; Fetter et al., 2004; Zelazny et al., 2007). PIP movement to the plasma membrane also requires physical interaction with SNAREs (soluble N-ethylmaleimide-sensitive factor protein attachment protein receptors), which are proteins involved in vesicle fusion with the target membrane. ZmPIP2;5 physically interacts with SYP121, a plasma membrane Qa-SNARE, regulating their abundance in the plasma membrane and possibly their activity (Besserer et al., 2012; Hachez et al., 2014a). Also, the post-Golgi trafficking of AtPIP2;7 involves specific interaction with both SYP121 and the trans-Golgi localized SNARE, SYP61 (Hachez et al., 2014a). ZmPIP2;5 also interacts with the vesicle-associated membrane protein (VAMP)-associated proteins (VAP27s) at ER-plasma membrane contact sites (EPCSs) and endocytic structures upon ionic stress (Fox et al., 2020). Whether this precise location allows a direct transfer of H_2O_2 from the apoplast to the ER or conversely is unknown and would be an interesting aspect for future investigation.

Other protein-protein interactions are important in abiotic and biotic stresses. Expression of the E3 ubiquitin ligase (Rma1) and the ubiquitin-conjugating enzyme UBC32, which are induced by drought stress, reduce AtPIP2;1 levels in the plasma membrane through a degradation process involving the proteasome, avoiding their deleterious impact on drought tolerance (Lee et al., 2009). Another way to regulate PIP abundance under stress is its physical interaction with the plant translocator protein (TSPO), a multi-stress regulator. Upon ABA treatment, TSPO physically interacts with AtPIP2;7 to reduce its plasma membrane abundance and membrane water permeability via the autophagy mediated degradation pathway (Hachez et al., 2014b; Jurkiewicz et al., 2020).

Phosphorylation is also involved in the alteration of PIP trafficking. Phosphorylation of the C-terminal Ser283 is required for AtPIP2;1 trafficking to the plasma membrane (Prak et al., 2008). Salicylic acid (SA), salt stress, and/or H_2O_2 treatment induced AtPIP2;1 internalization in root cells, a process dependent on the phosphorylation status of Ser208 and Ser283 (Boursiac et al., 2008; Prak et al., 2008). An increase in intracellular GFP-PIP2;1 labeling mutated these serine residues (S280A or S280A) in root epidermal cells, and this effect was enhanced upon salt treatment (Prak et al., 2008).

6. Plasma membrane AQPs and H_2O_2 signaling in plants

AQPs are found in all plant tissues and are significant players in maintaining water and nutrient balance. Therefore, they are key actors in many plant physiological processes including transpiration, seed germination, cell elongation, osmoregulation, plant development, etc. (for review, see Bienert & Chaumont, 2011; Chaumont & Tyerman, 2014; Heinen et al., 2009; Maurel et al., 2015). In this part, we will focus on recent physiological evidence for AQP-mediated H_2O_2 diffusion in plants and in signaling processes.

Upon biotic and abiotic stresses, H_2O_2 is produced by the activated NADPH oxidases, superoxide dismutases, and peroxidases in the apoplast and enters cells through AQPs to contribute to plant signaling mechanisms. This facilitated H_2O_2 diffusion through the plasma membrane was the first demonstration of a physiological role for AQPs as peroxiporins in plants, and was well documented for stomatal closure upon abscisic acid (ABA) treatment and plant defence upon pathogen interaction (Ding et al., 2022; Hirt, 2016; Qi et al., 2017; Rodrigues et al., 2017; Tian et al., 2016; Zhang et al., 2019).

6.1 Stomatal movement

Harsh environmental conditions such as cold, drought, and salt stress induce ABA accumulation, which further activates plant adaptive stress responses. ABA is important in stomatal signaling and closure (Hsu et al., 2021). In Arabidopsis, ABA binds to PYR/PYL/RCAR receptors leading to the inhibition of protein phosphatase (PP2C) activity and the activation of Snf1-related protein kinase 2 (SnRK2.6)/open stomata1 (OST1) kinase (Fig. 3). This activated SnRk2.6/OST1 kinase then phosphorylates several substrate proteins, including membrane transporters, which are involved in stomatal closure such as S-type anion channel SLAC1, NADPH oxidases, and PIP AQPs (Brandt et al., 2012; Geiger et al., 2009; Grondin et al., 2015; Hao et al., 2008; Kwak et al., 2003; Rodrigues et al., 2017; Sirichandra et al., 2009) (Fig. 3). Phosphorylation of the Ser121 of PIP2;1 enhanced its water channel activity, and it was proposed that this process facilitates water exit from the guard cells, resulting in stomatal closure (Grondin et al., 2015). Actually, while the aperture of stomata of *pip2;1* mutated plants is not affected after their transfer to light, their closing response after ABA addition is slower and

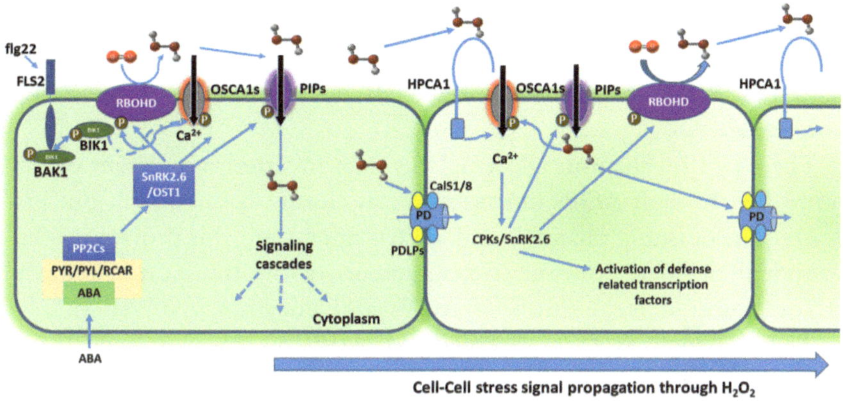

Fig. 3 AQP-mediated H_2O_2 membrane diffusion and its signaling roles. ABA activates the SnRK2.6/OST1 kinases, which phosphorylate several membrane proteins including RBOHD, calcium channels such as OSCA1s, and AQPs. Conversely, Flg22 is perceived by the FLS2 that later phosphorylates RBOHD with the help of BAK1 and BIK1 kinases. Wu et al. (2020) first identified HPCA1 as an extracellular H_2O_2 sensor in plants, which senses the apoplastic H_2O_2 that activates PM localized calcium channels (for instance OSCA1.3) and leads to cytoplasmic Ca^{2+} elevation that activates RBOHD, which contributes to produce apoplastic H_2O_2, which is further taken up to the cell via PIPs. Stresses induce elevated intracellular H_2O_2 that could activate several signaling cascades, such as activation of PM localized Ca^{2+} channels such as OSCA1s, and cell-to-cell communication through open plasmodesmata (PD) by activation of callose synthase genes (CalS1/8) (indicated as a light blue ball) and PD localized proteins (PDLPs) (indicated as yellow ball) at PD.

the stomata remain open compared to the wild-type stomata (Grondin et al., 2015). Also, the use of the genetic probe HyPer revealed that ABA triggers an accumulation of H_2O_2 in wild-type guard cells but not in an *Atpip2;1* KO mutant (Rodrigues et al., 2017), suggesting that AtPIP2;1 contributes to stomatal closure by also facilitating H_2O_2 entry, which in turn acts as a secondary messenger and is sensed by intracellular proteins, leading to activation of inward calcium channels and SLAC1 anion channel (Rodrigues & Shan, 2021).

The contribution of plasma membrane AQPs in facilitating the membrane diffusion of water and H_2O_2 in guard cells to control stomatal closure dynamics was also documented in maize plants over- and down expressing *ZmPIP2;5* (Ding et al., 2022). Upon mild-water deficit or low ABA concentration treatments, transpiration and stomatal conductance (g_s) decreased more in *PIP2;5* overexpressing (OE), and less in *pip2;5* KO lines, in comparison with WTs. In detached epidermis, ABA treatment induced faster stomatal closure in PIP2;5 OE lines compared to WT, while *pip2;5* KO stomata were ABA insensitive. These phenotypes were associated with guard cell ROS accumulation, detected by H_2DCFDA staining (Ding et al., 2022).

The exact mechanism allowing this dual inverted movement of water and H_2O_2 through AQPs during stomatal closure is unknown, but probably requires distinct channels and/or tetramers, and this process might be dependent on local concentration gradients of the transported molecules and/or clustered localization of AQPs and enzymes involved in H_2O_2 production, such as that observed for AtPIP2;1 in membrane microdomains (Cui et al., 2021; Wang et al., 2015). As various AQPs are expressed in the stomatal complexes, determining the contribution of each isoform to the stomatal dynamic behavior in relation to signaling molecules including H_2O_2 will be important for improving plant water use (Ding & Chaumont, 2020; Heinen et al., 2014).

6.2 Pathogen infection

Upon pathogen interaction, plants produce apoplastic H_2O_2, which induces an efficient intracellular immunity to overcome the infection. A role of AQPs in this signaling process was clearly demonstrated in several studies (Fig. 3). Upon bacterial infection of Arabidopsis leaves, an induction of *AtPIP1;4* gene expression is observed with concomitant H_2O_2 accumulation into the cytoplasm (Tian et al., 2016). This intracellular H_2O_2 accumulation is reduced or increased in *Atpip1;4* KO and *AtPIP1;4* OE plants, respectively, and triggers systemic acquired resistance (SAR) and pathogen-associated molecular patterns-triggered immunity (PTI). In wheat, fungal attack also induces apoplastic H_2O_2 production and its translocation to the cytoplasm through TaPIP2;10, resulting in a restriction of fungal attachment by activating PTI (Wang et al., 2021). *TaPIP2;10* overexpression enhances the resistance to powdery mildew and Fusarium head blight. In rice, OsPIP2;2 mediates the membrane diffusion of pathogen-induced apoplastic H_2O_2 into the cell to activate the MAPK cascade and downstream defence responses (Zhang et al., 2022). In response to the bacterial flagellin-derived synthetic peptide flg22 known to induce plant immune responses, the Ser125 of OsPIP2;2 is phosphorylated to activate H_2O_2 transport, and also triggers the translocation of the membrane-anchored transcription factor OsmaMYB into the nucleus to regulate gene expression and positively modulate plant innate immunity.

Actually, flg22 peptide is initially perceived by the receptor kinase flagellin sensing 2 (FLS2) in combination with BAK1 and BIK1 kinases (Mersmann et al., 2010), and activates RBOHD and apoplastic H_2O_2 production (Khokon et al., 2011). This process is important to induce cellular signaling and molecular responses and trigger stomatal closure, therefore preventing pathogen entry into the plant (Fig. 3). Rodrigues et al. (2017) showed that, similarly to ABA treatment, flg22-induced cell accumulation of H_2O_2 is dependent on AtPIP2;1 and its phosphorylation by BAK1.

AQP was also shown to be a plant-immunity-related gene. The conserved kinase CRN78 effector from *Phytophthora* promotes infection and physically interacts with NbPIP2;2 from *Nicotiana benthamiana* or GmPIP2;13 from soybean. This protein interaction mediates phosphorylation of the PIP C-terminus and PIP degradation via a proteasome-dependent mechanism (Ai et al., 2021). Interestingly, both NbPIP2;2 and GmPIP2;13 facilitate H_2O_2 membrane diffusion when expressed in *Nicotiana benthamiana* leaves. This effector-mediated mechanism, by targeting AQPs, might prevent H_2O_2 entry and the downstream signaling cascade leading to plant immunity (Ai et al., 2021).

6.3 Systemic signaling

An elegant study investigated the role of PIP AQPs in systemic ROS signaling in Arabidopsis (Fichman et al., 2021). In addition to inducing intracellular signaling, the apoplast-to-cytosol translocation of H_2O_2 is also responsible for cell to cell communication of the systemic signal, resulting in systemic acquired acclimation (SAA) throughout the whole plant (Fig. 3) (Miller et al., 2009). This involved the symplastic route through plasmodesmata, which connect the cytoplasm of adjacent cells. Furthermore, the apoplastic H_2O_2 produced by one cell can also move in the apoplastic space and enter neighboring cells to trigger systemic signaling, acclimation, and defence mechanisms. Using live imaging and ROS sensitive dye, Fichman et al. (2021) showed that, upon local high light stress application, RBOHD and plasmodesmata-localized protein 1 and 5 (PDLP1 and PDLP5) increase cell to cell transport and plasmodesmata pore size, and are required for the propagation of the systemic ROS signals and SAA (Fichman et al., 2021). Actually, *Atrobhd* and *Atpdlp5* KO plants display a decrease of carboxyfluorescein spread and plasmodesmata pore area size in response to local high light stress treatment, suggesting that RBOHD-generated ROS promote systemic cell to cell transport through a control of the plamodesmata pore size and, therefore, callose deposition in the plasmodesmata edge (Cui & Lee, 2016; Fichman et al., 2021).

The authors also tested the role of AtPIPs in this process. While *Atpip1;4* KO plants exhibited a systemic ROS response similar to wild-type plants, *Atpip2;1* KO plants were completely deficient in the initiation and propagation of this systemic response, and *Atpip1;2* KO enhanced the rate of ROS signal propagation. As mentioned above, AtPIP2;1 facilitates H_2O_2 diffusion (Dynowski et al., 2008; Rodrigues et al., 2017). Furthermore, grafting experiments suggested that AtPIP2;1 has a cell-autonomous function and is required for the amplification of the systemic signal in each individual cell along the path of the high light-induced systemic ROS wave (Fichman et al., 2021). The ROS entering cells through PIP2;1 is likely to be mobilized between cells through plasmodesmata. The different effect of *Atpip1;2* and *Atpip2;1* KO on the spread of systemic ROS signals in plants could be due to differences in their channel activity and/or regulation mechanisms such as heterotetrameric assembly, phosphorylation or other post-translational modifications, oxidative status and/or specific protein interaction known to modulate their subcellular localization, fate and activity (see above).

7. AQP-driven chloroplastic H_2O_2 signaling

The chloroplast membranes include the inner and outer envelopes and the thylakoid membrane containing the light-harvesting antenna and the photosynthetic electron transport chain. Oxygenic photosynthesis produces large amounts of ROS, including H_2O_2 produced by thylakoid-bound and stromal dismutases (Asada, 1999). H_2O_2 can move out of the chloroplast and trigger signaling to regulate energy metabolism and genetic and epigenetic responses, which contribute to plant adaptation (Fig. 1) (Foyer & Hanke, 2022). This process is associated with retrograde chloroplast-to-nucleus signaling. Different mechanisms are considered to explain this retrograde signaling: physical connections (stromules), organelle proximity, and formation of protein complexes between the organelles forming contact sites (Mittler et al., 2022; Mullineaux et al., 2020). This can also involve the ER membrane, which is contiguous with the nuclear envelope and forms physical interaction with chloroplasts (Anderson & Hetzer, 2007; Breeze & Mullineaux, 2022). While chloroplastic H_2O_2 movement toward the nucleus in stromules does not require crossing any membranes, this is not the case in organelle proximity or contact sites. Whether the latter requires AQP activity is still a matter of debate. The use of the Amplex red probe, which forms fluorescent resorufin when reacting with H_2O_2, showed that the fluorescence outside isolated chloroplasts is suppressed by 60% in the presence of the AQP inhibitor acetazolamide, suggesting that H_2O_2 can diffuse across the chloroplast envelope through AQPs (Borisova et al., 2012).

AQPs from the PIP and TIP subfamilies have been identified in chloroplast membranes. *Nicotiana tabacum* NtAQP1 (a PIP1 isoform) is present in both the plasma membrane and chloroplast inner envelope, and was shown to facilitate CO_2 diffusion into chloroplasts (Uehlein et al., 2003, 2008). However, to our knowledge, its contribution in H_2O_2 transport is still unknown. AQPs were also identified in proteomic analysis from Arabidopsis chloroplast envelopes and thylakoid membrane (Ferro et al., 2010; Simm et al., 2013). AtPIP1;2, AtPIP1;3, AtPIP2;1, AtPIP2;4, and AtPIP2;7 were detected in the envelope fraction, even if their presence could also be due to putative contamination as proposed for tonoplast AQPs (AtTIP1;1; AtTIP1;2 and AtTIP2;1) detected in the envelope (Ferro et al., 2010). AtPIP2;1, AtPIP2;4, and AtPIP2;7 were shown to be H_2O_2 channels when expressed in yeast and/or plant cells (Dynowski et al., 2008; Hooijmaijers et al., 2012; Rodrigues et al., 2017). Therefore, they could facilitate H_2O_2 transfer from the chloroplasts to the cytosol or to the associated organelles, and contribute to the retrograde ROS signaling.

AtTIP1;2 and AtTIP2;1 were specifically detected in the thylakoid fraction. The presence of AQPs in the thylakoid membrane has been considered for their water channel activity, and are proposed to regulate volume changes of the thylakoid lumen in light (Beebo et al., 2013). However, as H_2O_2 is also generated inside thylakoid membranes (Mubarakshina et al., 2006), a role in H_2O_2 membrane diffusion for the stroma is not excluded, and would need to be investigated in detail.

8. AQPs and their putative contribution in ROS signaling in mitochondria, peroxisomes, and ER membranes

ROS production in other organelles such as mitochondria, ER, and peroxisomes, and the subsequent signaling also imply H_2O_2 movement across membranes (Fig. 1). While no plant AQPs have been identified in peroxisomal membranes, proteomics analysis showed that six Arabidopsis AQPs (AtTIP1;1, AtTIP2;1, AtTIP5;1, AtPIP1;4, AtPIP2;2, and AtPIP2;7) localize to the inner mitochondria (Møller et al., 2020). AtTIP5;1 was also in the mitochondria of pollen tubes, but was proposed to be involved in nitrogen recycling (Soto et al., 2010). Human AQP8 was demonstrated to be a H_2O_2 pore in the inner mitochondrial membrane of different cell lines (Chauvigné et al., 2015; Krüger et al., 2021; Marchissio et al., 2012), making plant AQPs also plausible channels to regulate mitochondrial H_2O_2 fate and signaling.

AQPs are also found in the ER membrane. As previously explained, PIP1s expressed alone are retained in the ER (Zelazny et al., 2007) but are mainly found in the plasma membrane in intact tissues, due to their interaction with PIP2s. However, there is a subfamily of plant AQPs that are mostly present in ER, the SIPs (Ishikawa et al., 2005). These proteins are ubiquitously expressed in all plant tissues. Curiously, yeast strains expressing *Glycine max* GmSIP1;3, maize ZmSIP1;1, or ZmSIP2;1 are more resistant to increased H_2O_2 concentrations in the medium (Zhang et al., 2017; unpublished data). *GmSIP1;3* overexpression confers growth retardation with shorter root length. Also, *GmSIP1;3* OE seeds germinate earlier than wild type in the presence of H_2O_2, indicating that the protein confers tolerance to H_2O_2 and therefore plays a role in protecting plants from oxidative stress (Zhang et al., 2017). As mentioned above, the plasma membrane ZmPIP2;5 was also found in EPCS in direct interaction with the ER resident VAP27;1 (Fox et al., 2020). This localization was proposed to stabilize AQPs and guide their endocytosis in response to salt stress. However, we cannot exclude that this specific EPCS location also plays a role in the direct transfer of H_2O_2 from the apoplast to the ER.

9. AQPs at the crossroads of Ca^{2+} and H_2O_2 signaling

During stress conditions, a challenge for plants is to maintain water and solute homoeostasis (Mahdieh et al., 2008). Therefore, plants require elaborate adaptive responses to tackle water limitation in harsh environments. AQPs are certainly important actors in these responses and contribute in maintaining water and solute balance upon stress conditions (Chaumont & Tyerman, 2014). An important aspect is the multiple substrate channel activity of AQPs that led us to hypothesize that each function is interlinked. For instance, salt stress induces ionic, osmotic, and oxidative stress leading to H_2O_2 production in root tissues that functions as an intermediate between salt stress perception and regulation of root water transport by AQPs (Boursiac et al., 2008; Demidchik et al., 2003; Leshem et al., 2006). Also, AQP-mediated H_2O_2 membrane diffusion activates H_2O_2 dependent signaling pathways, which also affect AQP phosphorylation status (Prak et al., 2008; Rodrigues et al., 2017). Indeed, H_2O_2 treatment alters the AtPIP2;1C-terminus phosphorylation state (Prak et al., 2008). Also, receptor-like kinase (RLK) and FERONIA potentially physically interact with AtPIP2;1 in response to NaCl and H_2O_2 treatments to modify PIP phosphorylation status by unknown partners (Bellati et al., 2016). Furthermore, salt stress induces PIP internalization, which is mediated by H_2O_2 (Boursiac et al., 2008; Pou et al., 2016). Altogether, these observations suggest that H_2O_2 is a key signaling intermediate during salt stress that regulates PIP phosphorylation, activity, and subcellular dynamics to contribute to plant salt tolerance.

AQP-driven H_2O_2 transport possibly activates Ca^{2+} channels by indirectly modifying their phosphorylation status (Fig. 3). Activation of Ca^{2+} channels increases the Ca^{2+} influx into cells, which directly activates Ca^{2+} dependent kinases (CPKs), leading to the activation of RBOHD to continue the positive feedback loop (Dubiella et al., 2013). Besides, pathogen infection induces oxidative burst leading to transient Ca^{2+} increases, possibly by activating Ca^{2+} permeable channels. The plasma membrane localized AtOSCA1.3, a newly identified calcium channel, is rapidly phosphorylated upon flg22 perception (Thor et al., 2020). Possibly, this phosphorylation event activates OSCA1.3, triggers Ca^{2+} influx, and ultimately activates further RBOHD via another phosphorylation event (Fig. 3). Indeed Ca^{2+} cytosolic import triggers the activation of several membrane components including RBOHD and AQPs through CPK-dependent phosphorylation events to initiate the cytoplasmic import of H_2O_2 and activate PTI (Lee et al., 2020; Marcec et al., 2019).

Several studies proposed that intracellular Ca^{2+} and ROS signals are interlinked (Choi et al., 2014; Gilroy et al., 2016; Fichman & Mittler, 2020). During stress conditions, RBOHD generated H_2O_2 increased the Ca^{2+} wave by activating glutamate receptor-like Ca^{2+} channels (Gilroy et al., 2014). ROS waves are also involved in generation of electric signals (Suzuki et al., 2013). Mechano-activated Ca^{2+} channels were suggested to be involved in hydraulic waves initiation (Gilroy et al., 2016; Fichman & Mittler, 2020). Altogether, Ca^{2+}, ROS, and electric and hydraulic waves seem to interplay together. Also, electron transfer from the cell to the apoplast and RBOHD mediated H_2O_2 production could modify the pH and the membrane potential and regulate plasma membrane proteins such as H^+-ATPase (AHA1), AQPs, or other channels (Fichman & Mittler, 2020). However, several components of this complex signaling network still need to be experimentally validated. Furthermore, as explained above, AQP-driven cytoplasmic H_2O_2 entry might influence the cell-to cell connectivity by regulating plasmodesmata aperture (Fichman et al., 2021). Plasmodesmata are directly involved in cell to cell transport of plant nutrients and pathogens (Cui & Lee, 2016). For instance, plant viruses move from cell to cell by plasmodesmata, which could restrict the virus movement upon H_2O_2-mediated plasmodesmata closure (Tilsner et al., 2013; Wang, 2021). Systemic ROS signaling is also regulated by Ca^{2+} permeable channels (Glutamate receptor-like Ca^{2+} channels (GLR), Cyclic nucleotide gated channel (CNGC) and Mechano-activated Ca^{2+} channel (MSL) in addition to RBOHD, PLD1, PLD5, and PIP2;1 (Fichman et al., 2021). All these observations suggest that AQP-mediated H_2O_2 transport might play an important signaling intermediate in ROS and Ca^{2+} signaling.

10. Conclusion and perspectives

Current knowledge strongly suggests that AQPs are mediating facilitated membrane diffusion of H_2O_2 into and within plant cells, and that this activity triggers intra- and intercellular ROS signaling involved in many physiological processes, notably in relation to biotic and abiotic stress. This AQP-driven H_2O_2 transport activity seems to be an important process for initiating common stress signaling to the cell for plant acclimation, in interaction with other redox enzymes, Ca^{2+} channels, and other transporters. However, how all these actors are interconnected and the molecular mechanisms behind are still poorly understood. Recent technological advances such as genetically encoded ROS probes, super-resolution microscopy, and genome editing techniques will provide ways to explain the role of specific AQP isoforms and their potential partners in plant stress responses. A major difficulty will be to differentiate the H_2O_2 channel activity from the other substrate specificities of AQPs that might contribute to the observed phenotypes. Indeed, a plant AQP with a unique peroxiporin activity was not identified yet, and the H_2O_2 substrate specificity is expected to be very difficult to engineer. An interesting aspect to investigate is the contribution of AQPs in organelle to organelle ROS retrograde signaling (mitochondria, chloroplast, or ER to nucleus), which is an important component of plant response to environmental stimuli. While AQPs are found in the organellar membranes, their role as peroxiporins in these membranes was not demonstrated yet. Currently, the functional relationship between AQPs and H_2O_2 in plant immunity is well established. However, how pathogens escape plant immunity still needs to be investigated. It was recently proposed that pathogens target AQPs to inhibit their H_2O_2 channel activity and escape the host plant defence system (Ai et al., 2021), but this attractive mechanism would need additional experiments to be validated.

Overall, a better understanding of AQPs in mediating H_2O_2 cell autonomous and plant systemic signaling, together with characterization of molecular mechanisms/partners regulating their peroxiporin activity would be very useful to improve/select plants with better abiotic and/or biotic stress tolerance.

Acknowledgments

This work was supported by the Belgian National Fund for Scientific Research (FNRS, FRFC 2.4.501.06F), the "Communauté française de Belgique-Actions de Recherches Concertées" (grants ARC16/21-075). J.A. was supported by a research fellowship from the Fonds de Formation à la Recherche dans l'Industrie et l'Agriculture.

References

Ahmed, J., Mercx, S., Boutry, M., & Chaumont, F. (2020). Evolutionary and predictive functional insights into the aquaporin gene family in the allotetraploid plant *Nicotiana tabacum*. *International Journal of Molecular Sciences, 21*, 4743.

Ai, G., Xia, Q., Song, T., Li, T., Zhu, H., Peng, H., et al. (2021). A *Phytophthora sojae* CRN effector mediates phosphorylation and degradation of plant aquaporin proteins to suppress host immune signaling. *PLoS Pathogens, 17*, e1009388.

Aller, I., & Meyer, A. J. (2013). The oxidative protein folding machinery in plant cells. *Protoplasma, 250*(4), 799–816.

Almasalmeh, A., Krenc, D., Wu, B., & Beitz, E. (2014). Structural determinants of the hydrogen peroxide permeability of aquaporins. *The FEBS Journal, 281*(3), 647–656.

Anderberg, H. I., Kjellbom, P., & Johanson, U. (2012). Annotation of *Selaginella moellendorffii* major intrinsic proteins and the evolution of the protein family in terrestrial plants. *Frontiers in Plant Science, 3*, 33.

Anderson, D. J., & Hetzer, M. W. (2007). Nuclear envelope formation by chromatin-mediated reorganization of the endoplasmic reticulum. *Nature Cell Biology, 9*, 1160–1166.

Asada, K. (1999). The water-water cycle in chloroplasts: Scavenging of active oxygens and dissipation of excess photons. *Annual Review of Plant Physiology and Plant Molecular Biology, 50*(1), 601–639.

Azad, A. K., Ahmed, J., Alum, M. A., Hasan, M. M., Ishikawa, T., Sawa, Y., et al. (2016). Genome-wide characterization of major intrinsic proteins in four grass plants and their non-aqua transport selectivity profiles with comparative perspective. *PLoS One, 11*, e0157735.

Azad, A. K., Yoshikawa, N., Ishikawa, T., Sawa, Y., & Shibata, H. (2012). Substitution of a single amino acid residue in the aromatic/arginine selectivity filter alters the transport profiles of tonoplast aquaporin homologs. *Biochimica et Biophysica Acta (BBA)-Biomembranes, 1818*, 1–11.

Bansal, A., & Sankararamakrishnan, R. (2007). Homology modeling of major intrinsic proteins in rice, maize and Arabidopsis: Comparative analysis of transmembrane helix association and aromatic/arginine selectivity filters. *BMC Structural Biology, 7*, 27.

Beebo, A., Mathai, J. C., Schoefs, B., & Spetea, C. (2013). Assessment of the requirement for aquaporins in the thylakoid membrane of plant chloroplasts to sustain photosynthetic water oxidation. *FEBS Letters, 587*, 2083–2089.

Bellati, J., Champeyroux, C., Hem, S., Rofidal, V., Krouk, G., Maurel, C., et al. (2016). Novel aquaporin regulatory mechanisms revealed by interactomics. *Molecular & Cellular Proteomics, 15*, 3473–3487.

Berny, M. C., Gilis, D., Rooman, M., & Chaumont, F. (2016). Single mutations in the transmembrane domains of maize plasma membrane aquaporins affect the activity of monomers within a heterotetramer. *Molecular Plant, 9*, 986–1003.

Besserer, A., Burnotte, E., Bienert, G. P., Chevalier, A. S., Errachid, A., Grefen, C., et al. (2012). Selective regulation of maize plasma membrane aquaporin trafficking and activity by the SNARE SYP121. *The Plant Cell, 24*, 3463–3481.

Bestetti, S., Galli, M., Sorrentino, I., Pinton, P., Rimessi, A., Sitia, R., & Medraño-Fernandez, I. (2020). Human aquaporin-11 guarantees efficient transport of H_2O_2 across the endoplasmic reticulum membrane. *Redox Biology, 28*, 101326.

Bienert, G. P., Bienert, M. D., Jahn, T. P., Boutry, M., & Chaumont, F. (2011). Solanaceae XIPs are plasma membrane aquaporins that facilitate the transport of many uncharged substrates. *The Plant Journal, 66*, 306–317.

Bienert, G. P., Cavez, D., Besserer, A., Berny, M. C., Gilis, D., Rooman, M., et al. (2012). A conserved cysteine residue is involved in disulfide bond formation between plant plasma membrane aquaporin monomers. *Biochemical Journal, 445*, 101–111.

Bienert, G. P., & Chaumont, F. (2011). Plant aquaporins: Roles in water homeostasis, nutrition, and signaling processes. In M. G. A. K. Venema (Ed.). *Transporters and pumps in plant signaling* (pp. 3–36). Berlin-Heidelberg: Spinger-Verlag.

Bienert, G. P., & Chaumont, F. (2014). Aquaporin-facilitated transmembrane diffusion of hydrogen peroxide. *Biochimica et Biophysica Acta (BBA)-General Subjects, 1840*, 1596–1604.

Bienert, G. P., Heinen, R. B., Berny, M. C., & Chaumont, F. (2014). Maize plasma membrane aquaporin ZmPIP2;5, but not ZmPIP1;2, facilitates transmembrane diffusion of hydrogen peroxide. *Biochimica et Biophysica Acta (BBA)-Biomembranes, 1838*, 216–222.

Bienert, G. P., Moller, A. L., Kristiansen, K. A., Schulz, A., Moller, I. M., Schjoerring, J. K., et al. (2007). Specific aquaporins facilitate the diffusion of hydrogen peroxide across membranes. *Journal of Biological Chemistry, 282*, 1183–1192.

Bienert, G. P., Schjoerring, J. K., & Jahn, T. P. (2006). Membrane transport of hydrogen peroxide. *Biochimica et Biophysica Acta (BBA)-Biomembranes, 1758*, 994–1003.

Borisova, M. M. M., Kozuleva, M. A., Rudenko, N. N., Naydov, I. A., Klenina, I. B., & Ivanov, B. N. (2012). Photosynthetic electron flow to oxygen and diffusion of hydrogen peroxide through the chloroplast envelope via aquaporins. *Biochimica et Biophysica Acta (BBA)-Bioenergetics, 1817*, 1314–1321.

Boursiac, Y., Boudet, J., Postaire, O., Luu, D. T., Tournaire-Roux, C., & Maurel, C. (2008). Stimulus-induced downregulation of root water transport involves reactive oxygen species-activated cell signalling and plasma membrane intrinsic protein internalization. *The Plant Journal, 56*, 207–218.

Brandt, B., Brodsky, D. E., Xue, S., Negi, J., Iba, K., Kangasjärvi, J., et al. (2012). Reconstitution of abscisic acid activation of SLAC1 anion channel by CPK6 and OST1 kinases and branched ABI1 PP2C phosphatase action. *Proceedings of the National Academy of Sciences, 109*(26), 10593–10598.

Breeze, E., & Mullineaux, P. M. (2022). The passage of H_2O_2 from chloroplasts to their associated nucleus during retrograde signalling: Reflections on the role of the nuclear envelope. *Plants, 11*, 552.

Byrt, C. S., Zhao, M., Kourghi, M., Bose, J., Henderson, S. W., Qiu, J., et al. (2016). Non-selective cation channel activity of aquaporin AtPIP2;1 regulated by Ca^{2+} and pH. *Plant, Cell & Environment, 40*(6), 802–815.

Castro, B., Citterico, M., Kimura, S., Stevens, D. M., Wrzaczek, M., & Coaker, G. (2021). Stress-induced reactive oxygen species compartmentalization, perception and signalling. *Nature Plants,* 1–10.

Chaumont, F., Barrieu, F., Wojcik, E., Chrispeels, M. J., & Jung, R. (2001). Aquaporins constitute a large and highly divergent protein family in maize. *Plant Physiology, 125*, 1206–1215.

Chaumont, F., Moshelion, M., & Daniels, M. J. (2005). Regulation of plant aquaporin activity. *Biology of the Cell, 97*, 749–764.

Chaumont, F., & Tyerman, S. D. (2014). Aquaporins: Highly regulated channels controlling plant water relations. *Plant Physiology, 164*, 1600–1618.

Chauvigné, F., Boj, M., Finn, R. N., & Cerda, J. (2015). Mitochondrial aquaporin-8-mediated hydrogen peroxide transport is essential for teleost spermatozoon motility. *Scientific Reports, 5*, 1–10.

Chauvigné, F., Ducat, C., Ferré, A., Hansen, T., Carrascal, M., Abián, et al. (2021). A multiplier peroxiporin signal transduction pathway powers piscine spermatozoa. *Proceedings of the National Academy of Sciences, 118*(10), e2019346118.

Chevalier, A, Bienert, G. P., & Chaumont, F. (2014). A new LxxxA motif in the transmembrane helix 3 of maize PIP2 aquaporins is required for their trafficking to the plasma membrane. *Plant Physiology, 166*, 125–138.

Chevalier, A. S., & Chaumont, F. (2015). The LxxxA motif in the third transmembrane helix of the maize aquaporin ZmPIP2;5 acts as an ER export signal. *Plant Signaling and Behavior, 10*, e990845.

Choi, W.-G., Toyota, M., Kim, S.-H., Hilleary, R., & Gilroy, S. (2014). Salt stress-induced Ca^{2+} waves are associated with rapid, long-distance root-to-shoot signaling in plants. *Proceedings of the National Academy of Sciences, 111*, 6497–6502.

Cordeiro, R. M. (2015). Molecular dynamics simulations of the transport of reactive oxygen species by mammalian and plant aquaporins. *Biochimica et Biophysica Acta (BBA)-General Subjects, 1850*, 1786–1794.

Cui, W., & Lee, J. Y. (2016). Arabidopsis callose synthases CalS1/8 regulate plasmodesmal permeability during stress. *Nature Plants, 2*, 1–9.

Cui, Y., Zhao, Y., Lu, Y., Su, X., Chen, Y., Shen, Y., et al. (2021). *In vivo* single-particle tracking of the aquaporin AtPIP2;1 in stomata reveals cell type-specific dynamics. *Plant Physiology, 185*, 1666–1681.

Danielli, M., Marrone, J., Capiglioni, A. M., & Marinelli, R. A. (2019). Mitochondrial aquaporin-8 is involved in SREBP-controlled hepatocyte cholesterol biosynthesis. *Free Radical Biology and Medicine, 131*, 370–375.

Danielson, J.Å., & Johanson, U. (2008). Unexpected complexity of the aquaporin gene family in the moss *Physcomitrella patens*. *BMC Plant Biology, 8*, 45.

Daudi, A., Cheng, Z., O'brien, J. A., Mammarella, N., Khan, S., Ausubel, F. M., et al. (2012). The apoplastic oxidative burst peroxidase in Arabidopsis is a major component of pattern-triggered immunity. *The Plant Cell, 24*, 275–287.

Demidchik, V., Shabala, S. N., Coutts, K. B., Tester, M. A., & Davies, J. M. (2003). Free oxygen radicals regulate plasma membrane Ca^{2+}-and K^+-permeable channels in plant root cells. *Journal of Cell Science, 116*, 81–88.

Ding, L., & Chaumont, F. (2020). Are aquaporins expressed in stomatal complexes promising targets to enhance stomatal dynamics? *Frontiers in Plant Science, 11*, 458.

Ding, L., Milhiet, T., Parent, B., Meziane, A., Tardieu, F., & Chaumont, F. (2022). The plasma membrane aquaporin ZmPIP2;5 enhances the sensitivity of stomatal closure to water deficit. *Plant, Cell & Environment, 45*, 1146–1156.

Dubiella, U., Seybold, H., Durian, G., Komander, E., Lassig, R., Witte, C. P., et al. (2013). Calcium-dependent protein kinase/NADPH oxidase activation circuit is required for rapid defense signal propagation. *Proceedings of the National Academy of Sciences, 110*, 8744–8749.

Dynowski, M., Schaaf, G., Loque, D., Moran, O., & Ludewig, U. (2008). Plant plasma membrane water channels conduct the signalling molecule H_2O_2. *Biochemical Journal, 414*, 53–61.

Ferro, M., Brugière, S., Salvi, D., Seigneurin-Berny, D., Moyet, L., Ramus, et al. (2010). AT_CHLORO, a comprehensive chloroplast proteome database with subplastidial localization and curated information on envelope proteins. *Molecular & Cellular Proteomics, 9*, 1063–1084.

Fetter, K., Van Wilder, V., Moshelion, M., & Chaumont, F. (2004). Interactions between plasma membrane aquaporins modulate their water channel activity. *The Plant Cell, 16*, 215–228.

Fichman, Y., & Mittler, R. (2020). Rapid systemic signaling during abiotic and biotic stresses: Is the ROS wave master of all trades? *The Plant Journal, 102*, 887–896.

Fichman, Y., Myers, R. J., Grant, D. G., & Mittler, R. (2021). Plasmodesmata-localized proteins and ROS orchestrate light-induced rapid systemic signaling in Arabidopsis. *Science Signaling, 14*, eabf0322.

Fox, A. R., Scochera, F., Laloux, T., Filik, K., Degand, H., Morsomme, et al. (2020). Plasma membrane aquaporins interact with the endoplasmic reticulum resident VAP27 proteins at ER–PM contact sites and endocytic structures. *New Phytologist, 228*, 973–988.

Foyer, C. H., & Hanke, G. (2022). *Abiotic stress and adaptation of electron transport: Regulation of the production and processing of ROS signals in chloroplasts. Photosynthesis in Action.* Elsevier,85–102.

Froger, A., Thomas, D., Delamarche, C., & Tallur, B. (1998). Prediction of functional residues in water channels and related proteins. *Protein Science, 7*, 1458–1468.

Fu, D., Libson, A., Miercke, L. J., Weitzman, C., Nollert, P., Krucinski, J., et al. (2000). Structure of a glycerol-conducting channel and the basis for its selectivity. *Science, 290*, 481–486.

Geiger, D., Scherzer, S., Mumm, P., Stange, A., Marten, I., Bauer, H., et al. (2009). Activity of guard cell anion channel SLAC1 is controlled by drought-stress signaling kinase-phosphatase pair. *Proceedings of the National Academy of Sciences, 106*, 21425–21430.

Gilroy, S., Białasek, M., Suzuki, N., Górecka, M., Devireddy, A. R., Karpiński, S., et al. (2016). ROS, calcium, and electric signals: Key mediators of rapid systemic signaling in plants. *Plant Physiology, 171*, 1606–1615.

Gilroy, S., Suzuki, N., Miller, G., Choi, W. G., Toyota, M., Devireddy, A. R., et al. (2014). A tidal wave of signals: Calcium and ROS at the forefront of rapid systemic signaling. *Trends in Plant Science, 19*, 623–630.

Grondin, A., Rodrigues, O., Verdoucq, L., Merlot, S., Leonhardt, N., & Maurel, C. (2015). Aquaporins contribute to ABA-triggered stomatal closure through OST1-mediated phosphorylation. *The Plant Cell, 27*, 1945–1954.

Guo, H., Gu, L., Liu, F., Chen, F., Ge, F., & Sun, Y. (2019). Aphid-borne viral spread is enhanced by virus-induced accumulation of plant reactive oxygen species. *Plant Physiology, 179*, 143–155.

Gupta, A. B., & Sankararamakrishnan, R. (2009). Genome-wide analysis of major intrinsic proteins in the tree plant *Populus trichocarpa*: Characterization of XIP subfamily of aquaporins from evolutionary perspective. *BMC Plant Biology, 9*, 134.

Gustavsson, S., Lebrun, A. S., Norden, K., Chaumont, F., & Johanson, U. (2005). A novel plant major intrinsic protein in *Physcomitrella patens* most similar to bacterial glycerol channels. *Plant Physiology, 139*, 287–295.

Hachez, C., Besserer, A., Chevalier, A. S., & Chaumont, F. (2013). Insights into plant plasma membrane aquaporin trafficking. *Trends in Plant Science, 18*, 344–352.

Hachez, C., Laloux, T., Reinhardt, H., Cavez, D., Degand, H., Grefen, C., et al. (2014a). Arabidopsis SNAREs SYP61 and SYP121 coordinate the trafficking of plasma membrane aquaporin PIP2;7 to modulate the cell membrane water permeability. *The Plant Cell, 26*, 3132–3147.

Hachez, C., Veljanovski, V., Reinhardt, H., Guillaumot, D., Vanhee, C., Chaumont, F., et al. (2014b). The Arabidopsis abiotic stress-induced TSPO-related protein reduces cell-surface expression of the aquaporin PIP2;7 through protein-protein interactions and autophagic degradation. *The Plant Cell, 26*, 4974–4990.

Hao, F., Zhang, J., Yu, Z., & Chen, J. (2008). Involvement of NADPH oxidase NtrbohD in the rapid production of H_2O_2 induced by ABA in cultured tobacco cell line BY-2. *Progress in Natural Science, 18*, 267–271.

Hara-Chikuma, M., Satooka, H., Watanabe, S., Honda, T., Miyachi, Y., Watanabe, et al. (2015). Aquaporin-3-mediated hydrogen peroxide transport is required for NF-κB signalling in keratinocytes and development of psoriasis. *Nature Communications, 6*(1), 1–14.

Hedfalk, K., Tornroth-Horsefield, S., Nyblom, M., Johanson, U., Kjellbom, P., & Neutze, R. (2006). Aquaporin gating. *Current Opinion in Structural Biology, 16*, 447–456.

Heinen, R. B., Bienert, G. P., Cohen, D., Chevalier, A. S., Uehlein, N., Hachez, C., et al. (2014). Expression and characterization of plasma membrane aquaporins in stomatal complexes of *Zea mays*. *Plant Molecular Biology, 86*, 335–350.

Heinen, R. B., Ye, Q., & Chaumont, F. (2009). Role of aquaporins in leaf physiology. *Journal of Experimental Botany, 60*, 2971–2985.

Henderson, S. W., Nourmohammadi, S., Ramesh, S. A., & Yool, A. J. (2022). Aquaporin ion conductance properties defined by membrane environment, protein structure, and cell physiology. *Biophysical Reviews*, 1–18.

Henzler, T., & Steudle, E. (2000). Transport and metabolic degradation of hydrogen peroxide in *Chara corallina*: model calculations and measurements with the pressure probe suggest transport of H_2O_2 across water channels. *Journal of Experimental Botany, 51*, 2053–2066.

Henzler, T., Ye, Q., & Steudle, E. (2004). Oxidative gating of water channels (aquaporins) in Chara by hydroxyl radicals. *Plant, Cell & Environment, 27*, 1184–1195.

Heymann, J. B., & Engel, A. (2000). Structural clues in the sequences of the aquaporins. *Journal of Molecular Biology, 295*, 1039–1053.

Hirt, H. (2016). Aquaporins link ROS signaling to plant immunity. *Plant Physiology, 171*, 1540.

Hooijmaijers, C., Rhee, J. Y., Kwak, K. J., Chung, G. C., Horie, T., Katsuhara, M., et al. (2012). Hydrogen peroxide permeability of plasma membrane aquaporins of *Arabidopsis thaliana*. *Journal Plant Research, 125*, 147–153.

Hossain, M. A., Bhattacharjee, S., Armin, S. M., Qian, P., Xin, W., Li, H.-Y., et al. (2015). Hydrogen peroxide priming modulates abiotic oxidative stress tolerance: insights from ROS detoxification and scavenging. *Frontiers in Plant Science, 6*, 420.

Hove, R. M., & Bhave, M. (2011). Plant aquaporins with non-aqua functions: Deciphering the signature sequences. *Plant Molecular Biology, 75*, 413–430.

Hsu, P. K., Dubeaux, G., Takahashi, Y., & Schroeder, J. (2021). Signaling mechanisms in abscisic acid-mediated stomatal closure. *The Plant Journal, 105*, 307–321.

Ishikawa, F., Suga, S., Uemura, T., Sato, M. H., & Maeshima, M. (2005). Novel type aquaporin SIPs are mainly localized to the ER membrane and show cell-specific expression in *Arabidopsis thaliana*. *FEBS Letters, 579*, 5814–5820.

Israel, D., Lee, S. H., Robson, T. M., & Zwiazek, J. J. (2022). Plasma membrane aquaporin PIP2;3 facilitates hydrogen peroxide diffusion into Arabidopsis roots. Available at SSRN 4051184.

Johanson, U., Karlsson, M., Johansson, I., Gustavsson, S., Sjovall, S., Fraysse, L., et al. (2001). The complete set of genes encoding major intrinsic proteins in Arabidopsis provides a framework for a new nomenclature for major intrinsic proteins in plants. *Plant Physiology, 126*, 1358–1369.

Jurkiewicz, P., Senicourt, L., Ayeb, H., Lequin, O., Lacapere, J. J., & Batoko, H. (2020). A plant-specific N-terminal extension reveals evolutionary functional divergence within translocator proteins. *iScience, 23*, 100889.

Khabudaev, K. V., Petrova, D. P., Grachev, M. A., & Likhoshway, Y. V. (2014). A new subfamily LIP of the major intrinsic proteins. *BMC Genomics, 15*, 173.

Khokon, M. R., Okuma, E., Hossain, M. A., Munemasa, S., Uraji, M., Nakamura, Y., et al. (2011). Involvement of extracellular oxidative burst in salicylic acid-induced stomatal closure in Arabidopsis. *Plant, Cell & Environment, 34*, 434–443.

Kim, Y. X., & Steudle, E. (2009). Gating of aquaporins by light and reactive oxygen species in leaf parenchyma cells of the midrib of *Zea mays*. *Journal of Experimental Botany, 60*, 547–556.

Kirscht, A., Kaptan, S. S., Bienert, G. P., Chaumont, F., Nissen, P., De Groot, B. L., et al. (2016). Crystal structure of an ammonia-permeable aquaporin. *PLoS Biology, 14*, e1002411.

Kosinska Eriksson, U., Fischer, G., Friemann, R., Enkavi, G., Tajkhorshid, E., & Neutze, R. (2013). Subangstrom resolution X-ray structure details aquaporin-water interactions. *Science, 340*, 1346–1349.

Krüger, C., Waldeck-Weiermair, M., Kaynert, J., Pokrant, T., Komaragiri, Y., Otto, O., et al. (2021). AQP8 is a crucial H_2O_2 transporter in insulin-producing RINm5F cells. *Redox Biology, 43*, 101962.

Kwak, J. M., Mori, I. C., Pei, Z. M., Leonhardt, N., Torres, M. A., Dangl, J. L., et al. (2003). NADPH oxidase *AtrbohD* and *AtrbohF* genes function in ROS-dependent ABA signaling in Arabidopsis. *The EMBO Journal, 22*, 2623–2633.

Laloux, T., Junqueira, B., Maistriaux, L. C., Ahmed, J., Jurkiewicz, A., & Chaumont, F. (2018). Plant and mammal aquaporins: Same but different. *International Journal of Molecular Sciences, 19*, 521.

Lee, D., Lal, N. K., Lin, Z. J. D., Ma, S., Liu, J., Castro, B., et al. (2020). Regulation of reactive oxygen species during plant immunity through phosphorylation and ubiquitination of RBOHD. *Nature Communications, 11*, 1–16.

Lee, H. K., Cho, S. K., Son, O., Xu, Z., Hwang, I., & Kim, W. T. (2009). Drought stress-induced Rma1H1, a RING membrane-anchor E3 ubiquitin ligase homolog, regulates aquaporin levels via ubiquitination in transgenic Arabidopsis plants. *The Plant Cell, 21*, 622–641.

Leshem, Y., Melamed-Book, N., Cagnac, O., Ronen, G., Nishri, Y., Solomon, M., et al. (2006). Suppression of Arabidopsis vesicle-SNARE expression inhibited fusion of H_2O_2-containing vesicles with tonoplast and increased salt tolerance. *Proceedings of the National Academy of Sciences, 103*, 18008–18013.

Mahdieh, M., Mostajeran, A., Horie, T., & Katsuhara, M. (2008). Drought stress alters water relations and expression of PIP-type aquaporin genes in *Nicotiana tabacum* plants. *Plant and Cell Physiology, 49*, 801–813.

Marcec, M. J., Gilroy, S., Poovaiah, B., & Tanaka, K. (2019). Mutual interplay of Ca^{2+} and ROS signaling in plant immune response. *Plant Science, 283*, 343–354.

Marchissio, M. J., Francés, D. E. A., Carnovale, C. E., & Marinelli, R. A. (2012). Mitochondrial aquaporin-8 knockdown in human hepatoma HepG2 cells causes ROS-induced mitochondrial depolarization and loss of viability. *Toxicology and Applied Pharmacology, 264*, 246–254.

Maurel, C., Boursiac, Y., Luu, D. T., Santoni, V., Shahzad, Z., & Verdoucq, L. (2015). Aquaporins in plants. *Physiological Reviews, 95*, 1321–1358.

Maurel, C., Reizer, J., Schroeder, J. I., & Chrispeels, M. J. (1993). The vacuolar membrane protein gamma-TIP creates water specific channels in *Xenopus* oocytes. *The EMBO Journal, 12*, 2241.

Maurel, C., Verdoucq, L., Luu, D. T., & Santoni, V. (2008). Plant aquaporins: Membrane channels with multiple integrated functions. *Annual Review in Plant Biology, 59*, 595–624.

Mersmann, S., Bourdais, G., Rietz, S., & Robatzek, S. J. (2010). Ethylene signaling regulates accumulation of the FLS2 receptor and is required for the oxidative burst contributing to plant immunity. *Plant Physiology, 154*, 391–400.

Miller, G., Schlauch, K., Tam, R., Cortes, D., Torres, M. A., Shulaev, V., et al. (2009). The plant NADPH oxidase RBOHD mediates rapid systemic signaling in response to diverse stimuli. *Science Signaling, 2*, ra45.

Mittler, R., Zandalinas, S. I., Fichman, Y., & Van Breusegem, F. (2022). Reactive oxygen species signalling in plant stress responses. *Nature Reviews Molecular Cell Biology*, 1–17.

Møller, I. M., Rao, R. S. P., Jiang, Y., Thelen, J. J., & Xu, D. (2020). Proteomic and bioinformatic profiling of transporters in higher plant mitochondria. *Biomolecules, 10*, 1190.

Montiel, V., Bella, R., Michel, L. Y., Esfahani, H., De Mulder, D., Robinson, et al. (2020). Inhibition of aquaporin-1 prevents myocardial remodeling by blocking the transmembrane transport of hydrogen peroxide. *Science Translational Medicine, 12*, eaay2176.

Mubarakshina, M., Khorobrykh, S., & Ivanov, B. (2006). Oxygen reduction in chloroplast thylakoids results in production of hydrogen peroxide inside the membrane. *Biochimica et Biophysica Acta (BBA)-Bioenergetics, 1757*, 1496–1503.

Mullineaux, P. M., Exposito-Rodriguez, M., Laissue, P. P., & Smirnoff, N. (2018). ROS-dependent signalling pathways in plants and algae exposed to high light: comparisons with other eukaryotes. *Free Radical Biology and Medicine, 122*, 52–64.

Mullineaux, P. M., Exposito-Rodriguez, M., Laissue, P. P., Smirnoff, N., & Park, E. (2020). Spatial chloroplast-to-nucleus signalling involving plastid–nuclear complexes and stromules. *Philosophical Transactions of the Royal Society B, 375*, 20190405.

Murata, K., Mitsuoka, K., Hirai, T., Walz, T., Agre, P., Heymann, J. B., et al. (2000). Structural determinants of water permeation through aquaporin-1. *Nature, 407*, 599–605.

Navarro-RóDenas, A., Xu, H. A. O., Kemppainen, M., Pardo, A. G., & Zwiazek, J. J. (2015). *Laccaria bicolor* aquaporin LbAQP1 is required for Hartig net development in trembling aspen (*Populus tremuloides*). *Plant, Cell & Environment, 38*, 2475–2486.

Nyblom, M., Frick, A., Wang, Y., Ekvall, M., Hallgren, K., Hedfalk, K., et al. (2009). Structural and functional analysis of SoPIP2;1 mutants adds insight into plant aquaporin gating. *Journal of Molecular Biology, 387*, 653–668.

Pou, A., Jeanguenin, L., Milhiet, T., Batoko, H., Chaumont, F., & Hachez, C. (2016). Salinity-mediated transcriptional and post-translational regulation of the Arabidopsis aquaporin PIP2;7. *Plant Molecular Biology, 92*, 731–744.

Prak, S., Hem, S., Boudet, J., Viennois, G., Sommerer, N., Rossignol, M., et al. (2008). Multiple phosphorylations in the C-terminal tail of plant plasma membrane aquaporins: Role in subcellular trafficking of AtPIP2;1 in response to salt stress. *Molecular Cell Proteomics, 7*, 1019–1030.

Preston, G. M., & Agre, P. (1992). Appearance of water channels in *Xenopus* oocytes expressing red cell CHIP28 protein. *Science, 256*, 385.

Qi, J., Wang, J., Gong, Z., & Zhou, J. M. (2017). Apoplastic ROS signaling in plant immunity. *Current Opinion in Plant Biology, 38*, 92–100.

Rhee, J., Horie, T., Sasano, S., Nakahara, Y., & Katsuhara, M. (2017). Identification of an H_2O_2 permeable PIP aquaporin in barley and a serine residue promoting H_2O_2 transport. *Physiologia Plantarum, 159*, 120–128.

Rodrigues, C., Pimpão, C., Mósca, A. F., Coxixo, A. S., Lopes, D., da Silva, et al. (2019). Human aquaporin-5 facilitates hydrogen peroxide permeation affecting adaption to oxidative stress and cancer cell migration. *Cancers, 11*, 932.

Rodrigues, O., Reshetnyak, G., Grondin, A., Saijo, Y., Leonhardt, N., Maurel, C., et al. (2017). Aquaporins facilitate hydrogen peroxide entry into guard cells to mediate ABA- and pathogen-triggered stomatal closure. *Proceedings of the National Academy of Sciences, 114*(34), 9200–9205.

Rodrigues, O., & Shan, L. (2021). Stomata in a state of emergency: H_2O_2 is the target locked. *Trends in Plant Science, 27, 274–286.*

Sies, H. (2014). Role of metabolic H_2O_2 generation: Redox signaling and oxidative stress. *Journal of Biological Chemistry, 289*, 8735–8741.

Simm, S., Papasotiriou, D. G., Ibrahim, M., Leisegang, M. S., Müller, B., Schorge, T., et al. (2013). Defining the core proteome of the chloroplast envelope membranes. *Frontiers in Plant Science, 4*, 11.

Sirichandra, C., Gu, D., Hu, H.-C., Davanture, M., Lee, S., Djaoui, M., et al. (2009). Phosphorylation of the Arabidopsis AtrbohF NADPH oxidase by OST1 protein kinase. *FEBS Letters, 583*, 2982–2986.

Slesak, I., Libik, M., Karpinska, B., Karpinski, S., & Miszalski, Z. (2007). The role of hydrogen peroxide in regulation of plant metabolism and cellular signalling in response to environmental stresses. *Acta Biochimica Polonica, 54*, 39–50.

Smirnoff, N., & Arnaud, D. (2019). Hydrogen peroxide metabolism and functions in plants. *New Phytologist, 221*, 1197–1214.

Sorieul, M., Santoni, V., Maurel, C., & Luu, D. T. (2011). Mechanisms and effects of retention of over-expressed aquaporin AtPIP2;1 in the endoplasmic reticulum. *Traffic, 12*, 473–482.

Soto, G., Fox, R., Ayub, N., Alleva, K., Guaimas, F., Erijman, et al. (2010). TIP5;1 is an aquaporin specifically targeted to pollen mitochondria and is probably involved in nitrogen remobilization in *Arabidopsis thaliana*. *The Plant Journal, 64*, 1038–1047.

Sui, H., Han, B. G., Lee, J. K., Walian, P., & Jap, B. K. (2001). Structural basis of water-specific transport through the AQP1 water channel. *Nature, 414*, 872–878.

Suzuki, N., Miller, G., Salazar, C., Mondal, H. A., Shulaev, E., Cortes, D. F., et al. (2013). Temporal-spatial interaction between reactive oxygen species and abscisic acid regulates rapid systemic acclimation in plants. *The Plant Cell, 25*, 3553–3569.

Thor, K., Jiang, S., Michard, E., George, J., Scherzer, S., Huang, S., et al. (2020). The calcium-permeable channel OSCA1.3 regulates plant stomatal immunity. *Nature, 585*, 569–573.

Tian, S., Wang, X., Li, P., Wang, H., Ji, H., Xie, J., et al. (2016). Plant aquaporin AtPIP1;4 links apoplastic H_2O_2 induction to disease immunity pathways. *Plant Physiology, 171*, 1635–1650.

Tilsner, J., Linnik, O., Louveaux, M., Roberts, I. M., Chapman, S. N., & Oparka, K. J. (2013). Replication and trafficking of a plant virus are coupled at the entrances of plasmodesmata. *Journal of Cell Biology, 201*, 981–995.

Tong, H., Wang, X., Dong, Y., Hu, Q., Zhao, Z., Zhu, et al. (2019). A *Streptococcus* aquaporin acts as peroxiporin for efflux of cellular hydrogen peroxide and alleviation of oxidative stress. *Journal of Biological Chemistry, 294*, 4583–4595.

Tornroth-Horsefield, S., Wang, Y., Hedfalk, K., Johanson, U., Karlsson, M., Tajkhorshid, E., et al. (2006). Structural mechanism of plant aquaporin gating. *Nature, 439*, 688–694.

Tournaire-Roux, C., Sutka, M., Javot, H., Gout, E., Gerbeau, P., Luu, D. T., et al. (2003). Cytosolic pH regulates root water transport during anoxic stress through gating of aquaporins. *Nature, 425*, 393–397.

Tu, B. P., & Weissman, J. S. (2002). The FAD-and O_2-dependent reaction cycle of Ero1-mediated oxidative protein folding in the endoplasmic reticulum. *Molecular cell, 10*, 983–994.

Tyerman, S. D., McGaughey, S. A., Qiu, J., Yool, A. J., & Byrt, C. S. (2021). Adaptable and multifunctional ion-conducting aquaporins. *Annual Review of Plant Biology, 72*, 703–736.

Uehlein, N., Lovisolo, C., Siefritz, F., & Kaldenhoff, R. (2003). The tobacco aquaporin NtAQP1 is a membrane CO_2 pore with physiological functions. *Nature, 425*, 734–737.

Uehlein, N., Otto, B., Hanson, D. T., Fischer, M., Mcdowell, N., & Kaldenhoff, R. (2008). Function of *Nicotiana tabacum* aquaporins as chloroplast gas pores challenges the concept of membrane CO_2 permeability. *The Plant Cell, 20*, 648–657.

Varadaraj, K., & Kumari, S. S. (2020). Lens aquaporins function as peroxiporins to facilitate membrane transport of hydrogen peroxide. *Biochemical and Biophysical Research Communications, 524*, 1025–1029.

Verdoucq, L., Grondin, A., & Maurel, C. (2008). Structure–function analysis of plant aquaporin AtPIP2;1 gating by divalent cations and protons. *Biochemical Journal, 415*, 409–416.

Wallace, I. S., & Roberts, D. M. (2005). Distinct transport selectivity of two structural subclasses of the nodulin-like intrinsic protein family of plant aquaglyceroporin channels. *Biochemistry, 44*, 16826–16834.

Wang, A. (2021). Cell-to-cell movement of plant viruses via plasmodesmata: A current perspective on potyviruses. *Current Opinion in Virology, 48*, 10–16.

Wang, H., Schoebel, S., Schmitz, F., Dong, H., & Hedfalk, K. (2020a). Characterization of aquaporin-driven hydrogen peroxide transport. *Biochimica et Biophysica Acta (BBA)-Biomembranes, 1862*, 183065.

Wang, H., Schoebel, S., Schmitz, F., Dong, H., & Hedfalk, K. (2020b). Quantitative analysis of H_2O_2 transport through purified membrane proteins. *MethodsX*, 7, 100816.

Wang, X., Li, X., Deng, X., Luu, D. T., Maurel, C., & Lin, J. (2015). Single-molecule fluorescence imaging to quantify membrane protein dynamics and oligomerization in living plant cells. *Nature Protocols, 10*, 2054–2063.

Wang, X., Lu, K., Yao, X., Zhang, L., Wang, F., Wu, D., P, et al. (2021). The aquaporin TaPIP2;10 confers resistance to two fungal diseases in wheat. *Phytopathology, 111, 2317–2331.*

Waszczak, C., Carmody, M., & Kangasjärvi, J. (2018). Reactive oxygen species in plant signaling. *Annual Review of Plant Biology, 69*, 209–236.

Wu, F., Chi, Y., Jiang, Z., Xu, Y., Xie, L., Huang, F., et al. (2020). Hydrogen peroxide sensor HPCA1 is an LRR receptor kinase in Arabidopsis. *Nature, 578*, 577–581.

Ye, Q., Holbrook, N. M., & Zwieniecki, M. A. (2008). Cell-to-cell pathway dominates xylem-epidermis hydraulic connection in *Tradescantia fluminensis* (Vell. Conc.) leaves. *Planta, 227*, 1311–1319.

Ye, Q., & Steudle, E. (2006). Oxidative gating of water channels (aquaporins) in corn roots. *Plant, Cell & Environment, 29*, 459–470.

Yool, A. J., & Weinstein, A. M. (2002). New roles for old holes: Ion channel function in aquaporin-1. *Physiology, 17*, 68–72.

Zelazny, E., Borst, J. W., Muylaert, M., Batoko, H., Hemminga, M. A., & Chaumont, F. (2007). FRET imaging in living maize cells reveals that plasma membrane aquaporins interact to regulate their subcellular localization. *Proceedings of the National Academy of Sciences, 104*, 12359–12364.

Zhang, D., Huang, Y., Kumar, M., Wan, Q., Xu, Z., Shao, H., et al. (2017). Heterologous expression of GmSIP1;3 from soybean in tobacco showed growth retardation and tolerance to hydrogen peroxide. *Plant Science, 263*, 210–218.
Zhang, L., Chen, L., & Dong, H. J. (2019). Plant aquaporins in infection by and immunity against pathogens—A critical review. *Frontiers in Plant Science, 10*, 632.
Zhang, M., Shi, H., Li, N., Wei, N., Tian, Y., Peng, J., et al. (2022). Aquaporin OsPIP2;2 links the H_2O_2 signal and a membrane-anchored transcription factor to promote plant defense. *Plant Physiology, 188*, 2325–2341.

Kolibri 360 GmbH
Ferdinand-Jühlke-Straße 7
99095 Erfurt, Deutschland
produktsicherheit@kolibri360.de